Sensorimotor Life

Sensorimotor Life
An Enactive Proposal

Ezequiel A. Di Paolo

Thomas Buhrmann

Xabier E. Barandiaran

OXFORD
UNIVERSITY PRESS

OXFORD

UNIVERSITY PRESS

Great Clarendon Street, Oxford, OX2 6DP,
United Kingdom

Oxford University Press is a department of the University of Oxford.
It furthers the University's objective of excellence in research, scholarship,
and education by publishing worldwide. Oxford is a registered trade mark of
Oxford University Press in the UK and in certain other countries

Published in the United States of America by Oxford University Press
198 Madison Avenue, New York, NY 10016, United States of America

British Library Cataloguing in Publication Data

Data available

Library of Congress Control Number: 2017932492

ISBN 978-0-19-878684-9

Printed and bound by CPI Group (UK) Ltd, Croydon, CR0 4YY

Acknowledgments

This work is the result of years of collaboration, friendly conversations, and heated discussions. It would not have been possible without the support and criticisms of many people.

Many of the ideas in this book have been discussed at academic meetings, research seminars, and lectures. We are thankful to audiences and students who in many cases steered us in the right direction by asking very tough questions. We are grateful in particular to the following friends and colleagues: Eran Agmon, Miguel Aguilera, Malika Auvray, Michael Beaton, Manuel Bedia, Randall Beer, Fernando Bermejo, Leonardo Bich, Maggie Boden, Seth Bullock, Giovanna Colombetti, Elena Cuffari, Hanne De Jaegher, Matthew Egbert, Andreas Engel, Arantza Etxeberría, Tom Froese, Thomas Fuchs, Shaun Gallagher, Takashi Gomi, Inman Harvey, Matej Hoffmann, Phil Husbands, Daniel Hutto, Hiroyuki Iizuka, Takashi Ikegami, Eduardo Izquierdo, Peter König, Miriam Kyselo, Charles Lenay, Hugo Gravato Marques, Alexander Maye, Marek McGann, Alvaro Moreno, Erik Myin, Jason Noble, Susan Oyama, Rolf Pfeifer, John Protevi, Vasu Reddy, Andreas Roepstorff, Marieke Rohde, Kepa Ruiz-Mirazo, John Stewart, Evan Thompson, Steve Torrance, Nathaniel Virgo, Mike Wheeler, and Dan Zahavi. Our work over the years has benefited from the advice, criticism, and suggestions of these people and, in many cases, from the inspiration we have drawn from them as researchers and as persons.

Some people were kind enough to read chapters and sections of this book. We are grateful to Miguel Aguilera, Hanne De Jaegher, Matthew Egbert, Verena Fischer, and Marek McGann for their invaluable comments.

Much of the work presented here has been made possible by different sources of funding, in particular the project *eSMCs: Extending Sensorimotor Contingencies to Cognition (EU FP7-ICT-2009-6 no: 270212)*, which ran between 2011 and 2014. Other sources that contributed to our work are the *Basque Government Financing for Research Groups, IT590-13 (IAS-Research)* and the project *Interidentidad: Identity in Interaction, (MINECO, FFI2014-52173-P)* of the Spanish Ministry for Economy and Competitiveness.

Some of the chapters are based on work appearing previously in journal publications, although they have been substantially revised and extended.

Chapter 3 is based on Buhrmann, Di Paolo, and Barandiaran (2013). A dynamical systems account of sensorimotor contingencies. *Frontiers in Psychology*, Volume 4, p. 285.

Chapter 4 is based on Di Paolo, Barandiaran, Beaton, and Buhrmann (2014). Learning to perceive in the sensorimotor approach: Piaget's theory of equilibration interpreted dynamically. *Frontiers in Human Neuroscience*, Volume 8, p. 551.

Chapter 7 is based on Buhrmann and Di Paolo (2015). The sense of agency—a phenomenological consequence of enacting sensorimotor schemes. *Phenomenology and the Cognitive Sciences*, doi: 10.1007/s11097-015-9446-7. (Online first).

Some sections in Chapter 8 are based on Di Paolo (2016). Participatory object perception. *Journal of Consciousness Studies*, Volume 23, Issue 5–6, pp. 228–58.

The epigraph at the start of Chapter 2 was reproduced with kind permission of the Finnish Philosophical Society from Haugeland, J. (1995). Mind Embodied and Embedded. In: L. Haaparanta and S. Heinamaa (eds.), *Mind and Cognition: Philosophical Perspectives on Cognitive Science and Artificial Intelligence*, Acta Philosophica Fennica, Volume 58, pp. 233–267. Copyright © 1995 Finnish Philosophical Society.

The quotation from Bruner in Section 2.2 was reproduced with kind permission from ACTS OF MEANING: FOUR LECTURES ON MIND AND CULTURE by Jerome S. Bruner, Cambridge, Mass.: Harvard University Press, Copyright © 1990 by the President and Fellows of Harvard College.

The epigraph at the start of Chapter 4 was reproduced with kind permission of Taylor and Francis from Merleau-Ponty, M. (2012). *Phenomenology of Perception*. 2nd edition. (D. Landes, trans.). London: Routledge. (p.155). Copyright © 2012 Taylor and Francis.

The epigraph at the start of Chapter 6 was reproduced from Langer, Susanne K., *Mind: An Essay on Human Feeling*. 3 volumes. pp. 310–311. Reprinted with permission of The Johns Hopkins University Press.

The epigraph at the start of Chapter 7 was reproduced with kind permission of Taylor and Francis from Taylor, C. (2002). Foundationalism and the inner-outer distinction. In: Nicholas, H. S. (ed.), *Reading McDowell on Mind and World*. London: Routledge. Copyright © 2002 Taylor and Francis.

The epigraph at the start of Chapter 8 was reproduced with kind permission of Taylor and Francis from Merleau-Ponty, M. (2012). *Phenomenology of Perception*. 2nd edition. (D. Landes, trans.). London: Routledge. (p.155). Copyright © 2012 Taylor and Francis.

Finally, we are immensely indebted to someone none of us met in person but whom we had the fortune to know a little through the memories and contagious admiration of his many friends and colleagues: Francisco Varela. His influence on our work will be difficult to miss. He has been our imaginary interlocutor at various stages during the writing of this book. We hope that readers will perhaps encounter in it a trace of the spirit of his ideas.

Contents

Chapter 1

Introduction

I, for my part, cannot escape the consideration, forced upon me at every turn, that the knower is not simply a mirror floating with no foot-hold anywhere, and passively reflecting an order that he comes upon and finds simply existing. The knower is an actor, and co-efficient of the truth on one side, whilst on the other he registers the truth which he helps to create. Mental interests, hypotheses, postulates, so far as they are bases for human action—action which to a great extent transforms the world—help to make the truth which they declare. In other words, there belongs to mind, from its birth upward, a spontaneity, a vote. It is in the game, and not a mere looker-on.

—*William James (1878, p. 17)*

1.1 What the sciences of the mind have not told us

How accurate is the picture of the human mind that has emerged over the last decades from studies in neuroscience, psychology, and cognitive science? Anybody with an interest in how minds work, how we perceive and learn about the world, how we experience and remember people and events, how we make decisions, and so on, may at one point or another feel somewhat alienated by the answers contemporary science has to offer. The reason, often left implicit in these answers, could be summarized like this: "You are a machine. Biology and other people have programmed you. Moreover, you are in constant need of enhancement."

It is idle to abound in details here; all that the reader needs to do is look critically at the way scientific results are communicated to the public. Brains make decisions for us, they deceive us and undermine our resolutions, and they trick us into believing false things about the world and about ourselves. They do this to advance larger biological or ideological purposes and then they make us believe we are in control when in fact we are not. More recently, they also seem to demand maintenance, lifetime training, and pharmaceutical improvement. Our mental life is reduced to little more than an epiphenomenon, either an illusion or at most the effect of its underlying processes, but never a cause of anything.

These perspectives are likely to be dissonant with the way many people see themselves, how they experience their daily lives and relate to other people. They also seem to contrast sharply with what William James has to say about the mind in the epigraph. The mind, at least in the news coming from scientific circles, does not resemble the active, participative and world-creating picture James painted of it more than a century ago. Somewhere along the line the notion that we are agents of our lives seems to have been lost and replaced by the image of a quasi-automaton, driven by biological impulses and the external environment; a mere onlooker, not really in the game.

We could consider several reasons for this state of affairs, including the possibility that things really are the way we are told and that we are no more than self-deceiving machines. But perhaps the first question we should raise is whether the premises that drive some of these conclusions are sound.

Cognitive science has progressed on the basis of the computer metaphor. According to it, the mind is essentially a collection of integrated information processing functions and these functions explain how humans create meaning as they encounter the world. The processes of the mind, in this view, are those of a complex problem-solving computer, much more complex than the ones we can currently design, a computer built upon principles we do not yet entirely understand, but a computer nonetheless. With this premise, scientists can attempt to match the algorithmic structure of any given problem (say, how we weigh the pros and cons of a particular decision or how we anticipate someone's reaction to a piece of news or how we tie our shoelaces) to candidate mechanisms (typically in the brain) that could implement these computations. This is a successful formula for the practice of research, a powerful recipe for generating empirical hypotheses and testing them, accumulating results, and assembling theoretical frameworks. It is the way things are currently done in the sciences of the mind, the way they have been done for over six decades.

But we should be cautious. The assumptions that validate specific lines of investigation in this mode are not universal though they are treated as if they were, particularly the assumption that the mind works like a special kind of computer. A careful look at what we know about our bodies, about their biology, and the way they organize themselves into powers and sensitivities, a look at the way we experience the world as situated creatures in complex relations to other creatures tells a rather different story.

The computer metaphor imposes an implicit grammar; it promotes certain questions but not others. Attention is directed to *how things function*, more precisely, to how cognitive problems are or should be solved. But there are many important questions that cannot be answered solely, or at all, in this way. A theory of cognition, action, emotion, and perception, a theory of the mind, should also be able to tell us with some precision what a cognitive system *is* and which systems are not cognitive, no matter how complex. It should provide a good account of how an animal or a person is individuated as such, how the body is transformed through development and shaped by sociocultural influences. Fundamentally, a theory of the mind should be able to tell us in what sense we are autonomous meaning-making beings, what it means when we proclaim ourselves agents, and why our dealings with the world are experienced as imbued with significance and importance. Why does a living, cognitive system care about anything? *Why do we give a damn?*

There are *no* algorithmic answers to these questions because they are *not* questions about how a problem is solved or how some computation should be designed or implemented or which source of information is the most profitable. But this does not mean that there are no valid scientific answers we can consider. However, to see this wider and in many ways more fundamental set of issues, we must go beyond the grammar of the computer metaphor. To seek these answers should also be the task of cognitive science.

Without theorizing about these (and other) questions and testing these theories empirically, cognitive science cannot yet be said to cover properly the phenomena it is supposed to study. This should make us cautious about deriving sweeping conclusions about human lives and minds and spreading them as established, context-free facts.

1.2 Down the enactive path

This book belongs to a series of recent contributions that examine and expand the premises of the sciences of the mind. It aims to offer concrete answers to many of these open questions and in the process help develop an alternative picture of the human mind, one closer to people's daily experiences.

The work presented here is a development of ideas and proposals in the enactive approach, a way of thinking about embodied minds and lived experience that has seen important new articulations and applications over the last 25 years. The book is meant as a tributary to the flow of ideas first articulated by Francisco Varela, Evan Thompson, and Eleanor Rosch in 1991 in their book *The Embodied Mind*. In the first pages of that book the authors describe their project itself as a continuation, in a different register, of the philosophical and phenomenological research of Maurice Merleau-Ponty, whose thought demands a reconceptualization of the idea of the body, an idea that has been handed down to us by modernity (the body as machine). Merleau-Ponty invites us to look instead at our lived experience to see the ambiguities inherent in our embodied being and in the open materiality that constitutes our experience of the world and of ourselves. The enactive project begins with *these* premises and its goal is their scientific articulation.

Like a river that continues to add tributaries, enactive thinking also encourages us to look upstream and discover "new" predecessors. As with Kafka's literary precursors in Jorge Luis Borges' famous essay—whose condition as precursors is granted retroactively because their influence is only revealed once we have actually read Kafka—several philosophers and scientists (witness James's epigraph) may be called enactivists *avant la lettre*. Others have provided key elements and ideas that nourish enactive thought today. One case that merits particular mention is that of philosopher Hans Jonas (Varela read his work relatively late in his career and was impressed with the resonances with his own thinking). In a compelling series of essays collected in the book *The Phenomenon of Life* (published in 1966), Jonas expresses the conceptual and existential continuity between life and mind, a continuity that enactivists study today using the language of complexity science and dynamical systems theory. Jonas points in the direction of precise answers to some of the open questions we have referred to; Varela and many others provide elements for turning these answers into workable scientific proposals.

Within this broad context, and in continuation with the ideas of Merleau-Ponty, Jonas, Varela, and other "new" precursors we will encounter along the way, this book has one particular theme that relates to the open questions mentioned above. The theme is *sensorimotor life*.

A consequence of the increased attention to embodied aspects of cognition has been a move from an abstract conception of mental activity toward more concrete and situated, action-based practices and capabilities. Originally thought of as a "minor" set of problems in comparison with conceptual thought or language comprehension, the complexity involved in flexibly performing everyday tasks under a variety of unpredictable conditions is now appreciated as a major challenge for cognitive science. The realm of the concrete can be harsh and demanding. Performing tasks such as riding a bicycle in a city street, repairing the kitchen sink, getting on and off the bus during rush hour, and myriad other activities we go through daily—not to mention those activities proper of occupations demanding special training and skills—is still quite beyond the reach of the most sophisticated artificial intelligence systems and the most advanced theories of cognition. One of the reasons for this is that such tasks are highly context sensitive and full of constraints of all kinds, including, but not limited to time pressures at multiple scales, context-sensitive manipulation of tools and artifacts, the need to coordinate actions with variable objects and other people, dealing with conflicting goals and norms, constraints on the sensitivities to the risks and opportunities offered by the environment, and constraints on the capabilities and demands of the body. These constraints, in addition, are dynamic and can change radically and unexpectedly leaving us momentarily lost while we assess new situations and establish new objectives. Yet most people are able to cope with such breakdowns, most of the time, quite successfully.

This is one of the senses of the title of this book. Sensorimotor life is the life we live while we are engaged in doing things, in appreciating our surroundings, in organizing our activities, in executing adequate moves and correcting the wrong ones, in evaluating how things are going and what comes next, surfing opportunities and dodging risks. It is the life we spend perceiving, changing, and utilizing the world around us. This life occupies most of the hours in the day. It is the most concrete and recurring manifestation of our minds. We take sensorimotor life so much for granted that it is hard to see how much we are invested in it as embodied creatures. We care about what we do and commit to it. We expect things to go in certain ways and get frustrated when they do not. This is highly complex cognitive activity and it happens not in the remote seclusion of the brain as a series of calculations but in concrete movements, in the articulation of the joints, and in gross or subtle orientations of material flows. Sensorimotor life is, for the most part, for most people, and for most of the time, the ongoing bustle of animate embodied being.

1.3 Dimensions of embodiment

Of course, not all of our life is reducible to sensorimotor activities. In fact, most of these activities are always already embedded in a social and cultural world. Most of our concerns involve other people, what they do, what they think, and how we relate to them. Even the simplest practical sensorimotor activities are touched by social normativity. There is an intersubjective and participatory life that frames activities, and one could not hope to fully understand human embodiment if we did not acknowledge this crucial dimension.

Nor could one hope to understand embodiment in general, if we did not also take into consideration the fact that we are living creatures. The body is material, dynamic, and self-organizing. It can only partially be described in machine-like terms because it constantly fluctuates between stable and metastable states (critical states, rich in potentialities and sensitivities that may alter radically the course of the processes involved, and hence, the validity of any functional machine-like relation between them). The body is animated and affected by itself and by the world as it sets itself the restless task of remaining alive in the face of its own precariousness. For this reason, the body relates to the world in terms of needs and care. This never stops as long as we are alive. Sensorimotor life is thus also framed by the organic body, by its aches, by hunger and thirst, and by fatigue when the accumulated efforts of our activities become a burden as the day progresses. It is also framed by moments of joy and satisfaction when we prove ourselves efficacious, perhaps even dexterous in the way we achieve our goals.

These are the three mutually constraining and mutually enabling *dimensions of embodiment*, which Thompson and Varela have called the cycles of operation that constitute the life of an agent:

> Three kinds of cycles need to be distinguished for higher primates:
>
> (1) cycles of organismic regulation of the entire body;
>
> (2) cycles of sensorimotor coupling between organism and environment;
>
> (3) cycles of intersubjective interaction, involving the recognition of the intentional meaning of actions and linguistic communication (in humans)
>
> (Thompson and Varela 2001, p. 424).

Each of these dimensions of embodiment is intimately related to the other two. But at the same time, each of them establishes an autonomous domain of enquiry, in the sense that none of the dimensions is reducible to phenomena in the others. Each dimension involves cycles that are constrained but not fully determined by the others.

Of these three loci of research, the work in this book is focused on the second one, sensorimotor life, but we would do a poor job if we did not expressly link our arguments, proposals, and discussions to the other two dimensions. To the extent that it proves crucial to our arguments, we have done this, but of course, we have not been exhaustive in exploring all possible (not even many of these) connections.

Our objective, of course, is not to provide a detailed account of the entire range of human sensorimotor activities (for reasons we discuss later, this may actually be an impossible task, since these activities are open-ended, and new goals, new norms, new skills, and even new genres keep appearing in unforeseeable ways). We aim instead to develop a series of "mesoscopic" proposals that can be used to frame enquiries about sensorimotor life that are more specific. These proposals concern general questions about action and perception as the fundamental co-defined modes in which agents relate to the world. But to understand these modes, we must also address the question of *agency* itself.

1.4 **The nature of agency**

A concept of agency is already implied in some of the questions we have raised earlier. But what is this concept? Can it be articulated in scientific terms? Can agency be approached naturalistically and still be understood as embedded in the three dimensions of embodiment instead of reduced to its neurobiological underpinnings?

The nature of agency is studied in philosophical circles in connection with theories of intentional action. An agent is someone with the capacity to perform acts. How is this capacity exercised and in what senses can one attribute the ensuing consequences to the agent behind these acts? These are examples of the questions that occupy philosophers of mind and moral philosophers such as the nature of free will, mental causation, and intentionality. Agency is also a topic of research in ethics and theories of rationality. These important debates have significance for conceptions of justice and responsibility and implications for education science, economics, and political science. Roughly speaking, this work can be said to be concerned mostly with phenomena occurring in the third, intersubjective dimension of embodiment.

At the same time, the nature of agency, in terms of behavior, is also the concern of biology, from the regulation of chemical exchanges across the membrane of a unicellular organism, to the determinants of conducts in more complex animals. Some work in biology and philosophy of biology has begun to address exactly what distinguishes an agent from physico-chemical happenings, what is the nature of even minimal forms of life such that they can be considered to behave, and do so adaptively, according to vital norms. Such research could be said to be concerned with phenomena approximately located in the first cycle of embodiment, the processes involved in the construction and regulation of the organic body and its relations to the environment.

Agency in the second, sensorimotor, cycle of embodiment is also the object of copious studies, particularly in psychology, psychiatry, and neuroscience. There is special interest in studying not only how the manifestations of agency occur, but also how we perceive them in others and how we understand ourselves as being behind our own acts and intentions, as it were, in possession of them. Different models and theories look at possible neural mechanisms that could be involved in the sense and sensitivity for agency, and their pathologies.

However, at this level of enquiry, the implications of naturalized concepts of agency that we start to see in the case of philosophy of biology have not yet been properly expressed in ways that could inform psychology, psychiatry, and neuroscience. A naturalized theory of sensorimotor agency is missing from the sciences of the mind. Cognitive science, ruled by the computer metaphor, cannot provide this theory because it escapes the way questions can be formulated and answered in a functionalist framework. This is the reason, as we shall discuss later, why it can only theorize about how different sources of information could be used to distinguish between body movements that are self-initiated or not. But this distinction already assumes an implicit concept of agency, which is not explicated. In other words, not much can be said from this perspective about what it means to be a sensorimotor agent, with powers, skills, and trust in the world, who acts on her own behalf. A theory of sensorimotor agency and subjectivity is now long overdue.

Such a theory is one of the central contributions of this book. We make use of conceptual tools that have emerged when studying the first cycle of embodiment, such as the concepts of autonomy and operational closure that explain what it means for a living organism to self-individuate (i.e., to construct itself materially and make itself actively distinct from its environment) and the concept of sense-making that describes how living organisms relate to their world in terms of meaning. These tools can help us make sense of how biological agents, organisms that regulate their interaction with their environment, can sometimes also become *sensorimotor* agents, forms of life that are constituted as self-sustaining, habitual organizations in the structural and functional interrelations between their acts, skills, and dispositions. A good part of what we do when involved in concrete activities has to do not so much with biological or social aims, but with how acts and capabilities fit together with other acts and capabilities. The internal logic, the constraining relations, and the adaptive facilitation between acts can give rise to a sensorimotor normativity, even a sensorimotor style that develops into aspects of a personality. We propose that this is a form of life, enabled by, but not reduced to biological life: a *sensorimotor life*.

With the aid of this idea—the other meaning of our title—we explore questions of sensorimotor development and the organization of daily activities (and what happens when a breakdown occurs). We also explore what this concept means for understanding the experience of being an agent. We offer a new perspective on this issue that can account for the various phenomenological aspects of the sense of agency, and we formulate empirical predictions that differ from those of standard computational models. With the concept of agency at the sensorimotor level, we also explore in the last chapter of the book (Chapter 8) some open challenges for the enactive approach (and plausible responses), such as the question of virtual actions and the social origins of abstract perceptual attitudes.

1.5 **The sensorimotor substrate**

Before we dive into an enactive theory of sensorimotor agency, we must first familiarize ourselves with some of its elements. We have found that a fruitful way to do this is to discuss the sensorimotor approach to perceptual experience proposed by Kevin O'Regan and Alva Noë in 2001. This approach usefully integrates various motor-based theories of perception formulated throughout the twentieth century and before. It states that perceptual experience is constituted by a mastery of lawful sensorimotor relations proper of the mode of engagement with the environment currently enacted by an agent. One reason for focusing on this theory is that it is a well-known, elaborated, and much-debated body of work in which embodied action plays a fundamental role. This makes it highly compatible with much of what we want to say about sensorimotor agency. Another reason is that we want to advance an enactive, non-representational interpretation of this sensorimotor theory (as we shall discuss, one of the axes of the heated debates about this approach has been precisely whether it implies the use of internal representations; a question about which the proponents of the theory have remained a bit ambiguous).

To interpret the sensorimotor approach to perception in enactive terms necessitates the elaboration of certain specific proposals, starting with an operational definition of the concept of sensorimotor regularities or contingencies, which is central to the approach. Surprisingly, this idea, so amenable to a dynamical systems interpretation has not been explicitly defined in the original literature.

Another elaboration we require for our enactive interpretation is a concept of mastery that does not imply the need for internal representations. In order to do this, we propose a Piagetian approach to how sensorimotor capacities are organized via processes of differentiation and integration. Again, we formulate these ideas in dynamical systems language to arrive at a theory of sensorimotor learning that allows us to interpret mastery in enactive, non-representational terms.

With these elements we then proceed to our proposed theory of sensorimotor agency. But throughout the rest of the book, we keep coming back to the sensorimotor approach to perception. Our discussions about agency will also help us make better sense of why we perceive things differently according to the kind of activity we are engaged in. We will discuss how our embeddedness in concrete activities implies different ways in which we master the "virtualities" of action and perception. And we shall make better sense of the difference between the pragmatic and abstract attitudes with which we can perceive the world around us.

This is, broadly described, what this book is about, what we intend by its title. Before continuing, however, we should mention a couple of caveats.

This is not a book that will engage in detailed criticism of other approaches, but one that intends to show that a particular perspective, an enactive one, can provide useful explanations and suggest novel ideas. At various points, unavoidably, we will discuss other approaches critically, in particular representationalism (especially in Chapter 2. Representational pull, enactive escape velocity). However, these criticisms are only provided to clarify for the reader the motivations behind certain moves we make and also to illustrate an enactive way of thinking about certain problems. If we had to take these criticisms to the level of sophistication demanded by a full debate on these topics, we would have to write a different book: fortunately, an abundance of books and articles do precisely that. Our book is not one of them, even if at some points it may look like it is, because, after all, some debates are hard to resist. On the contrary, we hope that if this book will contribute to such debates it will do so mainly by showing that a non-representational theory of action and perception is viable and that it covers important phenomena that are often neglected or underplayed by representationalist theories. As we shall see, many arguments in defense of representationalism and the computer metaphor rely on there being no apparent alternative: that thinking of the mind in mediational/representational terms is the only game in town. We hope to show that this is no longer the case. This is non-representationalism at work.

We are also (sometimes painfully) aware that we do not cover in this book all the relevant ramifications of our proposals, nor all of their nuances. Many central themes related to sensorimotor life will remain undeveloped. In particular, we will not discuss in detail various topics such as perceptual modalities, illusions, various different genres of action such as locomotion, navigation, prehension, manipulation, production of sounds, etc., nor will we cover different sensorimotor pathologies. We also do not discuss in detail the neuroscientific implications of our proposals, although we often make use of neuroscientific evidence and theories to illustrate or support our points. We hope, however, that the central theoretical lines of the book will prove sufficiently open for developing these connections in future work.

Unavoidably, as we have said, the three dimensions of embodiment are deeply intertwined. Nothing we say in the book is entirely and strictly limited to the sensorimotor realm and not touched by the other two. Nevertheless, this is often the way we shall express our ideas, which at various points are technically complex in themselves. We will not help the readers (or ourselves) if we try to cover all of the angles at once. Moreover, sensorimotor life is precisely the dimension that has received less attention from enactivists in the past, a gap that deserves to be filled in.

A particularly important point where this may create some confusion has to do with the way we present the idea of sensorimotor agency as constituted as a set of processes involving a single cognitive individual. We clarify that many of the processes involved would not be feasible, or even possible, in the absence of a social world, but these connections are not explicitly developed in this book (with the exception of the last chapter). They are merely signaled here and there in the hope of creating points of coupling with future work. This strategy may give the reader the impression that we are following an individualist methodology. This impression will be understandable, but it is not what we intend. Our ideas—this applies to the whole epistemology of enactivism—are part of ongoing epistemic cycles. These cycles parallel the dimensions of embodiment already mentioned. We have seen enactive theories of minimal agency and of sense-making as grounded in the organization of living systems. That is one cycle. Some of these concepts have been applied to understanding the intersubjective world. That is another cycle. Now we contribute work that explicates further the sensorimotor dimension of the living, social body; yet another cycle. Each of these cycles—this is the main point—*will keep turning*. Our formulations in this book will surely be refined as we consider more social and biological elements. We think these refinements will not contradict our main messages but rather allow us to express and understand them better.

Our proposal for a theory of sensorimotor life may itself contribute to another turn of the epistemic cycles of enactivism. We hope, in any case, that it will make for a less partial view of the mind, a view better attuned to the experiences of people who live, work, love, struggle, and age, thrown into a world of meaningful relations they help create, always and inevitably, through their bodies.

Chapter 2

Representational pull, enactive escape velocity

As our ability to cope with the absent and covert, human intelligence abides in the meaningful—which, far from being restricted to representations, extends to the entire human world. Mind, therefore, is not incidentally but *intimately* embodied and *intimately* embedded in its world.

—*John Haugeland (1995, p. 267)*

2.1 **The body knows**

You may be tying your shoes incorrectly. Many people do. It only takes one inverse crossing of the laces and, even if the final knot looks right, even if it holds for a while, the result will not last long. About 30 percent of the population does it "wrong." If you are among those who are tying your shoes incorrectly, it may be hard to notice. You may have compensated by doubling the knot to be sure the laces do not get untied during the day, or you might have got used to tying your shoes 2 or 3 times a day. You have incorporated a practice, you have accommodated, and it feels right to you. There is a subtle difference between the balanced shoelace knot and the granny knot, as it is called. You can check online tutorials and discover which way you have been doing it all your life. Now, if it is not the way you have done it in the past, you might consider the advantages of the balanced shoelace knot and tie your shoes balanced (for the first time!) and experience the difference. You may feel a bit embarrassed, surprised, even irritated, and then finally satisfied to have found the "right" way. What is perhaps more surprising is how long it can take you to relearn the way to tie your shoelaces. Procedurally speaking it is a very small difference; it would take a minimal change in a computer program to "fix the mistake." Yet, relearning something you have done differently for most of your life can take some time. You may not remember it, but it probably took much longer for you to learn this skill as a child. Teaching how to do it (ask any parent) is also a complex process, it involves all kinds of techniques, including direct physical guidance, demonstrations, singing mnemonics, and just plain trial and error. As a learner, the task slowly gets easier and more natural with practice. Once you incorporate it, each step calls for the next, one hand pushes against the other; you are not done with the first knot when your fingers start with the bow. Your body *knows*. And it knows in a sense that is very different from what you can make explicit by verbalization, by writing down the procedure or making a drawing of it. Try to explain how to tie your shoes and you will struggle with language while you find yourself prefiguring the movements. You may even be able to feel the movements of your hands as you try to think about it.

In a sense, as we are about to see, the study of the mind is caught in the tension between the way in which we communicate experience and knowledge through language (the structure of logic, composition, reasoning, symbolization, etc.) and the way in which we exercise, experience, and live life as an embodied practice.

Tying your shoelaces, making breakfast for the kids before they go to school, finding your way around the central station during a busy commute, these are all instances of concrete cognitive activities that involve sophisticated abilities, sensitivities, affect, effort, and the power to manage and regulate your relations to the world and to others. They take time to learn and effort to alter. The intelligence, adaptivity, adequacy, and complexity of these activities remain, for the most part, tacit—we know more than we can tell, as Michael Polanyi (1966) famously put it. Attempts to render our know-how explicit always miss something that cannot be expressed as the summation of several rules, objective standards and precise norms.[1] These embodied activities are performed fluently, efficiently, and safely without much concern for the fact that they cannot be easily described in explicit procedural terms, not just by those who perform them, but often by those who study them. Conversely, enacting small procedural differences like the one between the balanced knot and granny knot can take disproportionate amounts of time and effort to learn.

Can cognitive science explain how we achieve these and more complex things every day? Can it provide an account of embodied know-how from its prevailing assumption about cognition as the processing of information? This is the notion that has made contemporary cognitive science possible since its beginnings in the 1950s. Information processing has been the unifying idea behind the diversity of cognitive phenomena, from the sophisticated navigational capacities of a wasp to the quiet strategic planning of a chess player. But it has also biased cognitive explanations toward the disembodied and intellectualist end of the spectrum, the kind of explanations that, to many people, do not match well the situated and richly context-dependent experiences and activities they enact every day. Such an explanation would predict that it should be rather easy to learn the right way to tie your shoelaces.

Increasingly, over the last three decades, some explanations in cognitive science have begun to move away from this intellectualist bias, in some cases even plainly rejecting the idea of cognition as computation. More attention is being paid to dynamical engagements

---

[1] Tacit knowledge is often overlooked by the pervasive drive in modern Western societies to over-standardize professional practices in terms of rules, guidelines, benchmarks, performance targets, and protocols. Standardization is meant to make activities fit together rationally, but it often ignores the subtle embodied skills and communication links between practitioners who in the concrete embeddedness most of the time find it more efficient to reinterpret rules or work around them so as to get the job done.

between embodied creatures and their complex environments, to lived experience, to the organization of animate movement, and to the structure of sensorimotor engagements. This is happening in different areas. It is happening in situated artificial intelligence, in autonomous robotics, and in cognitive linguistics. Complex cognitive capabilities such as language use, mental imagery, and decision making are now found to depend deeply on embodied, affective, historical, and dynamical aspects of the agent in action. Such aspects used to be no more than implementation details, a mere question of proper interfacing between the *really* cognitive, algorithmic mechanism for problem solving and the real world. But the tables have started to turn. It is a move from an abstract and intellectualist picture of the mind to a concrete and embodied one. A re-enchantment of the concrete, as Francisco Varela once put it.

Have the tables turned completely? Are we now living in the golden age of embodied cognitive science? The answer must be a qualified "no." It has not happened yet. Even though there has been a proliferation of terms like "embodiment," which is now frequently found in book and conference titles, academic articles, research labs, and funded projects, much of the work in cognitive science remains well within the assumption of the mind as an information processing machine. In fact, the proliferation of embodied terminology may be part of the problem. There is no overarching agreement of how "embodied" criticisms of traditional, disembodied approaches should be interpreted. Many interpret them as amendments, improvements to the computational metaphor of the mind, but not as alternatives to it. Others see a different path, one that requires rethinking basic assumptions about what it means to be a living embodied mind in the world.

Among the reasons for this increasingly apparent split within the embodied camp—which we briefly explore in this chapter—we find a deeply rooted commitment to what counts as a scientific explanation of cognition. This commitment is expressed in the belief that no such explanation is possible if we do not provide an account of how facts about the outside world get internalized in the form of knowledge inside a cognitive agent. This is the single, most influential assumption in the contemporary sciences of the mind, and it inevitably leads to some version of cognition as internal computations, typically occurring in the brain. From within this picture, there seems to be little choice other than staying trapped by its frame. For this reason, as we shall see, arguments by default ("there is no alternative") abound in debates about embodiment.

But there *are* alternatives if we look for ways of escaping the picture of cognition as internalization and seek to replace it with a picture of the living cognitive agent as moving outward into the world and navigating its possibilities. In this chapter we discuss what such a positive story requires in the context of embodied theories of action and perception.

2.2 **From abstract thought to embodied action**

Steadily as the movement of tectonic plates, cognitive science has been shifting from abstract cognitivism to more dynamical, embodied, embedded, extended, and situated approaches.

The term *embodiment* is the most encompassing of the labels associated with these changes. But it also hides a considerable amount of ambiguity (Wilson 2002). And it is only recently that researchers have begun to question explicitly whether with the embodied turn we are witnessing the formation of a new continent or no more than a rearrangement of existing geographical features. Let us briefly review the context of these developments.

In the early years of cognitive science[2] (the 1950s), the drive was to bring the notion of the mind back into scientific discourse. Jerome Bruner, one of the key participants in the Cognitive Revolution, recounts its objectives as profound:

> to discover and to describe formally the meanings that human beings created out of their encounters with the world, and then to propose hypotheses about what meaning-making processes were implicated ... Its aim was to prompt psychology to join forces with its sister interpretive disciplines in the humanities and in the social sciences.
>
> (Bruner 1990, p. 2)

Soon, however, the attention shifted from meaning construction to information processing. The Cognitive Revolution crystallized into a powerful research program: cognitivism. Its commanding metaphor was the computer. Mainstream investigations targeted the various forms of human abstract cognition (conceptual reasoning, problem solving, language production and comprehension, etc.). In his classical text on *Cognitive Psychology*, Ulric Neisser defined cognition as referring "to all the processes by which the sensory input is transformed, reduced, elaborated, stored, recovered, and used" (1967, p. 4). The assumption was that such processes were fundamentally like computer programs and that the mind bore to the brain the same relationship as software does to hardware. This turn, according to Bruner, was a loss of the original target, since meaning in its nuances, vagueness, polysemy, and connotations could not be reduced to bits of information. And, with the exception of some branches of linguistics and analytical philosophy, cognitive science in this period did not quite join forces with other interpretive disciplines in the humanities and social sciences, a disconnection attributable to the reduction of the forms of meaning-making studied by these disciplines to information processing.

The software/hardware divorce between the domain of the mental and that of the underlying physical/biological substrate and medium rested on the assumption that mental states could be properly identified by their functional role (i.e., by their relation to other mental states and their role in the production of behavior). These mental states (beliefs, desires, perceptions, etc.) were assumed to be implemented as states of the brain, which in turn led to approaching the study of this biological organ in terms of the computations that could be instantiated by it. Functional units within the brain were assumed to be nearly decomposable (i.e., their interactions do not alter the way they work); time was

---

[2] For a rich and illuminating exposition of the history of cognitive science, its main characters and events, and its relation to other disciplines, see Margaret Boden's (2006) account of the study of minds as machines. For the fascinating story of the cybernetic roots of cognitive science and artificial intelligence see Dupuy (2000) and a compilation of articles and interviews edited by Husbands, Holland, and Wheeler (2008).

paced by an independent external clock and uniformly imposed on the system; inputs and outputs were coded in a processor-readable manner. This is the architecture that permitted to "freely" compose and manipulate propositional structures in a manner presumed analogous to human abstract thinking.

At the start, there were few criticism of this cognitivist program. Hubert Dreyfus (1965, 1967, 1972) drew attention to the insufficient grounding of the assumptions of cognitivism in concrete lived experience. The proposed explanations simply did not match the phenomenology and instead tended to rely overly on intellectualist assumptions about how a cognitive problem *should* be solved if it was in the hands of a clever computer programmer. A worry, in some ways related, expressed by John Searle's (1980) Chinese Room thought experiment, was that formal operations on "symbols" could never produce a theory of meaning. These criticisms made it quite apparent that the cognitivist programme, based as it was entirely on disembodied syntactic manipulations, had not clearly resolved the semantic question of how any sort of personal meaning could emerge out of such calculations.

Little changed in view of these general worries, even though cracks in the cognitivist framework steadily deepened. George Lakoff and Mark Johnson, for example, questioned the way we think about meaning and concept formation: human language reveals a deep entanglement between embodied, spatial, and temporal experience and the meaning of even the most abstract concepts (Lakoff and Johnson 1980; Lakoff 1987; Johnson 1987). Far from being a detail to be abstracted away when it comes to understanding the computational functions that presumably define the mind, the body is itself a source of meaning and concept articulation. Lakoff and Johnson argued that the primary tools for thinking are metaphors, as opposed to logical inferences, and it is through reframing problems and situations in an embodied context that we normally find their solutions.

Around the same time, cognitivism came under a series of criticisms from the fields of autonomous robotics (Brooks 1991; Beer 1990; Maes 1991; Steels 1993; Harvey, et al. 1996; Pfeifer and Scheier 1999) and situated artificial intelligence (Agre 1988; Suchman 1987; Winograd and Flores 1986). When putting cognitivist principles to the test in the design of intelligent applications, user interfaces, or efficient robots that must deal with the real world, it became apparent that more attention needed to be paid to the body in direct physical interaction with the environment. Behavior in a robot was modeled by following repetitions of a "sense-model-plan-act" flow from sensory information to activation of effectors. The linear conception was an obstacle for designing robots capable of even "simple" real-world interactions with environments involving noisy and partial information and tight constraints at different timescales. These environments are the rule rather than the exception. Some roboticists envisioned an alternative bottom-up approach in which behavior resulted from the combination of several functional layers of control. They started from the simplest layers (functional loops such as stop advancing and turn around if the proximity sensor is high resulting in collision avoidance) and from there they scaled up to more complex behavior emerging from the interaction between several robust sensorimotor layers and the direct engagement with the environment. This move toward situatedness was met with (sometimes fierce) opposition, often acknowledging the success of simple robots but considering it not so relevant for understanding human cognition (Kirsh 1991; Vera and Simon 1993; more recently, Sloman 2009).

"Simple" as the behavior of these autonomous robots was (robust navigation, obstacle avoidance, and picking up objects), the bottom-up strategy behind their design indicated the need to rethink cognition in general. Human level cognition, it was argued, emerged from a long history in which complex forms of adaptive behavior evolve from and reuse simpler ones. The key to modeling human intelligence is, therefore, the understanding of the relatively simple, yet robust and efficient, adaptive behavior it is built upon. One of the broader lessons from autonomous robots was that intelligence should not anymore be conceived as essentially a process of abstract symbol manipulation, but rather as fluid and flexible sensorimotor couplings. In such dynamic couplings, the body plays as central a role as brains and control circuits (Beer 1990; Webb 1995). In some cases, lifelike sophisticated control, for instance, the one required for balanced bipedal walking, was shown to be almost exclusively a question of having the right body. Passive-dynamic walkers (McGeer 1990; Raibert 1986) are machines able to walk down a slope using only the dynamics of the their legs. They are a striking demonstration of how much can be achieved by providing a robot with the appropriate bodily structures (see Pfeifer and Scheier [1999] for other examples making the same point). Another important insight that came from robotics was the emphasis on approaching the study of cognition by understanding the continuities and changes in the natural history of the mind and not by relying solely on introspective intuitions about how cognitive problems should be solved.

The concept of emergence is central in this brand of robotics. Behavior cannot be broken down into suboperations that play well-delimited functional roles. Rather, it arises out of nonlinear loops between brain and body components and between the agent and the environment (Steels 1993; Hendriks-Jansen 1996). Dynamical systems theory (Beer 1990; Kelso 1995; Van Gelder and Port 1995) was found better suited than computational routines for capturing the emergent nature of adaptive behavior. These new tools for thought allowed a treatment of the body and the environment as part of a dynamic continuum out of which coherent behavior could emerge and where no privileged status could in principle be attributed to anything like a "central processing unit."

The understanding of cognitive systems as embodied and situated in an environment, and of cognition as the result of complex, dynamical, and emergent interactive patterns, led to another insight that began to draw the attention of cognitive scientists: the primacy of action.

This insight is manifested in proposals regarding the sensorimotor constitution of perceptual experience (Noë 2004; O'Regan and Noë 2001). And also in the call to see cognitive mechanisms not as mirroring an objective external world but instead as generating effective action based on partial, noisy, egocentric information (Clark 1997; Wheeler 2005). More broadly, the insight fits a conception of cognition as subserving action and being grounded in sensorimotor coupling (Engel et al. 2013). Such calls for a pragmatic, action-oriented perspective resonate with an initial tenet of the enactive approach according to which:

> (1) perception consists in perceptually guided action and (2) cognitive structures emerge from the recurrent sensorimotor patterns that enable action to be perceptually guided.
>
> (Varela, Thompson, and Rosch 1991, p. 173)

Originally the two bread slices of the "sandwich model" of the mind (Hurley 1998), action and perception, are increasingly recognized as intertwined, co-dependent, and co-determined. For traditional cognitivism the direction of influence is predominantly from perception to action. Behavior is regulated by the perceived features of the environment. In the 1970s William T. Powers proposed an overturning of this idea in his Perceptual Control Theory (Powers 1973). Accordingly, perception is regulated by action; in other words, an agent behaves so as to compensate for deviations from expected perceptual goals, which are themselves set up hierarchically (a dynamical forerunner of predictive coding approaches, Clark 2013; Friston 2010). Instead of either arc, perception to action or action to perception, being taken as primordial, some approaches, particularly dynamical ones, propose to consider a continuous, closed sensorimotor circle. Cognition, in this case, is the way an agent regulates these entanglements between action and perception. This is a rediscovery of old but not often heeded advice to see action and perception as part of an ongoing loop. It is worth quoting at length a couple of well-known but apposite passages in this respect:

> Upon analysis, we find that we begin not with a sensory stimulus but with a sensori-motor coordination, the optical-ocular, and that in a certain sense it is the movement which is primary, and the sensation which is secondary, the movement of body, head and eye muscles determining the quality of what is experienced. In other words, the real beginning is with the act of seeing; it is looking, and not a sensation of light. The sensory quale gives the value of the act, just as the movement furnishes its mechanism and control, but both sensation and movement lie inside, not outside the act.
>
> (Dewey 1896, pp. 358–9)

> The organism cannot properly be compared to a keyboard on which the external stimuli would play and in which their proper form would be delineated for the simple reason that the organism contributes to the constitution of that form ... Doubtless, in order to be able to subsist, it must encounter a certain number of physical and chemical agents in its surroundings. But it is the organism itself—according to the proper nature of its receptors, the thresholds of its nerve centers and the movements of the organs—which chooses the stimuli in the physical world to which it will be sensitive.... This would be a keyboard which moves itself in such a way as to offer—and according to variable rhythms—such or such of its keys to the in itself monotonous action of an external hammer.
>
> (Merleau-Ponty 1942/1963, p. 13)

In contemporary terms the sensorimotor loop is manifested in the way in which motor variations induce (via the environment) sensory variations, and sensory changes induce (via internal processes) the agent to change the way it moves. The regularities that emerge from recurrent sensorimotor cycles are constitutive both of action and perception. These are also the regularities that the cognitive system is sensitive to and attempts to manage.

Analysis of the coupled agent-environment system, typically through dynamical systems techniques, is the prevalent research method for studying the emergent regularities in closed sensorimotor loops. This is clearly the case in situated robotics (Beer 1990; Pfeifer and Scheier 1999). Much in the same line but somewhat parallel are the dynamical approaches coming from the tradition of ecological psychology (Chemero 2009; Turvey and Carello 1995). The agent-environment coupling is so dynamically rich, according to this tradition, that it is deemed sufficient for the agent to perceive information directly without the need for internal reconstructions of the world. Typically the regularities in this coupling take the form of meaningful opportunities for engagements, or affordances, which are directly perceived by the agent.[3]

With notable influence, Kevin O'Regan and Alva Noë (2001) have presented a view of perceptual awareness in which the body plays a central, action-oriented role. Like researchers in autonomous robotics before them, they note that, when a body moves through the world, certain sensorimotor regularities occur: for instance, sound gets louder as you move toward its source but stays constant in volume when you travel at a fixed distance around it, different parts of an object come into view in a regular order as you move around it, and so forth. The way in which sensory activity changes varies in a lawful way as one moves around in the world.

Building on this insight, O'Regan and Noë argue that perceptual experience is constituted by the mastery of these sensorimotor regularities or contingencies. For instance, visual perceptual experience depends on sensorimotor regularities that affect a creature with eyes moving around a three-dimensional space. Tactile perception, in turn, involves the mastery of the changing touch sensations that occur as one moves one's skin in contact with objects, and so on. As they would put it, one cannot perceive unless one masters these changes in sensory stimulation as a result of movement. Objects in the world are

----

3 A full treatment of the similarities, complementarities, and opposing views between ecological psychology and the enactive approach (described later in this chapter) is beyond the scope of this book. As proof of how close the approaches can be in concrete cases, we make use throughout this book of work originating in the Gibsonian tradition (Gibson 1979). This tradition usually supplies some of the clearest examples of how dynamical engagements and bodily synergies can be explanatorily powerful. But the relation between the schools of thought is one of strange familiarity, as if their respective practitioners were staring at each other across an uncanny valley. It is true that both approaches overlap in their rejection of representationalism, but this does not mean they are necessarily rejecting the same thing. As we will discuss further in this chapter and demonstrate in the following ones, the enactive perspective rejects a functionalist general approach to cognition, whereas ecological psychology rejects the assumption of the poverty of environmental information. These are not the same thing. For Gibsonians, perception is still about information pickup, but not for enactivists, who conceive of perception as an aspect of sense-making, a concept that is explicitly grounded in the notion of autonomous agency. This is one of the reasons for the admittedly rather quick dismissal of ecological psychology by Varela, Thompson, and Rosch (1991), who saw it only capable of providing a theory of cognition on the side of the environment. In later years, there have been many attempts at bringing the two traditions closer to each other. We think the dialog should continue so as to work out (or at least to clarify) the differences. In this respect we should mention the valuable work of people like Anthony Chemero, Erik Rietveld, and Marek McGann, who have provided insights into how the links between the two approaches may be better articulated (see, e.g., Chemero 2009; McGann 2014; Rietveld 2008a).

patterns of regularity between sensory and motor activity. To perceive an object is to correctly grasp that these regularities are present here and now and in one's capacity to move and in the expected changes in sensorimotor activity. The key to understand perception for the sensorimotor approach is not in the explicitation of how an agent builds an internal representation of the world, but in how it masters sensorimotor regularities.

The sensorimotor perspective puts the whole embodied agent, rather than just the brain, at the center of the story. Perception and other cognitive states are not states of the brain; they are not even in the head, but rather they emerge from the sensorimotor process itself. For even the simplest movements (moving an eye or adjusting the position of the head) involve the whole agent and are norm governed (they can be right or wrong within the context of the overall activity; they can work or fail to work). The sensorimotor contingencies approach does not, however, answer the question of how these norms arise or what distinguishes body movements from acts. It has no concept of self or agency and thus fails to explain fundamental aspects of perceptual experience. It also presupposes the presence of a body (and a subject, see Thompson [2005]), because it takes its structure and possibilities for interaction with the world as its starting point. But, despite these problems, it is one of the most articulated expressions of embodied approaches that take action as primary.

2.3 **A forking path**

[handwritten margin notes: "How to", "what are the reactions to the "reappraisal of the body"?"]

Broadly speaking there have been two different reactions to the reappraisal of the body that have taken place in cognitive science. One reaction is to maintain the core functionalist principles that informed cognitivism and to extend these ideas beyond the skull by taking instead the brain, the body, and the environment as the combined "hardware" whose job it is to sustain the fundamentally software-like functions of the mind. This attitude—embodied functionalism, see e.g. Rupert (2016)—is best illustrated by the extended-mind hypothesis (Clark and Chalmers) 1998). The paradigmatic example in this case is that of a notebook becoming, under certain conditions, a physical part of the mental state of its user. Another way of interpreting this manner of looking at embodiment is to say that we can "offload" computations onto the body and the world, freeing the brain from having to compute everything and allowing for other structures, perhaps better attuned to environmental opportunities and regularities, to process information. In short, the body and its extensions are part of the implementational resources that sustain computational functions. Other variants of embodied functionalism include approaches in which the body provides a frame for "body-formatted" neural representations (Goldman and de Vignemont 2009) or situates and "grounds" brain activity, which involves body-based simulations as opposed to amodal computations (Barsalou 2008). The extended mind hypothesis and other variants of embodied functionalism—with characteristic wit Shaun Gallagher (2015a) refers to them as *"the body snatchers"*—are elaborations and updates of cognitivism in the sense that while in practice the body matters, it still

[handwritten margin note right: "extended mind hypothesis"]

[handwritten note at bottom: "what is body snatching?"]

plays second fiddle to the algorithmic nature of cognition, which remains its undisputed essence.[4]

A different reaction to the lessons of "embodied" criticisms involves the suspicion that something more fundamental is at stake than merely a reform of functionalism. Admitting of a more central role for the active, living body is a way of questioning the supposed informational essence of cognition. It is also an attempt to undo the steps that, according to Bruner, reoriented the original Cognitive Revolution from a science of meaning toward a science of information. This alternative take on embodiment goes deeper into the nature of the body, its organismic and ecological roots, and its self-organizing properties, in order to project a new foundation for thinking about the mind. In the words of John Haugeland in the epigraph to this chapter, minds are embodied and embedded in a meaningful world in an intimate, not incidental, sense. The enactive approach is one of the perspectives that illustrates this alternative path.[5] It entails a conception of the body not as a machine, but as a precarious self-individuating network of processes involved in multiple cycles of regulation. The enactive approach also questions the basic ontological and epistemological assumptions of functionalism in ways that bear consequences for understanding meaning, norms, and experience, as well as the nature of the self and the mind.

As introduced in cognitive science by Francisco Varela, Evan Thompson, and Eleanor Rosch (1991) and in its more recent articulations (Di Paolo, Rohde, and De Jaegher 2010; Thompson 2007a), the enactive approach provides an attempt to ground cognition using categories that describe the organization of living systems. Instead of understanding cognition as a *computer-like* process, enactivism starts by considering it as a *lifelike* process anchored in the living body. Instead of being merely a surrogate for the mind, a container, vehicle, or instrument controlled by the brain, even one that could implement clever computational functions, the body—the living and the lived body, the material and the experiencing body—is conceived as the source of all that the mind is and can be.

---

[4] The majority of examples of embodied functionalism happen also to be representational approaches. We should note that it is conceivable for an embodied functionalist theory to be formulated in non-representational terms, for instance, in terms of dynamically instantiated functions (as when we describe the function of the heart as that of pumping blood). Some dynamical systems explanations in cognitive science could be construed as non-representational forms of embodied functionalism, although many of them aim instead at full operational descriptions.

[5] Several closely related approaches share this view of the living body with enactivism. They may not call themselves enactive but the convergences are more than superficial. Unfortunately, we cannot discuss all of them here with the depth they deserve. We should first mention Alicia Juarrero's dynamical explanation of intentional behavior (Juarrero 1999). She proposes that intentions involve the modulation of emergent dynamical constraints in far-from-equilibrium self-organizing systems (such as living systems). This is close to some of the ideas we will discuss in Chapter 5. Another important contribution is Anthony Chemero's development of a radically embodied and dynamical elaboration of ecological psychology, which he brings close to enactive ideas (Chemero 2009). Like enactivists, he is critical of representationalism and advocates what in this book we call a world-involving perspective.

The enactive conception of the body would be of limited interest if it were only a collection of speculative arguments. But the enactive approach has articulated these ideas in specific domains in terms that allow for them to be tested, improved, and, if necessary, rejected by scientific standards. There are currently several enactive "takes" on a variety of questions often calling for a rethinking of the basic assumptions of computationalism. Apart from the work covered in the following chapters, there are, to name only a few, enactive perspectives on emotion and affectivity (Colombetti 2014); on intersubjectivity and social cognition (De Jaegher and Di Paolo 2007, 2010); enactive hypotheses on the social brain (Di Paolo and De Jaegher 2012); on sociocognitive development (e.g., Fuchs 2012; Gallagher 2015b); on language and narrativity (Cuffari, Di Paolo, and De Jaegher 2015; Caracciolo 2012; Popova 2015); on education (Maiese 2015); on music cognition and education (Matyja and Schiavo 2013; van der Schyff 2015); on the interactive factors affecting imitation (Froese, Lenay, and Ikegami 2012); enactive, person-based approaches to autism (De Jaegher 2013); organism-based theories of color vision (Thompson, Palacios, and Varela 1992); enactive perspectives on brain function (Fuchs 2011; Gallagher et al. 2013; Varela 1995; Varela et al. 2001); on perceptual modalities (McGann 2010); on speech coordination (Cummins 2013), on mental imagery (Thompson 2007b); on schizophrenia (Kyselo 2015), on artificial intelligence and robotics (Vernon 2010; Froese and Ziemke 2009); and on neurobiological and bodily factors in prehistoric art and material culture (Froese and Ikegami 2013; Malafouris 2013). There are also enactive reflections on ethics (Varela 1999a; Urban 2014; Colombetti and Torrance 2009) and consciousness (Varela 1996; 1999b; Thompson 2007a, Thompson and Varela 2001). The epistemological attitude of the enactive view—exemplified in most of this work and throughout the book—is one of always attempting to ground the key building blocks of cognitive science in naturalistic terms.

This also applies to the concept of the body itself. According to the enactive view, the body is not a label of convenience. It is not taken for granted or thought of only along the lines inherited from studies of anatomy or social convention. Rather, the constitution of the body, its identity, is closely tied to the autonomous processes of material self-individuation that occur at different levels and that become interlinked with what the body does in the world. These processes include metabolic and other physiological processes (the activities of the nervous system, the lymphatic system, the immune system, etc.), as well as the historically situated, self-sustaining loops of interaction with the world (habits,

Robert Hanna and Michelle Maiese advance the thesis that minds are essentially embodied; by this they understand rooted in the self-organizing properties of the living organism (Hanna and Maiese 2009). Influenced by enactive thinking they elaborate a theory of action and mental causation in resonance with some of our proposals in Chapters 7 and 8. Several other authors who have looked closely at the relations between the activity of the lived/living body and the phenomenological and psychological aspects of meaning making, including its pathologies, have done so in ways that avoid representational discourse and resonate with enactive ideas. Such authors include Frederik J. J. Buytendijk (1958, 1970), Eugene Gendlin (1962/1997), Mark Johnson (2007), Helmuth Plessner (1970), Matthew Ratcliffe (2008), Richard Shusterman (2008), Maxine Sheets-Johnstone (2011), and Erwin Straus (1966).

sensorimotor scheme networks, social relations, etc.). And, in the case of human beings, these constitutive processes also include language-mediated engagements and sensitivities that literally transform the human body into a linguistic body (Cuffari, Di Paolo, and De Jaegher 2015).

According to enactivism, the body counts as a cognitive system because it is possible to deduce from processes of precarious, material self-individuation, the concept of *sense-making* (i.e., a perspective of meaning on the world invested with interest for the agent itself). With the individuation of an autonomous identity there arises (in history, but also at each moment) an intrinsic norm aimed at securing the systemic identity, a basic dimension of care, and thus a subjective viewpoint from which interactions with the world are evaluated and thus become meaningful (Di Paolo 2005; Barandiaran and Moreno 2006; Jonas 1966; Thompson 2007a, Weber and Varela 2002).

We can begin to appreciate the differences between this path and the embodied functionalist path. For enactivism, once we clarify the concept of sense-making, minds *cannot* be conceived without bodies.

It is clear then that the claim that cognition is embodied is not a sufficiently discriminating calling card (the schematic position of the different approaches mentioned can be appreciated in Figure 2.1). It does not separate two quite different interpretations of the criticisms of the cognitivist framework that we have briefly overviewed.

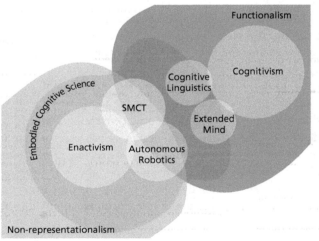

Figure 2.1 A schematic for contemporary cognitive science. The diagram depicts the relative positions of the different approaches discussed in this chapter (SMCT = sensorimotor contingencies theory). Anything lying outside the "Non-representationalism" set is by default a form of representationalism. Some versions of ecological and dynamical systems approaches may be considered functionalist in the broad sense of approaching cognition in terms of functional architectures but without appeal to full-blown representational explanations, hence the overlap between the two sets.

For embodied functionalists, it is important to note that minds are empirically implemented in bodies and this cannot be ignored since it bears important consequences for how the architecture of information processing is organized. For enactivists, minds are inherent attributes of bodies in the world, they could not happen other than as bodies in relation. Often these two interpretations get confused in part because there is indeed quite a lot they agree on, especially in terms of the rejection of abstract disembodied information processing. But what embodied functionalists and enactivists reject are in fact different things. Embodied functionalists reject what is "abstract and disembodied" about information processing. And here enactivists will agree. But functionalists gladly keep the idea that information processing is the core assumption of what is constitutively cognitive. And here enactivists disagree. Like with the way we tie our shoelaces, there is not one but two different paths for embodied cognitive science. One is less secure than the other. The sharpest contrast between these views always appears at the inevitable point when someone drops the word representation into the conversation.

2.4 **Trapped by a representational pull**

Western philosophical tradition since Descartes has been haunted by a pervasive mediational epistemology: the widespread assumption that one cannot have knowledge of what is *outside* oneself except through the ideas one has *inside* oneself. Charles Taylor describes this mediational epistemology as "an understanding of the place of mind in a world such that our only knowledge of reality comes through the representations we have formed of it within ourselves" (2006, p. 26; see also Dreyfus and Taylor [2015] for an expanded development of this idea). No matter how many transformations and refinements the notion of a representation as the guarantor of access to knowledge has undergone, its mediational structure remains intact. "Knowledge of things outside the mind/agent/organism only comes about through certain surface conditions, mental images, or conceptual schemes within the mind/agent/organism. The input is combined, computed over, or structured by the mind to construct a view of what lies outside" (Taylor 2006, p. 27). One cannot escape this pervasive image simply be declaring it incorrect or unsupported by facts. It has the status of a framework, of a common sense. The pull of representations is like the gravitational pull that backdrops our everyday activities. It has provided "the necessary irreplaceable context for all thinking about these matters, hence not something one would ever need to examine. In this way, it worked insidiously and powerfully" (Taylor 2006, p. 28).

In the sciences of the mind, the overarching mediational epistemology takes the shape of an inescapable appeal to internal representations. Few ideas have caused as much controversy as this one. Initially aligned with the concept of mental models introduced by Kenneth Craik (1943) and Edward C. Tolman (1948) as a resource for vicarious activity, the notion of internal representations has been articulated in many, sometimes contradictory manners. It has significantly flourished with the development of the information processing metaphor and the rise of computationalism (e.g., Newell and Simon 1976; Dretske 1981; Fodor 1981; Marr 1982). Its use continues to be widespread within neuroscience, cognitive psychology, cognitive science, artificial intelligence, and robotics. Yet, from phenomenology to robotics, the idea of internal representations has been criticized since its origins.

It is not our purpose to review these criticisms in detail. We shall only summarize some salient and well-known points. To begin with, we must remark on the operational[6] ambiguity of the term "representation." The word is used in wildly different manners ranging from the philosophical sense of intentional aboutness to just that of an "internal state," from a "stand-in" for a state of affairs to a mere "correlation" with it. In practice, the circulation of the word representation is as self-validating as the circulation of a subprime mortgage. It is used, passed around, accepted, and re-packaged with the tacit agreement that nobody should ask the embarrassing question of how to cash it out.

At the intersection of several of its many meanings, there is the idea that an internal representation is what carries cognitive *content*. This is its functional role. Yet few people attempt to spell out this strange power invested in neural activity, or even try to express it in naturalized terms. In fact, it is difficult to pin down what the idea of content adds to the concept of statistical correlation. And once correlations are explained it can be argued that the notion of content provides no additional explanatory value (Hutto and Myin 2013). For some, representations would then be explanatorily epiphenomenal once cognition is naturalized (Keijzer 2001; Calvo 2008). If that is the case, each single use of the term belies how much we actually ignore about cognition.

Even when representations are well defined in the case of specific models (say, as probability density functions describing the likelihood of a sensorimotor event given a current sensorimotor state) or correlated with precise brain activity or regions, representationalism faces the problem of homuncularity: the need to postulate an internal agent that interprets and uses representations. As a result, representations, instead of explaining cognition, simply displace the explanatory burden further up the "processing hierarchy," since to interpret a representation is itself a cognitive act. This is inherent in the grammar of the term. A representation not only establishes a relation between two entities (the represented and the representing) but it *always* includes at least one subject who bears

---

6 Throughout this book, the term *operational* is used in two senses, neither of which corresponds exactly to the meaning given to this term in experimental psychology. The latter is merely a synonym of *measurable*. In our case, one meaning of the term refers to the possibility of a noncircular grounding for a given concept (this is almost contrary to some misuses of the term in psychology, which can sometimes be circular as in, e.g., intelligence is what is measured by intelligence tests). Typically, an operational definition is one that is grounded naturalistically (i.e., it involves a description of the multiple elements that constitute a particular concept, and these elements are themselves grounded operationally). A consequence of having a naturalistic grounding is that in principle at least, an operationally defined concept will also lead to ways of measuring a particular phenomenon, though not necessarily straightforwardly. A different sense of operationality, used in some of the following chapters, is applied to processes, whose unravelling over a course of time is described as their *operation*. The two different senses should be clear from the context.

the burden of interpreting this relation and one subject (who could be the same) that establishes it.[7]

The problem of homuncular regress is manifested in the ambiguous stance of representational explanations with respect to the personal/subpersonal divide. Literally speaking the activity of representing is a personal level activity, and like many other mereological fallacies (Bennett and Hacker 2003), it becomes mysteriously transformed into the activity of subpersonal mechanisms (along with "predictions," "simulations," and "inferences," to name a few) in a confusion of *explanans* and *explanandum* that can sometimes be kneaded into the shape of an explanation, provided nobody looks too closely.[8]

As we have said, for embodied functionalism the notion of representation is not only acceptable despite its problems, it is an essential part of the framework: cognition *is* information processing that takes place as the manipulation of representations, whether within a skull or extended into notebooks. To be fair to the complexity of the embodied turn in functionalism, the concept of representation has been transformed in various ways that incorporate the dynamic, situated, and pragmatic aspects of embodied criticism. For instance, Wheeler (2005) acknowledges that some embodied-embedded activity in situated cognitive agents may not require any kind of representational explanation. But, like others (e.g., Clark 1996, 1997; Dennett 1993) he sees the dynamical-coupling sort of explanations typical of autonomous and evolutionary robotics as fundamentally limited. According to Daniel Dennett's review of Varela, Thompson, and Rosch (1991): "The trouble is that once we try to extend Brooks' [1991] interesting and important message

---

[7] Many have said similar things concerning the (mis)use of representational language, but someone who, over the years has pointed this out, explicitly, repeatedly, and vehemently to the authors of this book has been evolutionary and adaptive systems researcher Inman Harvey at the University of Sussex (see e.g., Harvey 2008).

[8] Homuncular explanations are not a problem in themselves; they can help us understand complex systems in familiar terms and regulate our dealings with them. When a car mechanic tells the driver that the car engine "does not like" his jerky driving style or that the broken thermostat keeps "telling the automatic choke that the engine is hot enough" when it is not (Wheeler 2005, p. 256), he is using homuncular explanations. The car engine and its parts are given the attributes of agents, and the explanation is helpful precisely because there is no chance of misunderstanding these parts as real agents. The problem comes when this kind of "as-if" narrative is adopted as the systematic basis for a general scientific framework. The situation only gets worse if homuncular stories are meant to explain agency itself. The assumed solution to the problem of infinite regress is to declare that a regress may exist but that it is finite: at some sufficiently low level, an agent-like, homuncular description will be like that of the car engine, merely a metaphor, (see the discussion in Wheeler [2005]). The (slightly perverse) strategy relies on using the everyday intuitive understanding of what an agent is or is not (grammatically sustained by the use of certain intentional vocabulary) to break down "real" agency into "quasi-agent" components. However, what makes these components agent-like or not is the same intuitive understanding. In other words, there is no chance that an explanatory strategy of this kind would ever make us question or refine our intuitions about agency because these are adopted without examination. An alternative approach to agency will be introduced in Chapters 5 and 6, which does not face such problems of circularity.

beyond the simplest of critters (artificial or biological), we can be quite sure that something awfully like representation is going to *have* to creep in like the tide, in large waves," (Dennett 1993, p. 126).[9] One wonders where this certainty comes from, especially since representationalist explanations of the human mind have so far been found wanting. Clark comes close to making it a matter of definition by following Haugeland's (1991) first requirement for a system to be considered representational, viz, that it must behave in coordination with aspects of the environment that are "not always reliably present" (Clark 1997, p. 144; see also discussion in Chemero [2009]). The reasoning goes something like this: As soon as the "target" that is tracked by cognitive activity is not available in the dynamical coupling—as when, say, riding on the bus I make plans for an important work meeting later that day—a representational story must come back into the picture because I must somehow operate on information which is not in my coupling with the environment (we come back to this sort of argument by default in Chapter 8). From an embodied functionalist perspective, however, this need not be the kind of representational story that ignores bodily constraints or attempts to mirror agent-independent states of the world. The story could well involve, Wheeler suggests, action-oriented representations (i.e., representations framed in egocentric, body-relevant coordinates that are context- and activity-dependent). Modernized or not, representations are therefore necessary for a complete embodied functionalist approach even if it is admissible that some performances by cognitive agents do not require them.

The enactive approach, in contrast, is non-representational. This is not only because the criticisms of representationalism are too many and too serious to be ignored, but also because representationalism is not the only game in town (as the rest of the chapters in this book will attempt to show) and so argumentations by default ("there is no alternative") do not work (see, also, Haugeland [1995] for plausible shapes non-representational explanations can take). The enactive approach seeks to provide a solid conceptual grounding for the key terms of the sciences of the mind (meaning, agency, identity, interaction, environment, norms, action, perception, etc.). For many of the reasons already mentioned (phenomenological dissonance, lack of operationality, mix-ups between personal and subpersonal levels, and homuncular vicious circles) a naturalistic conceptual grounding is not possible for the notion of internal representation. Instead, enactivists see agents as making sense of their environment by coupling precarious processes of self-individuation (at different levels) with environmental dynamics. These processes are imbued with value and sensitivities to the potential effects on the maintenance of the agent's identity. The concept of autonomy at the heart of this activity cannot be approached in functionalist terms because it relies on the precarious metastability of constitutive processes, a negative

---

[9] As Dennett (2013) himself recommends, one should always check for the "surely operator" in order to spot a weak point in an argument—that moment when, instead of evidence or a demonstration supporting a point, an author offers only a bald statement. Surely, in Dennett's own statement about the inevitability of representations "we can be quite sure that" must count as "surely."

aspect of materiality that puts limits to the validity of positive functional descriptions (Di Paolo 2009).

This does not mean that functional descriptions should be discarded altogether (enactivists frequently use them), but that they are inherently limited to contingent conditions and so cannot be the basis for a general framework for understanding the mind. Any functional description relies on certain stationarity of background conditions. For instance, sensory activity can be considered as "informative" provided it statistically stands out against a background distribution of sensor states. But in order to say this we must make some assumptions about expectations, probability distributions of states must be known, remain unchanged or change lawfully, and so on. Provided these conditions are stable enough, a specfic functional-informational description can work. For example, one can estimate the information-transfer capacity of the optic nerve but only under assumed known distributions of neurophysiological states and stimulation statistics. However, outside the lab the conditions that grant validity to such descriptions are in constant flux. Indeed, they change in particular ways as the agent moves from one activity to the next, or as activities face breakdowns. Enactivism is concerned with explaining precisely these critical transitions between particular conditions that sometimes afford different functional descriptions and those "in-between" dynamics that (re)constitute these or novel conditions. By definition, those in-between, open-ended transitions cannot be described in informational terms, or they would not be open-ended. They lie exactly at the blind spot of functionalism.

In the enactive view, therefore, meaningful, cognitive activity does not depend on' vehicles storing information, but on the coordination of dynamic processes at various scales by an autonomous agent. Internal representations (in the strong sense of internal states bearing cognitive content) are therefore rejected by the enactive approach. The advantage of this rejection is that the notion of *representing*, as a kind of cognitive activity, demands an explanation, although not in terms of elements that are themselves representational as it is a good idea to avoid mixing *explanans* and *explanandum*. And this is the reason why enactivism is *non*-representational instead of *anti*-representational: representing as a family of complex, late-arriving, and mostly socially mediated cognitive activities does exist and deserves to be explained, but not in terms of other representations.

In fact, the pull of representationalism, the sustained pervasiveness of the mediational epistemology, may in part be attributable to the ubiquitous amount of and symbols that populate our habitat. Behind each of these images and symbols lie acts of producing representations and interpreting them. In fact, representations are nothing but the reification of these acts. There is something inherently familiar about the act of representing. We do this all the time. We invoke the features, voice, and attitude of a friend when she is away; we describe events to other people who were not there to witness them for themselves; we make use of maps, shopping lists, and books. Early on, children engage in pretend play. And as adults we continuously play infinite variations of the representing game. Moreover, we live surrounded by increasingly networked informational gadgets. We copy and paste chunks of information, images, text, and sounds, we carry them around in storage devices or upload them to clouds, we download, share, and capture information on smartphones, digital cameras, and so on. Perhaps this game is so inescapable that it is difficult to conceive the whole of cognition if not as a variation of it.

But do these acts of producing and interpreting representations imply that something of the same mediational nature must be happening inside our heads, inside our bodies?

This question is often answered in the affirmative by invoking imagination and such "higher mental" capacities as sufficient evidence for the presence of internal representations. Planning, strategizing, thinking, imagining a tree all seem to imply that I need to have at my disposal a stock of objects to manipulate (schedules, maps, propositions, images) and to represent the world and its possibilities. But if I do not have such objects at hand, how is it that I can still (to some extent) perform these activities? If I can imagine a tree, so the intuition goes, surely I need to have an internal image of the tree, perhaps like an internal photograph stored in an appropriate format somewhere in my brain. Well ... maybe not.

There is nothing in these cases like going back to the experience itself. Take your time and imagine you are seeing a tree. If you pay attention to how you do this you may realize that a clear, richly detailed image of a tree does not simply pop up in your mind. It is something you have to achieve. Some people have more difficulty than others in visual imagination. Perhaps, instead of immediately and vividly visualizing a tree, you are actually wondering about where this line of argumentation is going or are being distracted by noises or the weight of the book in your hands. So first you have to clear some "mental space" before you start seeing something like a tree. Is it a full tree seen from afar, or a trunk climbing next to you and lots of thick branches seen from below? Why one and not the other? Why does it have a point of view at all? At this point, your body is quite active fending off distracting stimuli and you might even feel the saccadic eye movements that you enact when you imagine observing the tree from different vantage points. Does the experience resemble that of looking at an image as when you look at real photograph of a tree? Instead of activating some internal image-like neural pattern stored in your brain and then "looking at it," you are closer to acting or, as in a theatre play, *enacting* the visual experience as if the tree were manifested at a possible moment of viewing it. It is an *act* of presenting something to yourself again, or *re*-presenting. "Imagery experience is not a species of picture-viewing. In visual imagining, one apprehends an object not by means of a phenomenal mental picture, but by re-presenting that object as given to a possible perceptual experience" (Thompson 2007b, p. 408). The same goes for thinking about how you tie your shoelaces: you almost need to position yourself in the right manner to start imagining it. You probably cannot imagine the middle steps of the process on their own, without going through the whole thing from the start. It is hard. Revealingly, much harder than simply doing it. You are performing to yourself, for yourself, enacting, acting *as if*. This may even involve many of the elements that confront a real actor in an actual theatre play, including the potential spectators and the norms of a good performing style.

The question that arises when paying careful attention to the experience of concrete thinking or concrete imagination is not how one internalizes a token that stands in for something out there, but rather, how is it possible to act upon something, to incorporate something into your acts as when using a tool, or to pretend play together with somebody, so that these acts, or similar ones, may still occur in the absence of people and things. Something even more fundamental needs to be clarified in advance: what is it *to act* in the first place, how to distinguish action from other physical happenings or an "incidental" movement of a limb. Surprisingly, we are much more familiar with the structure of an object than with the structure of an act. Yet, the object and its properties (three-dimensional shape, rigidity, color, etc.) are perceptual and cognitive achievements in themselves. It takes hundreds of microscopic and macroscopic acts to generate stable units of perception, to elaborate and constitute a perceived object. Moreover it takes years of socially mediated training to discriminate, name, combine, and abstract those objects and their properties, eventually to be able to represent them, describe them, draw them on a piece of paper, and so on. How can all or most cognitive phenomena be explained by appealing to representational capacities that presuppose and demand, before it is even possible to approach the act of representing, so many other prior cognitive capacities?

We still need to provide proper answers to these questions from an enactive perspective. However, one thing we can say is that the concrete embodied experience of representing undermines the functionalist argumentation by default in favor of representations. One cannot claim that there is no alternative to explain cognition when representationalism itself is not an alternative in the first place.

2.5 The sensorimotor approach as playing field (or battleground)

It remains to be seen whether the enactive approach is able to provide an alternative to representationalism as an explanatory framework. This book is intended as a contribution in the affirmative to this question, but it will not settle the matter once and for all. The strategy is to start asking more systematically some of the questions we have raised in the previous section and apply enactive concepts, models, and ways of thinking to a new embodied account of action and perception.

Some groundwork has already been laid down by the sensorimotor approach to perceptual experience (O'Regan and Noë 2001; Noë 2004). Properly interpreted (and we will need to spell out what that means) this is a good locus to move beyond the criticism of representationalism and into concrete positive proposals for an alternative account. The sensorimotor approach provides us then with a suitable frame or playing field for testing various non-representational proposals. But the resulting enactive story developed throughout the book will be broader than the sensorimotor approach to perceptual experience and will connect with other embodied theories of action and perception and their development.

2.5.1 **A sensorimotor approach without representations?**

In contrast to embodied functionalism and enactivism, the attitude toward representationalism implied in sensorimotor contingency theory is ambiguous. Ostensibly, this approach claims to categorically reject the need for internal representations. In their emphasis on the world as an outside memory, O'Regan and Noë propose that "visual experience does not arise because an internal representation of the world is activated in some brain area. On the contrary, visual experience is a mode of activity involving practical knowledge about currently possible behaviours and associated sensory consequences. Visual experience rests on know-how, the possession of skills" (O'Regan and Noë 2001, p. 946). They claim that all that is needed to explain perceptual experiences are methods for probing the three-dimensional world itself, methods that involve a mastery of sensorimotor contingencies. In this view, the sensorimotor approach presents itself as consisting of the positive claim that perceiving consists in the exercise of this mastery, and the negative claim that representations are unnecessary (at least for perception).

However, the sensorimotor approach is as easily trapped by the representational pull as other embodied functionalist theories. O'Regan and Noë seem to accept that the visual system extracts, stores, and categorizes information about the environment in one form or another, and makes use of it to influence current or future behavior. Also, mastery of the laws of sensorimotor contingency is itself a form of knowledge that presumably would need to be stored somehow. The authors are happy to label such stored information representations, in apparent conflict with their claim that there are no representations of the world in the brain (O'Regan and Noë 2001, pp. 950, 1017). However, a closer look reveals that the claim is in fact not that representations do not exist, but rather that it is not the brain's role in vision to recreate a detailed, pictorial, three-dimensional representation of the world from retinal images. We do not have such representations, nor need them to explain conscious experience.

In support of this claim, O'Regan and Noë discuss the phenomenology of change blindness, perceptual completion, visual inversion, and color perception as showing that we do not have (access to) detailed internal representations of the world. Regarding the necessity for such representations, according to sensorimotor contingencies theory, we can have "flawless, unified and continuous experience" without having to build internal representations with the same qualities from distorted retinal images that are punctured by blind spots and interrupted by saccades. This is because seeing does not consist in some part of the brain reflecting upon an internally represented image, the suggestion of which would be to commit the homuncular fallacy. Seeing is rather an act in the visual modality, such as visual exploration of one's environment.

Though in essence a sensori*motor* account of perceptual experience, the approach does not claim that all experiencing depends on action. Rather, action is necessary to acquire knowledge of the ways movement affects sensory stimulation. Making use of this knowledge then is required for experience, but not action itself. This implies that neural activity may sometimes be sufficient for some forms of experience (dreaming, mental imagery). Similarly, though dependent on capacities for action, it is not the claim of this approach that seeing, that is, exercising visual sensorimotor contingencies, is always *for* the guidance of action. Hence, the sensorimotor approach is compatible with the idea that sometimes what we see does not guide our action directly, but might instead impinge upon our desires or beliefs (e.g., when looking at art in a gallery). This explanation may bring some coherence to the approach; however, it does so by putting the approach strictly within the representationalist camp and through a "there-is-no-alternative" argumentation by default, no less: "our view relies on the existence of representations. Knowledge of the laws of sensorimotor contingency themselves *must surely* be represented. We readily grant this" (O'Regan and Noë 2001, p. 1017, our emphasis).

In short, many of the claims of the sensorimotor approach are easily subsumed by a representationalist interpretation, which is arguably the one the authors seem to hold themselves. But does it have to be like this? Some researchers claim that the only way the claims of the approach can gain coherence is by going in the opposite direction (i.e., by abandoning representationalist interpretations and by bringing the sensorimotor approach closer to enactivism).

Making use of Gilbert Ryle's (1949) distinction between knowing-that and knowing-how, Daniel Hutto (2005) sees a missed opportunity for the sensorimotor approach to break free of representationalist orthodoxy. Given the reliance on practical engagements with the world, the sensorimotor approach does not need to appeal to notions of mastery of sensorimotor contingencies if what is meant by this is some sort of stored knowledge in the brain. For a body (not just a brain) to master the regularities of its engagements with the world all that is needed is that it is able to operate in the world in a robust, adaptive, skilful manner. This may take the form of plastic equilibrations in the face of perturbations and uncertainties (see Chapter 4) the consequence of which, on the body side, is the simple retuning of operational parameters (muscle tone, tendon elasticity, timing and coordination of moves, coherence of neural oscillations, etc.). This process of attunement and its results do not even need to be reflected as structural/functional changes in the brain at all, let alone take the shape of represented knowledge.

The lesson to be drawn from Hutto's criticisms is that enactivists would do well in better explicating Ryle's concept of know-how. In fact, one of the reasons that the claims of the sensorimotor approach can be interpreted in representationalist terms is that several of its key concepts lack operational specificity. The idea itself of sensorimotor contingencies needs some clarification (which we aim to provide in Chapter 3). And the notion of mastery may allow the kind of non-representational interpretation that Hutto advocates not only in some, but in all possible cases, yet this still needs to be properly articulated. (Chapter 4 contains our own attempt.)

The same is the case regarding Thompson's (2005) worries about the sensorimotor approach lacking a proper account of agency and selfhood. For him, while the approach can offer a rich account of several properties of perceptual experience, it does not account for its subjective aspects, such as pre-reflective bodily self-consciousness or the sense of agency (these worries will be the topic of Chapters 5, 6, and 7).

Not having concrete, positive proposals to address these matters makes sensorimotor contingencies theory an easy target for the there-is-no-alternative crowd. Ironically, some of these gaps are also challenging for embodied functionalism. So it is not as if turning to representationalism will solve the open problems of the sensorimotor approach.

Let us overview some of the areas where the sensorimotor approach would need operational specificity, as well as some of the ideas that could help us provide an enactive interpretation.

2.5.2 Sensorimotor contingencies, what are they?

If not representations, what could be the appropriate unit of analysis for an enactive interpretation of the sensorimotor approach? We could consider in the first instance the notion of sensorimotor contingencies itself. This idea (and related ones: sensorimotor coordination, regularities, etc.) is not meant to describe a way in which the agent internalizes information about the environment. Instead, it takes the raw and quantifiable variation of sensory and motor surfaces of the organism as a departure point. The primary correlation that is available to an agent is the manner in which the sensory stream changes as a function of its own actual movement and its possibilities and dispositions for movement. Moreover, what the scientific observer first gets to grasp, analyze, and infer from recorded data about the behavior of an agent is that some coordination is taking place, that different kinds of coordination follow each other, and in between there are periods of coordination breakdown, metastability, and openness. These coordinations can be bodily coordinations, coordinations between environmental processes and the agent, or between agents in interaction. Importantly, taking sensorimotor contingencies as a basic notion compels us to always understand the environment from the perspective of the agent and not as a set of objective properties available to the external observer. Sensorimotor contingencies are always the result of a "dialog" between agent and environment and are therefore neither "subjective" nor "objective" in the traditional senses of these terms. From this perspective, agency is about enacting effective sensorimotor relations according to the agent's norms and goals. These are the relations that the agent helps create and which are immediately available to it. Nothing so far in this picture is required to undergo a sort of outer-to-inner transformation and recording as demanded by the picture of mediational epistemology.

There are many important precursors to the idea of perception as dependent on the mastery of sensorimotor contingencies in the history of psychology (see Scheerer [1984] for a more general historical review of motor theories of perception and cognition). Helmholtz (1867) provided one of the earliest scientific expressions of the idea that perception is underpinned by "unconscious inferences" that take into account the regularities of muscle activity and the changes in sensory stimulation they induce. Thus, the visual perception of the world can be relatively stable in spite of the constant, ongoing movements of the retina, because their effects are compensated by a kind of reafference principle (von Holst and Mittelstaedt 1950; Scheerer 1984). Versions of this idea underwent various elaborations and refinements in the second half of the 20th century, often expressed in terms of constraints and contingent expectations as an agent samples a visual scene, rather than inferences (e.g., Hochberg 1968; MacKay 1962; Rock 1983).

But there is also another sense in which sensorimotor activity can show regularities. And that is in how different possible actions solicited by the current situation interact and influence each other, even if they are not enacted. In the early 20th century Margaret Floy Washburn (the first woman to obtain a PhD in psychology) expressed this idea in her theory of how tentative movements and incipient motor responses (muscle group activations associated with an action without actually proceeding to a full enactment) influence perceptual consciousness and other mental processes (Washburn 1914, 1916). Attention for Washburn was itself a motor process, and sensations can be discriminated not only because they involve action on particular sense organs (modalities), but also because they activate motor responses that are peculiar to a particular stimulus or situation. These need not be enacted, but they nevertheless influence how a perceiver perceives and acts. This perspective allowed Washburn to interpret Gestalt phenomena in sensorimotor terms: perceptual forms and groupings are organized in relation to the action possibilities they solicit (Washburn 1924).

In his 1932 essay on *The Physical Thing*, George H. Mead offers a similar description of action and perception in terms of the organization of sensorimotor couplings, tendencies to act, and the inhibitions of some of these tendencies, which, precisely because they are actively held back, influence the way behavior is enacted. In many ways Mead and Washburn prefigure the subtler points of the sensorimotor approach.

> … the resistance which the volume of a body offers to the hand, or to any surface of the body, and the tendencies to manipulate it when seen at a distance, are organized in various ways. There is, for example, the tendency to pick up a book on a distant table…. My thesis is that the inhibited contact responses in the distance experience constitute the meaning of the resistance of the physical object. They are, in the first place, in opposition to the responses actually innervated or in prospect of being innervated. They are competitors for the field of response. They also within the whole act fix the conditions of the actual response…. If I see a distant book an indefinite number of manipulatory responses are aroused, such as grasping it in a number of ways, opening, tearing its leaves, pressing upon it, rubbing it, and a host of others. One, picking up the book, is prepotent and organizes the whole act. It therefore inhibits all others. The tendencies to perform these others involve the same resistance of manipulation, and are now in direct opposition to the prepotent response; but while in opposition they provide the conditions for the exercise of the prepotent response. The feel of the book if one rubbed it, the contours if one passed one's hands about it, the possibility of opening the book, etc., determine the form that the grasping and lifting up of the book will take. In general what one does not do to the book, in so far as this calls out the same resistance as that given in actually manipulating the book, and in so far as it is inhibited by what one does do to the book, occupies in the experience the "what the book is" over against the response which is the expression of the act. Inhibition here does not connote bare nonexistence of these responses, for they react back upon the prepotent response to determine its form and nature…. The act is a moving balance

within which many responses play in and out of the prepotent response. What is not done acts in continual definition of what is done.

(Mead 1932, pp. 127–8)

It seems, therefore, that there exist many levels in which sensorimotor contingencies are mastered, from lawful correlations to interactions between actual and potential sensorimotor flows. These possibilities demand some clarification. Surprisingly little effort, however, has been directed toward formalizing the notion of sensorimotor contingencies in detail. O'Regan and Noë, and other colleagues, provide various examples that serve to pin down the notion intuitively, but not to fully define it. Some cases involve *general* regularities in the mapping from motor activity to sensory activity like the transformations undergone by the projection of a horizontal line on the retina as the eyes move in the vertical or in the horizontal directions. Other cases involve *particular* regularities created by specific, agent-controlled activity, as when the softness of a sponge is perceived by pressing it a few times between the fingers or the smoothness of a table is perceived as we slide the hand across its surface.

These cases are not typically distinguished. Sensorimotor contingencies often describe the manner in which sensory stimulation changes as result of motor movement, (i.e., the structure of the mapping M→S). But it is also the case that the agent guides its movements in relation to sensations and in a sense that it "chooses" to engage in specific forms of S→M. This results, in a circular manner, in specific agent-driven M↔S regularities. In this way, one thing is the existence of general invariant or lawful relations for a given perceptual modality. For instance, opaque objects will occlude each other when light is sensed from a small, point-like sensor in space (any animal eye). A different thing is the specific enactment of sensorimotor patterns for a given sensorimotor embodiment. For instance, the visual system in several species of crabs distinguishes between two broad stimulus categories, those on the horizon (typically other crabs) and those above the horizon (typically potential predators). And the particular use of any of these regularities by a given agent according to context and motivation, and the mutual influence between various viable sensorimotor possibilities as described by Mead, is yet another thing. All of them, at one point or another, have been called sensorimotor regularities. Attempting to disentangle these various meanings is the goal of Chapter 3.

2.5.3 Mastery and world-involvement

As we have said, the concept of mastery of the laws of sensorimotor contingencies is one of the key elements of the sensorimotor approach, and one that invites representational interpretations even by its main proponents. There are two broad ways of interpreting the idea of mastery: as conceiving mastery as *in the head* or as *not just in the head*. We will refer to the latter kind throughout the book as *world-involving* perspectives. In-the-head approaches claim that mastery can be sufficiently pinned down to states in an agent's functional architecture. They are representational, whether the relevant computations are instantiated inside a skull or not, *pace* the label. World-involving interpretations, in contrast, claim that

What are representational accounts of mastery deemed inevitable by the in-the-head... [handwritten marginalia at top]

processes both in the agent and in the environment are constitutive of mastery and they make appeals to non-representational forms of know-how to account for it.

Representational accounts of mastery are deemed inevitable by in-the-head approaches because they accept the functionalist partition between agent and environment as two systems that affect each other fundamentally via informational inputs and outputs. For this separation to work, the assumption of near-decomposability (Simon 1969) must be applicable. This assumption implies that inputs received from the environment do not fundamentally alter how the agent's internal functions operate. From this perspective, mastery can only be a functional evaluation and regulation of the relation between input and output streams. And since the operation of computational processes in the agent (and even just in the brain) is assumed to be independent of the environmental input, the only way that mastery can be achieved is by fine-tuning (within pregiven parameters) an internal functional representation within a set of pre-existing possibilities (typically expressed by learning rules). Such is the case, for instance, with Anil Seth's (2014) account of sensorimotor mastery in terms of hierarchically organized predictive models where the world only contributes to an error-driven update of probability distributions. A model like this remains entirely *in the head* of the agent as do other versions of predictive models (see Di Paolo 2014; Froese 2014; Froese and Ikegami 2013).

What does empirical evidence suggest? [handwritten marginalia in left margin]

Contrary to this, empirical evidence suggests that across a range of timescales neurophysiological, sensorimotor, and environmental processes form a dynamic entanglement, especially at the moments when agents are involved in complex activities. This evidence negates the near-decomposability assumption and suggests that the relation between the agent's internal dynamics and the dynamics of the environment cannot be fully captured by the input/output metaphor (see, e.g. Aguilera et al. [2013] for a demonstration of this claim). This is in line with calls to investigate the braided coordination of neural, behavioral, and social processes in social neuroscience (Dumas, Kelso, and Nadel 2014; Hari and Kujala 2009). It also coheres with cumulative evidence of the brain-body as an *interaction-dominant* system (the opposite of a nearly decomposable one), based on findings of correlations of neural and behavioral variability across a wide range of timescales (Kelso et al. 2013; Van Orden, Kloos, and Wallot 2003). Interaction-dominant systems are characterized by the causal inextricability of the various processes involved, as well as the unpredictability of the behavior of the whole from knowledge of the parts in isolation. Evidence of interaction-dominance has also been found to involve extraneural factors: for example, in agent–tool systems (Dotov, Nie, and Chemero 2010) and during social interaction (e.g., Bedia et al. 2014; Shockley, Santana, and Fowler 2003; Riley et al. 2011).

This evidence suggests that instead of an in-the-head account of mastery we need a world-involving one. In such an account, the environment is not an input to a perceptual system, but is fundamentally involved in the realization of meaningful sensorimotor performance. Jumping on a springboard alters how high we can reach and playing in a swimming pool removes the need for bipedal balancing; being underwater reduces our visual acuity, slows downs several movements, and frees up others. Processes in the world contextually co-create the possibilities for action and perception together with the agent, and play a central role in their actualization, their success or lack of it. No amount of computational input can ever generate the fundamental material conditions for enaction, nor play constitutive roles in it, as the coupling between agent and environment does. Throughout development, the agent-environment interaction shapes both the agent and its surroundings and can do so beyond pregiven learning rules and parameter tuning.

In consequence, a world-involving conception of mastery sees it not as a functional representation of the relations between sensory and motor activities, but as an emergent property of a whole embodied agent in interaction with the environment. It may involve the diversification of sensorimotor schemes, the appearance of totally new skills, new functional spaces to explore, and the ongoing integration of the old and the novel into bundles of concrete, embodied activities. In bodily terms, mastery may take the form of loosely assembled synergies between muscle groups attuned to the current situation of the agent (Bernstein 1967; Latash 2008; Turvey 1990; Turvey and Carello 1996). An example of bodily mastery is the capability of compensating interaction torques in multijoint movements. For instance, when we move our arms activating both the elbow and shoulder joints, each of these joint movements influences the other in complex nonlinear ways that depend on the direction and body orientation. Yet we do actively and contextually compensate for these nonlinear deviations from a desired trajectory. It can be shown that no internal model is required for this compensation and that, instead, it is sufficient for simple descending signals to specify the desired movement kinematics, while spinal feedback mechanisms are responsible for the appropriate creation and coordination of dynamic muscle forces (Buhrmann and Di Paolo 2014a).

Attunements of this kind are typically quick, robust, and unconscious. They may be manifested by the appearance of poised critical states of neural and bodily activity that permit a fast, but still adaptive, response to environmental perturbations. Thus, cases of ultrafast cognition in go/no-go tasks (Wallot and Van Order 2012) show precisely that the mastery of a sensorimotor situation cannot involve internal functional calculations as the agent's brain simply has no physical time to implement any. Yet these responses are quickly adaptable and not automatic as shown when a perturbation of the jaw is induced at the moment that a phoneme is pronounced (Kelso et al. 1984). Some of these cases may involve a kind of anticipation but of a non-representational sort, or strong anticipation according to Stepp and Turvey (2010), which involves a whole system/environment attunement. The body-in-action as a whole possesses mastery in these cases, and this property can be measured for instance by investigating long-term correlations in sensorimotor variability. Notably, skilful, fluent performance is strongly consistent with certain distributions of correlations across timescales, typically described as 1/f or pink-noise. This is found, for instance, in tool use (Dotov, Nie, and Chemero 2010), reading fluency (Wallot and Van Order 2012; Wijnants et al. 2012), precision aiming (Wijnants et al. 2009), and the maturation of gait in children (Hausdorff et al. 1999)—see Van Orden et al. (2011) for further discussion.

In Chapter 4, we will elaborate a dynamical systems interpretation of Piaget's theory of sensorimotor development that does not assume near-decomposability (what distinguishes the agent from the environment is discussed below and further elaborated in Chapters 5 and 6). This interpretation builds on our dynamical concepts of sensorimotor contingencies and relates to other approaches, such as Esther Thelen and Linda Smith's (1994) framework for the embodied dynamics of development. Mastery is the ongoing process by which an agent continuously adapts to the challenges of a changing world. In our proposal mastery consists in the refining and acquiring of new sensorimotor responses and their integration into an existing repertoire. It is not the storage of information resulting from the accumulated evidence or facts about one's sensorimotor regularities. It is instead the condition of a sensorimotor repertoire that is in equilibrium with the agent's world. This condition results from concrete enactments and ongoing attunements. Mastery corresponds to the agent having the right response at hand and her sensorimotor schemes being internally and relationally coherent.

2.5.4 Agency, subjectivity, and self

The sensorimotor approach proposes that perceiving is a form of acting: I visually perceive, for example, when I make use of my practical knowledge of the sensorimotor regularities involved in exploring the world with my eyes. But although the theory emphasizes the act as primary, it does not provide an explanation of what distinguishes acts from other movements. As a result, it also lacks a principled approach to determine if and how robots and other artifacts making use of "sensorimotor" regularities in their basic functioning may be considered perceivers, and if so, how they differ from, say, humans in this respect.

In everyday use we consider acts or actions to be performed by agents in their own interest. In other words, they are movements performed by an agent according to his or her own norms (which does not preclude that they may also adopt and follow external norms as well). We do not usually consider machines to be agents, nor their movements to be actions. Does that mean that such systems should not be considered capable of perceiving either? The sensorimotor approach itself has nothing to say about agency, or the origin of the norms that govern an agent's behavior. The theory is thus unable also to describe what kinds of systems are potential subjects of experiences. As Evan Thompson points out, the sensorimotor approach addresses the question of how experiences in different modalities differ from each other but it does not explain why our activities are accompanied by any experience at all, nor does it account for the subjective, first-personal character of such experiences. In order to fill this gap, according to Thompson, the theory "needs to be underwritten by an enactive account of selfhood or agency in terms of autonomous systems. Second, it needs to enrich its account of subjectivity to include prereflective bodily self-consciousness" (Thompson 2005, p. 417).

In Chapters 5 to 7, we aim to do exactly this. Chapter 5 introduces an enactive and non-representational approach to agency that seeks to define in operational terms the organizational properties required of a system to be considered an agent. We propose that living agents are systems that produce and distinguish themselves from their environment and adaptively regulate their interactions with the world. Systems of this kind evaluate their interactions in terms of what is good or bad for their continued, active self-maintenance, and this, we argue, is a precondition for having a subjective outlook on the world. The idea is closely related to other interesting proposals for naturalizing agency in the self-sustaining organization of living systems and self-organized processes (Kauffman 2003; Moreno and Etxeberria 2005; Barandiaran 2008; Barandiaran, Di Paolo, and Rohde 2009; Skewes and Hooker 2009; Kelso 2016). Recent overviews of these proposals can be found in Moreno and Mossio (2015) and Rosslenbroich (2014). Perhaps because of the complexities involved, these ideas have yet to become mainstream in cognitive science.

But how exactly can an explanation in terms of self-organization explain action, and more generally cognition? An advance on the dilemma of mental causation (how can a mental state called "intention" be the cause of an action) can be made as soon as we stop thinking of intentions as causes of the billiard-ball kind. Instead, by expanding the vocabulary using concepts from dynamical systems theory, some researchers explore the notion of constraints and control parameters to circumvent the paradoxes of seeing intentions as cases of efficient causation (Kelso 1995; Van Orden et al. 2011; Wallot 2015). Alicia Juarrero develops a perspective on intentions as dynamic constraints that she uses to scientifically reinstitute other forms of causation (formal and final). Consider a complex self-organizing system, like an autocatalytic set of chemical reactions (we discuss this example further in Chapter 5). In this set some reactions are catalyzed by the products of other reactions, forming a network of mutual constraints between chemical processes. According to Juarrero, this emergent organization imposes a top-down, global to local kind of causation—the kind that is not supposed to exist in the billiard-ball world:

> As a distributed whole, a self-organized structure imposes second-order contextual constraints on its components, thereby restricting their degrees of freedom. As we saw, once top-down, second-order contextual constraints are in place, energy and matter exchanged across an autocatalytic structure's boundaries cannot flow any which way. The autocatalytic web's dynamical organization does not allow any molecule to be imported into the system: in a very important feature of self-organizing dynamical systems, their organization itself determines the stimuli to which they will respond. By making its components interdependent, thereby constraining their behavioural variability, the system preserves and enhances its cohesion and integrity, its organization and identity. As a whole it also prunes inefficient components. Second-order contextual constraints are thus in the service of the whole. They are, also therefore, the ongoing, structuring mechanism whereby Aristotle's formal and final causes are implemented.
>
> (Juarrero 1999, p. 143)

Seeing global dynamic constraints as enabling and restricting the possibilities for "lower-level" dynamics, which in turn sustain these global constraints is a powerful concept that can help solve dilemmas typical of reductionist perspectives on the mind. This relation between formal and efficient causes is at the core of the concept of emergence adopted by enactivists (Di Paolo, Rohde, and De Jaegher 2010; Thompson and Varela 2001; Thompson 2007a).

But Juarrero describes more than the emergence of global organizational constraints. She also points to the individuation of a system, or more precisely of an agent (see also Juarrero Roqué 1985). In this, her proposal comes closer to those mentioned above that link agency to the organization of living systems. Agency from these perspectives is not merely an ascriptional, *as-if* property. Instead, it relies on an ongoing process of self-individuation as a constitutive aspect, inextricably linked with the normativity and goal-directedness of action. This provides several clues as to how biological, organic agency may be naturalized.

But not all our activities are governed by our needs as organisms. In Chapter 6 we ask whether agency can also be found in levels that are to some extent decoupled from basic biological viability. We propose a positive answer and suggest that a network of mutually supporting sensorimotor schemes may indeed become individuated and adaptively self-regulating. This proposal resonates with classical views in psychology before the advent of behaviorism and cognitivism. For instance, William James (1890) saw animals as "bundles of habits." Habits being self-affirming sensorimotor structures, their integration into a network links them structurally and functionally and frees them thus from their "compulsive" re-enactment. We may propose that a self-sustaining organization of "habits" can emerge from these networked relations. If so, then at some point we may speak of a sensorimotor subject, whose richer world of interactions opens up an equally richer domain of new sources of normativity. A sensorimotor subject's activities become meaningful not only in virtue of their contribution to biological survival, but also in virtue of their contribution to the stability and coherence of a sensorimotor repertoire. The kind of meaningful experiences such a subject may enjoy depends on the myriad ways in which she reaffirms her sensorimotor selfhood through action. Precursors for similar views include Merleau-Ponty's description of the corporeal schema "as 'conducts' at work in the world, as ways of 'grasping' the natural and cultural world surrounding us" (1964, p. 117). Susanne Langer (1967) also offers an account of the individuation of an agent as constituted by relations between acts, and similarly, Christine Korsgaard (2009) proposes that it is the efficacy and autonomy manifested in each action that grant these same properties to the agent herself (i.e., her self-constitution and autonomy as a person). Our non-representational account of sensorimotor agency is largely compatible with these views.

Finally, Chapter 7 focuses on one particular aspect of the bodily selfhood that emerges at the sensorimotor level: the sense that we are the agents of our actions. Current explanations of this sense of agency involve the use of internal models aimed at comparing sensory and other nervous signals in order to determine whether a movement is self-initiated or not (Feinberg 1978; Frith, Blakemore, and Wolpert 2000a, b). Such representational models have some empirical support (Frith 2005; David, Newen, and Vogeley 2008; Synofzik, Vosgerau, and Newen 2008a) but have been criticized because by themselves they are insufficient to explain the full phenomenology of the sense of agency (e.g., Synofzik, Vosgerau, and Newen 2008a, 2008b). We will also argue that they are not necessary either.

Further elaborating on the idea of a network of sensorimotor schemes, we propose an alternative account in which the different pre-reflective aspects in the sense of agency can be explained in terms of processes underlying the selection and control of sensorimotor schemes in a particular situation. In this model each act reaffirms its relation to other schemes (relations of mutual structural support, inhibition, and other functional relations). These relations are constitutive of sensorimotor agency. The processes of equilibration between sensorimotor schemes through which this reaffirmation of agency occurs correlate well with the phenomenology of the sense of agency.

2.6 Toward a world-involving account of action and perception

If sensorimotor contingencies theory provides a good basis for elaborating a theoretically loaded enactive approach to perception and action, addressing the lacks in this theory, in particular those that easily tend to be filled in with representational discourse, is the essential first task of enactivism. This is the boost required to free ourselves from the representational pull and achieve enactive escape velocity. Starting with Chapter 3 we will address the issues raised in the previous section from an enactive (non-representational, world-involving) perspective. This will lead to some novel proposals, which are supported by evidence and phenomenological insights, but which nevertheless will call for further articulation, refinement, and testing.

At the end of this process we will be in position to use various intermediate results with the aim of looking at open challenges for an enactive theory of action and perception (Chapter 8). These challenges include theorizing the problems of virtual actions and long-range sensitivities, as well as evaluating the social factors that influence perception, in particular the constitutive social skills that underlie abstract perceptual attitudes (e.g., contemplating an object noninstrumentally). These final sketches will be more speculative, but in themselves, they serve to illustrate further a point already implied by the rest of the book: That it is possible to think scientifically about the mind outside the mediational picture that leads inevitably to representationalism. In other words, there *are* alternatives.

Chapter 3

Structures of sensorimotor engagement

... the reflex arc idea, as commonly employed, is defective in that it assumes sensory stimulus and motor response as distinct psychical existences, while in reality they are always inside a coordination and have their significance purely from the part played in maintaining or reconstituting the coordination;

[...]

What we have is a circuit, not an arc or broken segment of a circle. This circuit is more truly termed organic than reflex, because the motor response determines the stimulus, just as truly as sensory stimulus determines movement. Indeed, the movement is only for the sake of determining the stimulus, of fixing what kind of a stimulus it is, of interpreting it.

—*John Dewey (1896, pp. 360, 363)*

3.1 Perceptually guided action

You open the door that leads into a familiar room where you must look for something you have forgotten. You are in a hurry. Yet, if we were to slow down time and focus on the fractions of second it takes you to reach for the door handle, turn it down, and push the door open, what would we find? Clearly, your attention is elsewhere, and still you perform a small series of coordinated eye, arm, and hand movements quite flawlessly in the act of opening the door. You probably do not need to search for the door handle. Some broad peripheral awareness confirms that it is to be found where it has always been. In fact, your body knows already how to proceed. You slow down your walking or even stand still in front of the door. Your shoulder and elbow joints adopt adequate angles so that a slow approach brings your open hand right up to the handle. Upon contact, your fingers close around the handle, the smooth metal surface feeling slightly cold. The strength of your grip increases until it reaches a comfortable point, a safe sensation. You feel most of the handle in contact with your palm and fingers. It is not too thin or too thick. Then keeping a firm grip, you initiate an angular movement of your forearm and wrist, bringing the handle down. The resistance it offers is just right. It is neither too slack nor unmovable. You feel that the resistance is not uniform as you push the handle down. You sense a click as the latch bolt retracts. After that, you stop pushing down but keep the handle in that position where you still feel it pushing upward. You pull the door open.

What goes on in these fleeting moments is an example of your body's attunement to the world. You may not have had your attention focused on this action—your mind is on finding what you need in the room—and yet your activity was perceptually guided. Any unexpected deviation, the door handle is stuck, or its temperature strangely warm, or the door does not move when you push, would be immediately noticed. Yet, what you perceive during these events are not the properties of the door or the handle as external objects. You perceive what happens as they react to the directed movements of your body. The comfort of the grip is a property of the encounter between the enclosing hand and the handle. So is the resistance it offers as you push it down an encounter between your gyrating forearm and the springs and mechanism of the lock. These regularities emerge from the meeting of your active body and the world. And they make up your perception of the door-opening act as a whole.

Action in the world is always perceptually guided. And perception is always an active engagement with the world. The situated perceiver does not aim at extracting properties of the world as if these were pregiven, but at understanding the engagement of her body with her surroundings, usually in an attempt to bring about a desired change in the relation between the two. To understand perception is to understand how these sensorimotor regularities or contingencies are generated by the coupling of body and world and how they are used in the constitution of perceptual and perceptually guided acts. According to Varela, Thompson, and Rosch (1991, p. 173), the task of the enactive approach is therefore "to determine the common principles or lawful linkages between sensory and motor systems that explain how action can be perceptually guided in a perceiver-dependent world."

An examination of this notion of lawful linkages between sensory and motor systems—or, as they are also called, sensorimotor contingencies (SMCs)—is a first step toward making the idea of perceptually guided action operational, in other words, describable in scientific terms. Far from being a straightforward idea, we will see that the concept of sensorimotor dependencies admits different interpretations and refinements. In fact, we will suggest that there is some gain in distinguishing between different usages of this notion and that they deserve different names.

To show this, we adopt a dynamical systems perspective in order to lay the foundations for a formalization of the sensorimotor approach to perception. This task will continue in Chapters 4, 5, 6 and 7, where we investigate the concepts of mastery, sensorimotor agency, and sense of agency, using the notions introduced here.

At this stage, we will not yet formulate a distinctly enactive theory of action and perception. We must first lay some foundations. To start with, we must attempt to operationalize some of the basic ideas in order to avoid confusion. In a way, this means that the notions of SMCs we will propose are more "neutral" than other concepts elaborated later in the book. They, in principle, could be adopted by a non-enactive perspective, even by approaches that are more traditional, provided they accept the language of dynamical systems theory. And since the concepts here introduced do not make any assumption about the kind of agent under consideration, they are also equally applicable to human beings and animals, as well as to robots. If the corresponding ideas of sensor and effector

activities are properly defined, some of these concepts may even apply to non-animal forms of life, like plants and bacteria.

When we say that we want to operationalize the notion of SMCs, we simply mean that we aim at a formulation that describes this concept in terms of lawful relations within the general constraints of scientific naturalism. On the one hand, this means not invoking superfluous entities or relations for which there is no scientific basis: no magic allowed. On the other, it also means not accepting at face value basic terms that themselves lack a proper and generally accepted operational grounding. This second requirement is more subtle and difficult to accomplish because we must start somewhere and take certain basic concepts as generally understood and not deserving further scrutiny.

Some of the notions we will use here can be said to be like this, especially the dynamical systems vocabulary of variables, parameters, mappings, couplings, etc. We will not question them further. However, in the chapter we also rely on some terms like agent, environment, sensors, motors, etc., which for the moment, we simply use in the accepted, intuitive meanings they possess in the context of psychology, cognitive science, and neuroscience. They are not yet presented as operationally grounded and they *do* deserve further scrutiny. We will come back to reflect on these ideas in the following chapters. For the moment, we take out a conceptual loan on these ideas and promise to repay it using precisely the concepts of SMCs we develop in the following sections.

3.2 What exactly is a sensorimotor contingency?

We have discussed some general aspects of motor theories of perception and cognition in Chapter 2. As we noticed, these theories tend to rely on the coordination of sensory and motor signals—both as they actually occur and as they could occur—in order to generate the properties of perceptual experience, in particular, those properties that allow the agent to perceive the environment as external (objects, spatial relations, and so on). The sensorimotor approach to perception has made explicit the connection between perceptual experience and the way an agent masters such sensorimotor correlations, or SMCs.

But what are we talking about when we talk about SMCs? The notion seems to point in an unproblematic manner to regularities in the sensorimotor field: predictable or "lawful" co-variations of sensory stimulation, neural, and motor activity. For instance, the projection of a horizontal line onto the retina changes from a straight line to a curved arc as one shifts the eye's fixation point from the line itself to points above or below it. In contrast, if the focus is moved along the line, no such transformation takes place. The geometry of the viewed object, the morphology of the retina, and the particular movement pattern employed all determine these regularities in the sensory stimulation pattern (O'Regan and Noë 2001, p. 941). Such regularities, in principle, could be described if enough detail is known about the sensory system and the environment.

However, what counts as a sensorimotor dependence can change depending on whether we decide to focus on all possible sensorimotor scenarios given the general properties of an agent's body and of its surroundings, or if we study the agent as the active creator of such regularities. After all, the agent may decide to move the eyes in the vertical or horizontal direction or in a particular combination of both when looking at a line. The notion of regularity also changes in terms of what we want to use it for. For instance, if we consider different task-oriented situations with different salient sensorimotor patterns we could be interested in particular regularities that contribute to the agent's performing a task such as recognizing a complex object visually. The idea of SMCs also suggests the relevance of regular structures in these dependencies, but what counts as regularity can also depend on the scale of observation, on whether we make purely dynamical considerations or whether the focus is on the functional organization of the task, and so on.

Which of these various possibilities is relevant for an enactive theory of perception? The answer is all of them. But we should perhaps first look at how the term has been used in order to start clarifying these ideas. O'Regan and Noë's original formulation describes SMCs as follows:

> ... *the structure of the rules* governing the sensory changes produced by various motor actions, that is, what we call the *sensorimotor contingencies* governing visual exploration.
>
> (O'Regan and Noë 2001, p. 941; emphasis in the original)

There are no other explicit definitions in this work but we can illustrate the way in which the idea is used:

> According to this theory, seeing is a skilful activity whereby one explores the world, drawing on one's mastery of the relevant laws of sensorimotor contingency.
>
> (O'Regan and Noë 2001, p. 966)

> Now, perceiving, according to the sensorimotor contingency theory, is an organism's exploration of the environment that is mediated by knowledge of sensorimotor contingencies
>
> (Myin and O'Regan 2002, pp. 33–4, emphasis removed)

> ... to *see* something is to interact with it in a way governed by the dynamic patterns of sensorimotor contingency characteristic of vision, while to *hear* something is to interact with it in a different way, governed by the different patterns of sensorimotor contingency characteristic of audition.
>
> (Hurley and Noë 2003, p. 146, emphasis in the original)

Apart from expressions like these, there have been a few attempts at formalizing the idea of SMCs. Some of the existing models, for instance, (Philipona, O'Regan, and Nadal 2003; Philipona and O'Regan 2010) focus on the problem of extracting general properties about the environment in which a robot operates, for instance the number of spatial dimensions. The authors make use of a notion of SMCs as revealed by invariants in the set of movements executed by an agent. The resulting lawful relations become clear after doing the math. Implicitly, such models already lean on a particular interpretation of the notion of SMCs and not all of the above quotations fit this interpretation. The existence of invariant and lawful relations in sensorimotor patterns is one thing; the *use* of such relations by the agent is another. In these models, one can claim that the interaction between agent and world is "governed by dynamic patterns of sensorimotor contingency" as Hurley and Noë suggest. But one cannot quite say that the agent is "drawing on its mastery of relevant laws" or acting in a way that is "mediated by knowledge of SMCs."

Is this a problem only with the above quotations? No, it is more general. For instance, many of examples of SMCs employed in the literature have in common that they describe the static sensory consequences of arbitrary motor changes, i.e. the immediate, local effect on sensory surfaces of an agent's change from one state to another. This is so in the example of the change in curvature of a line projected onto the retina as the eye moves, or in the lawful change in the properties of light reflected from an object's surface as the observer or the object move around (O'Regan and Noë 2001, p. 942; Philipona and O'Regan 2006), or the set of distortions that a shape undergoes as its position relative to the observer changes (O'Regan and Noë 2001, p. 942).

Other examples, in contrast, acknowledge the active role of the agent's ongoing engagement with the environment in the creation of invariant sensorimotor structures. For example, O'Regan and Noë (2001) refer to sensory substitution experiments by Lenay, Canu, and Villon (1997), in which the distance of a light source can be estimated using only the vibration created in response to a single on/off photosensor fixed to the hand. This is possible not only because the person is moving, but also because some movements allow a person to coordinate with the environment and others do not. The effective coupling occurs only when participants perform movements along an arc, such that elbow and shoulder angles vary together according to a fixed relation, while the light remains fixated. The relevant contingency is not found in the instantaneous consequences of arbitrary observer motion, but is rather created through a time-extended sensorimotor engagement that couples specific movement patterns with sensory feedback.

All of this suggests that there may in fact be not one, but several useful interpretations for the notion of SMCs, from broad descriptions of lawful structures in an agent–environment interaction to coordinated structures that are strategically deployed by the agent in a goal-oriented or normative manner.

3.3 **Several kinds of sensorimotor regularities**

In order to appreciate the fact that sensorimotor regularities come in different kinds, let us imagine a situation that can help us see these differences more clearly.

One of the permanent attractions that welcome the visitor to the Guggenheim Museum in Bilbao is Richard Serra's massive installation *The Matter of Time*. It consists of eight colossal structures in the shape of tall curving walls made out of reddish weathering steel (Figure 3.1). They form circles, spirals, and wavy corridors. As the spectator walks around, within, and in between these sculptures, she experiences smooth but powerful spatial changes in the atmosphere. The effects of the unbroken reshaping of the spatial surroundings can be dizzying and oppressive, but they are interspersed with sensations of release as a large area suddenly opens wide at the interior of a sculpture or as the walker comes to the end of a long corridor.

Let us use this setting for an imaginary situation. Picture yourself as a solitary spectator exploring this installation on a quiet evening. Unexpectedly, as you are exploring one of the most vertiginous spirals, the lights in the museum go out. A power failure. You are in the dark and decide to make your way out of spiraling walls. Once outside the spiral you have lost orientation, and there is no obvious way to find the exit. You decide to start

Figure 3.1 A setting for sensorimotor exploration. Richard Serra's *The Matter of Time, 1994–2005*. Eight sculptures, weathering steel, variable dimensions. Guggenheim Bilbao Museoa.
Copyright © ARS, NY and DACS, London 2017, with permission. Photograph reproduced courtesy of Ezequiel A. Di Paolo.

walking toward one of the ends of the long room, but which one? Let us imagine for the sake of the example that you remember seeing a large round sculpture when you entered the room, but at the other end, before the lights went out, you managed to see a long wavy one. So you decide to try to find the exit by locating a wall curving smoothly into a circle, without being able to see it, using only your walking and your extended arms to touch the walls.

How can you tell if the steel wall you are touching is a large quasi-cylinder or a long wavy corridor? It does not seem too difficult. You decide to walk while keeping one hand on the wall, essentially being guided by its shape. The contact with the wall is insufficient by itself to help you find your goal as the local curvature can be ambiguous. But as you continue walking, you become confident that you are in contact with a circular wall. How? Perhaps you noticed that you kept turning into the same direction for some time. If so, you have made use of kinesthetic and proprioceptive cues, bodily sensations generated in this case by some regularities in your own wall-guided walking. After this you can hear some voices and guide yourself toward the exit.

This situation is not unlike the one faced by a small wheeled robot designed by Rolf Pfeifer and Christian Scheier in the mid-1990s. Without any complex visual system, equipped only with infrared distance sensors, the robot lives in a plane environment full of small and large cylindrical objects (Figure 3.2). Its task is to find the small cylinders, grab them with a gripper, and move them to a predetermined location (Pfeifer

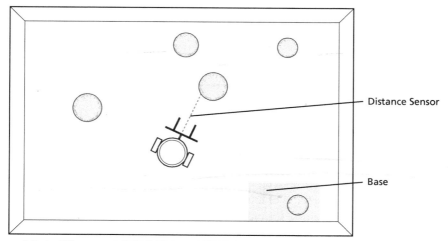

Figure 3.2 A different setting for sensorimotor exploration. A miniature two-wheeled robot (with a gripper) in an environment with cylinders uses sensorimotor coordination to categorize the cylinders into large and small.

and Scheier 1997). After some learning, the robot uses a strategy that, with few modifications, could also help it find its way out of Richard Serra's installation. The robot approaches an object (remember that it can only sense its proximity; the object could be a large or small cylinder or a wall) and starts moving alongside the object, trying to keep its sensor activation constant. This makes the robot turn around the cylinders and follow walls. If a cylinder is small, the robot's angular velocity will be higher than if the cylinder is large. The robot can use this sensorimotor coordination between constant sensor activity self-generated movement, and resulting angular velocity to discriminate between large and small cylinders and then proceed to move away or grab as required.

It may not be immediately clear, but in both examples, there is more than one concept of sensorimotor regularities at play.

One kind of regularity depends mainly on the relation between properties of the environment and properties of the body. The installations in the museum are curved in particular ways, and as the visitor moves her hand along the walls, the curvature can be concave or convex, slanting upward or downward as the case may be. Because the walls do not move, at different given spots, a displacement of one step produces a change in tactile sensation particular to that spot. This relation is one kind of sensorimotor regularity, perhaps the most general one, since it depends only on the relation between bodily and environmental structures.

A different kind of regularity however emerges out of the visitor's and the robot's self-generated activity. It is different from the structural correlations between body and environment because it depends on how the body moves as a response to the changing sensations. Sensorimotor activity is closed in a loop, and we cannot say what comes first, the sensation that affects the movement or the movement that alters the sensation. In both examples a new kind of regularity emerges out of this sensorimotor loop, one that relates properties of environmental structures to invariant properties of proprioceptive bodily changes; the feeling the visitor has of walking in a circle or the regular relation between the left and right motor activity in the wheeled robot. If the visitor or the robot tried moving differently, perhaps other regularities would emerge too.

But there is more. In both cases, the emergent regularities are put to use. The emergence of active sensorimotor invariants depends on given structural body-environment regularities and, in turn, it enables the recognition of environmental properties that exceed the direct discrimination capabilities of touch and infrared sensors. The museum visitor can figure out the rough shape of a sculpture too large to discriminate by moving the hand without walking. The robot can discriminate different sizes of cylinders without a visual system to compare them. These sensorimotor regularities are used to coordinate action. And moreover, they are also linked to larger schemes involving other instances of sensorimotor coordination, which themselves may show regularities in the way they combine, while aiming at a general goal, finding the exit in one case, deciding whether to pick up an object in another.

So we have at least four related but different notions of sensorimotor regularities. They move from the abstract to the concrete and from the more objective end to the more subjective. We first have the regular changes in sensation that occur in relation to arbitrary activity in a motor system (e.g., taking a step while touching the wall at a particular spot). Then we have regularities that are generated by the agent's own time-extended activity (e.g., invariant bodily relations provoked by walking at a fixed distance from the wall). The agent uses these emergent regularities; they form part of an activity that they help coordinate (e.g., discriminating the shape of a large structure). Finally, they combine in sequence or in parallel with other instances of coordination to form schemes (e.g., finding the exit, keeping the other arm outstretched in the direction of movement to avoid bumping into something or someone). Each new kind of sensorimotor regularity seems to rely on and make use of the former.

We elaborate these distinctions next.

3.4 Four kinds of SMCs

Let us describe these different senses of sensorimotor regularities more formally. We consider a generic agent (an animal, a person, a robot, etc.) in ongoing interaction with a generic environment. One possible relation between the sensory and motor activity of the agent simply describes how sensory input changes with induced motor activity in an open-loop fashion. It is equivalent to asking the question: what would be the change in the sensory activity if we were to alter the motor activity so that the body configuration of the agent changes in such-and-such a way? This relation between motor and sensory activities depends only on the embodiment of the agent and on the environment. A different sensorimotor relation looks at co-variations that obtain once the loop is closed by taking into account the agent's internal activity and responsiveness to sensory changes (i.e., when the agent is allowed to move on its own). The next sensorimotor relation is more specific and looks at the coordination patterns that contribute to the performance of a task that the agent is currently engaged in. Finally, another sensorimotor relation indicates how such coordination patterns may be organized according to some norm to distinguish between levels of skillfulness, efficiency, stability, etc.

One way to shape these intuitions is to formalize the sensorimotor coupling of an agent with the environment using a dynamical systems approach. For this, we will assume that the agent and its environment can be sufficiently approximated as well-defined systems, at least in terms that are relevant to our investigation. Much of the ideas in this and the following chapters will make extensive use of dynamical systems concepts. We will present these key concepts by keeping the mathematical notation to a minimum.

Making some simplifying assumptions, we may describe the situation of our abstract agent using the diagram in Figure 3.3 made famous by Randall Beer's minimal cognition research (see Box 3.1). We consider three dynamical systems, each defined by a set of variables and equations or rules of transformation that describe how these variables evolve over time. These systems are the agent's body, from which we distinguish its nervous system as a separate system (here this separation is merely a matter of convenience), and the environment. The agent's body and nervous systems are coupled, and so are the environment and the body. This is indicated by the arrows between the systems in Figure 3.3. Two systems are said to be coupled when some of the parameters governing the equations in one of the systems are dependent on the state of the variables of the other. Coupling can be unidirectional or bidirectional.

We represent the state of the environment by a vector of variables **e**. These contain any relevant, measurable variable that may have an effect on the agent (movements of objects, lighting conditions, changes in temperature, and so on). Similarly, we use the vector **p** to represent the state of the agent's body (position and configuration, the activity of effectors such as muscles, and so on). Since what happens in the environment depends on its own current state and on what the agent is doing we have that:

$$\frac{d\mathbf{e}}{dt} = E(\mathbf{e}, \mathbf{p});$$

[handwritten annotations: "state of env." pointing to e; "position of agent's body" pointing to p]

in other words, the derivative or rate of change over time of the environmental variables **e**, is a function (E) of those same variables and the state of the agent's body **p**.

In the case of the nervous system (or the "controller," say, in a robot) we distinguish three sets of variables: those describing the state of the sensor activity **s** (peripheral stimulations of organs sensitive to the state of environmental variables), those describing the state of the motor activity **m** (the outflowing movement-producing signals closest to the periphery), and those describing the internal neural dynamics **a**. The change in the activity of the

[handwritten note at bottom: "What are the variables in the nervous system? ① sensor activity ② motor activity ③ internal neural dynamics"]

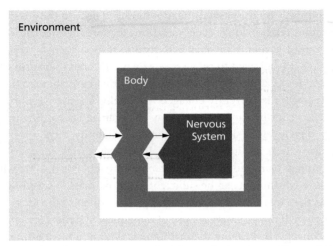

Figure 3.3 Agent and environment as coupled dynamical systems. The world is partitioned into components describing the dynamics of the environment and the agent and their interactions (arrows indicating mutual influences). For convenience, the agent is further subdivided into body and nervous system.

Copyright © 2017 Ezequiel A. Di Paolo, Thomas Buhrmann and Xabier E. Barandiaran, with permission.

sensors **s** depends on what goes on in the environment (**e**). It also depends on modulatory effects produced by the neural dynamics (**a**). In turn, changes in the internal neural activity (**a**) depend on the sensor activity and on the neural activity itself. And the changes in motor activity (**m**) depend on the internal neural activity (motor activity depends on the state of the sensors, as well, but we describe this dependency here indirectly through the mediation of internal neural activity). These dependencies are described by the functions S, A, and M:

$$\frac{d\mathbf{s}}{dt} = S(\mathbf{e}, \mathbf{a}),$$

$$\frac{d\mathbf{a}}{dt} = A(\mathbf{a}, \mathbf{s}),$$

$$\frac{d\mathbf{m}}{dt} = M(\mathbf{a}).$$

And finally, the interaction loop is closed by noticing that changes in the configuration of the body (**p**) depend on the current state of the body, on the activity of the motors, and on the state of the environment through a function B.

$$\frac{d\mathbf{p}}{dt} = B(\mathbf{p}, \mathbf{m}, \mathbf{e}).$$

Box 3.1 Minimal Cognition

The terms *minimal cognition* or *minimally cognitive agent* were introduced by Randall Beer (e.g., 997; 2003) to describe a style of modeling using simulated agents (or real robots) with minimal motor and sensor capabilities, in ongoing coupling with the environment, which are required to perform a task deemed of sufficient cognitive interest. Minimal cognition models involve modeling a full agent by including all the arrows in Figure 3.3. In other words, instead of simplifying the mutual coupling between nervous systems, body, and environment to focus on the minute details of a specific subsystem, minimal cognition models massively simplify most aspects of these systems but maintain their mutual coupling. Similar work has been done under the names of computational neuroethology (Chiel and Beer 2009; Cliff 1991) and evolutionary robotics (Harvey et al. 1996; 2005; Nolfi and Floreano 2000; Vargas et al. 2014).

The "minimally cognitive" problems studied with these models are often less trivial than they might seem. They include problems of absolute and relational categorization (Beer 2003; Beer and Williams 2015), associative learning (Izquierdo and Buhrmann 2008; Phattanasri, Chiel, and Beer 2007; Yamauchi and Beer 1994), action selection and behavioral choice (Agmon and Beer 2014; Iizuka and Di Paolo 2007), and social coordination (Di Paolo 2000a; Quinn 2001). Although there is a variety of modeling choices, there are a few constants in most of these models. (1) The agent's sensors, effectors, and neural controllers tend to be as simple as possible, thus encouraging the emergence of active strategies for solving the task. (2) The agent–environment sensorimotor loop is modeled as an ongoing dynamical coupling. (3) No strong assumptions are made as to how the task must be solved; usually once the problem is described, solutions are explored by artificial selection methods. (4) As a consequence of the last point, agents capable of the desired performance must still be analyzed; this is done using dynamical approaches and looking at controlled interactions between neural dynamics, body activity and environment.

Minimally cognitive agents follow the tradition of synthetic models utilized to help us think about systems engaged in complex interactions, where intuitions often fail. The use of such models saw a boom in the mid-20th century with the advent of cybernetics—a well-known example from that era that is still today conceptually very useful is W. Ross Ashby's Homeostat (1960), a machine capable of various kinds of adaptive responses and learning. Another example is W. Grey Walter's (1950; 1953) "turtles": light-seeking, obstacle-avoidance robots capable of a variety of behaviors using a simple controller. Before the availability of transistor electronics, such models were rare and limited to the purely conceptual sphere. One good example of these is the "schematic sow bug" designed by Edward C. Tolman (1939; 1941). This artificial creature could orient itself and move in a field of stimulation, while learning about its environment and achieving internally imposed goals. A constant feature of synthetic systems like these is that their construction is relatively simple, and their interactions with the world relatively complex. From mere observation of their behavior, it is difficult to guess how these systems work. One diligent explorer of such synthetic systems, Valentino Braitenberg, condensed this relation into a formula: the "law of uphill analysis and downhill invention" (1984, p. 20). The important lesson here should be clear: complex behavior does not always result from equally complex mechanisms: hence, the importance of testing ideas using minimal models. As stated by Randall Beer (2003), minimally cognitive agents provide concrete examples to explore contentious notions and stretch intuitions about mechanisms: a kind of mental calisthenics. Minimal cognition models are relatively simple but they make us contemplate significantly more sophisticated explanations than we normally could think of without their aid.

The details of these equations depend on the specific case under consideration and the level of relevant information. The important point is that for every agent–environment system, we can choose some variables to describe the environment and others to describe the agent. For the moment, we take the choice as conventional and the distinctions could be drawn differently depending on the focus of interest. This picture could be refined in many ways. For instance, we could have made explicit the distinction between physiological and mechanical bodily variables if we needed to. Or in some cases, we may not wish to define sensory activity as an explicit variable but would be happy to describe the agent as a whole system coupled with the environment (with sensor inputs as parameters as in, e.g. Beer [2003]). In other occasions we may require to divide the environment into kinds of systems, for instance, if we wish the make explicit the activity of other agents. And so on. Such refinements, however, would simply add or remove equations, and this may serve a purpose in some cases. Once these decisions have been made and the necessary distinctions have been drawn, whether by convenience or as a matter of principle, it is important that such distinctions be maintained. As long as this rule is observed, the following analysis will remain valid.

Using this abstract description[1] of the agent and environment as coupled systems we can now identify the four notions of SMCs suggested earlier.

---

[1] The expression of this formalization corresponds to an abstract case. Specific details depend on the case under consideration: the equations may or may not depend explicitly on time and they may exhibit different degrees of nonlinearity, piecewise, or conditional functional relations, and dependence on various parameters. Some of the equations may be valid only within a timeframe. In a system capable of plastic adaptation, some functions will change over time. It is possible to formalize some of these changes through the inclusion of rules of plasticity. Other changes are harder to capture, such as the emergence of new relevant variables through development or during the acquisition of novel perceptual systems. We want to emphasize that this general formal description is not necessarily complete, but it is sufficient for our current purposes, and as we shall see, once introduced, the new kinds of SMCs can be investigated by a range of other methods that do not demand the level of detail of a full dynamical description.

3.4.1 **Sensorimotor environment**

Many examples of SMCs in the literature refer to the instantaneous sensory consequences of given changes in the agent's perspective or specific movements of the sensor organs. Such is the case of the regular deformation of the stimulated area of the retina as we scan a horizontal line in the vertical direction that we mentioned earlier. These regularities do not depend on the origin of these movements. They can be described without considering how the movements themselves relate to sensory feedback; in other words, they regard motor changes from an open-loop perspective. Movement of the body leads, for example, to lawful instantaneous changes in the optic flow on the retina (expanding when moving forward, contracting when moving backward, etc.). Equally, rotations of the head lead to a lawful change in the temporal asynchrony between sounds received in the left and right ear. These lawful relations are independent of how such movements originate or combine with other movements. They depend solely on the sensorimotor apparatus of the agent and the structures of the environment in which it is situated.

In terms of our formal abstract agent, we can capture this kind of SMCs by considering how the sensor values **s** change in relation to given motor states assuming that the motor command **m** varies freely—in other words, **m** is taken as an independent variable, decoupled from the agent's internal variables **a**. The sensorimotor loop is opened by removing the equation $d\mathbf{m}/dt = M(\mathbf{a})$. These motor-induced sensory changes can depend on several environmental and bodily factors; for instance, the position/configuration of the body **p** (technically considered a parameter dependent on **m** in this open-loop version of the model). In the most general terms, this is expressed as an implicit function involving sensor and motor values: $f(\mathbf{s}, \mathbf{m}) = 0$, where **m** is the independent variable. In some cases, this relation can be expressed explicitly as $\mathbf{s} = g(\mathbf{m})$. Mathematically, using these mappings we can capture relevant aspects of this relationship between sensor and motor activity, whenever possible and at least locally, by the partial derivative $\partial\mathbf{s}/\partial\mathbf{m}$, that is, the change in sensor values **s** resulting from changes in the independent variable **m** while all other variables are held constant.

We call this functional relation (f or g) the *sensorimotor environment*. It is a relation between sensors and motors specific to agents that share the same body structure and environment, but it is independent of the agent's internal state and the actual actions it performs. The SM environment[2]—not to be confused with the environment *for* the agent—constitutes the set of all possible sensory dependencies on motor states (**s**, **m**) for a particular type of agent and a particular environment. Whatever specific behavior the agent exhibits, its sensorimotor projections will always be found within this set, which constitutes the most abstract sense of sensorimotor dependencies.

Identifying the SM environment can be very useful, as we shall see. There are different practical ways in which this mathematical relation can be approximated in empirical

---

2 Throughout this chapter and in some tables in the book when referring to one of the four notions of SMCs, we will use the abbreviation SM to stand for "sensorimotor."

cases. The SM environment will have several properties (e.g., of smoothness, dimensionality, and symmetry) that constitute the most general constraints to any actual sensorimotor trajectory, including both successful and unsuccessful strategies for solving a given task. The properties of the SM environment are the most general kind of regularities or "laws" of SMCs. They are shared by all agents with sufficiently similar bodies in a given environment.

3.4.2 **Sensorimotor habitat**

Freely behaving animals and persons are not open-loop systems. They live in an ongoing sensorimotor coupling with their environment, as indicated by the arrows in Figure 3.3. This means that there may be other kind of sensorimotor regularities apart from those identified by studying the SM environment. New regularities can be created through patterns of agent–environment interactions (i.e., in the time-extended closed loop between motor variations and sensory feedback).

Such is the case, for instance, in the example of sensory substitution experiment described earlier, in which the distance to a light source can be determined by performing a certain class of arm movements while keeping the light source fixated (Lenay, Canu, and Villon 1997). Other examples include perceiving the softness of a sponge (we will come back to this), or the strategies employed by baseball outfielders, who move in such a way as to create and exploit certain sensorimotor invariants in order to catch fly balls without having to predict where and when the ball will land (Chapman 1968; Sugar, McBeath, and Wang 2006).

In our formal framework, we can identify regularities of this kind by examining how an agent actually "navigates" the SM environment, what it actually does. We close the sensorimotor loop that we had opened to define the SM environment and now take into account the agent's internal states and their influence on the effectors. The result is that the regularities we find will be dependent on the internal activity **a** of the agent (typically its neural activity).

We use the term *sensorimotor habitat* to refer to the set of all sensorimotor trajectories (i.e., movements in sensorimotor space) that can be generated by the closed-loop system in a given situation. For this, we need to take into account the evolution of the internal states **a**. This set describes how an agent "moves" within the SM environment. In the SM habitat, we should be able to identify various patterns: the different instantaneous tendencies, the regions of sensorimotor space that are most likely to be visited, the temporal patterns of these trajectories, and other regularities. The SM habitat inherits some constraints from the SM environment. But it is not solely determined by it. In fact, as a space, it is likely to be of a higher dimensionality because of the addition of internal dynamics. In other words, although the SM environment limits the possible habitats, there are still an infinite number of ways in which an SM environment can be "inhabited." (It is simply a question of varying the function A that controls the evolution of the internal activity **a**.)

Formally, the SM habitat corresponds to the set of actual sensorimotor trajectories traveled by the closed-loop system (given by the equations described earlier) for a range of values of relevant parameters (initial positions, initial states, environmental parameters, etc.). Whenever possible, local information about the SM habitat can be captured by the total derivative $ds/d\mathbf{m}$, using the full set of equations, where $\mathbf{m}$ is not any more a free variable but is influenced by the agent's internal states $\mathbf{a}$ and, of course, the coupling of the agent with its environment. As is the case with the SM environment, the SM habitat will have certain properties of smoothness, dimensionality, symmetries, and so on, to which we can now add dynamical properties such as attractors, regions of (meta)stability, etc. All of these structural and dynamical regularities will characterize a particular type of agent in an active engagement with its environment.

It is important to say once again that neither the SM habitat, nor its properties can, in principle, be fully deduced from the SM environment. It is true that whatever the SM habitat, its existence must be a possibility in the SM environment. And, in specific cases, properties of the former might reflect properties of the latter. But the agent's internal dynamics also create completely new behavioral constraints and new ways in which sensors and motor can be coupled in a time-extended manner. A metaphor might be to think of the SM environment as the room within which a person is doing some activity. The walls of the room constrain the possible behaviors but cannot determine whether the person will sit in meditation, lie down, or do exercise, activities that would be described by the SM habitat. In other words, the "laws" of the SM environment constrain but do not fully determine the regularities of the SM habitat.

This point is important because noticing a sensorimotor dependency that belongs to the SM environment does not have the same explanatory status as noticing a dependency in the SM habitat. Both kinds of regularities may help explain a specific instance of action or perception, but their contributions are different. Contingencies belonging to the SM environment will in general play the role of constraints to an explanation (as we shall see later in an example). These constraints can often facilitate the explanatory task. For example, an animal's bilateral symmetry often means that we can expect any sensorimotor regularities in the SM environment also to be bilaterally symmetric (under appropriate reflections of the environmental situation), and so if we are able to formulate an explanation of behavior, the default expectation (unless other factors intervene) is that the explanation will also work if we reflect the situation around the axis of symmetry of the body. But this knowledge does not tell us which of the innumerable bilaterally symmetric behaviors will be engaged by the agent. For this the regularities of the SM habitat are more informative. Still, these regularities will not tell the whole story. For that we need to look closer at the role played by *patterns* in the SM habitat as the agent is engaged in real actions, in other words, within a functional context of attempting to complete a task or reach a goal.

3.4.3 Sensorimotor coordination

So far, we have considered the agent outside of any functional context, for instance, the performance of a specific task. Within the SM habitat, we may find certain regularities as we have said. Some dynamical patterns may be repeated for a large set of parameter values; there may be stable trajectories, and even transients may occur reliably for a set of circumstances, things we would externally describe as what an agent *typically* does. Some such regular patterns are often found to be crucial for task performance. In the area of autonomous robotics, coordination between sensor and motors plays crucial roles, as we have seen in the example of the robot that must discriminate between cylinders of different sizes (see, also, the work of Beer and colleagues, for other examples of complex behavior using simple sensors). Following the usage in this context, we call any such reliable sensorimotor pattern a *sensorimotor coordination* if it contributes functionally to the performance or goals of the agent.

For example, the presence or absence, depending on the environment, of a stable oscillatory attractor within the SM habitat could lead the agent to perform a categorical discrimination (e.g., Beer 2003). Or, as we have seen, a correlation between self-sensed left and right angular velocities of the wheels in a robot while keeping a proximity sensor activity nearly constant can be used to discriminate between cylinders of a large or small diameter by indirectly "measuring" their size (Pfeifer and Scheier 1997). Or in the now classical example, the softness of a sponge can be determined by squeezing it between the fingers, the quality of interest resulting from a specific correlation between applied pressure and felt resistance. Or in the situation of the minimal sensory substitution experiment by Lenay and colleagues in which distance to a remote source and its orientation can be estimated by local tactile stimulation and by co-varying the movements of the arm joints. All of these are typical cases where SMCs are invoked as part of an explanation of behavior (often of a perceptual task) and so they deserve to be treated separately from the general mappings described in the first two kinds of SMCs. All of these are examples of SM coordination, that is, specific SMCs found within a SM habitat and described by co-dependencies between **s** and **m** that reliably contribute to functionality.

SM coordination patterns are specific, often local sensorimotor co-dependencies that are dynamically organized in time in the context of a task. Dynamically, they can be stable sensorimotor patterns but not always necessarily so. Transients, as long as they are "used" reliably, be explanatorily linked to functionality and so count as instances of SM coordination (we show this in the model discussed later in the chapter).

In practical terms, a SM coordination pattern is determined by a dynamical analysis of the agent within the context of specific action and perception. These patterns inherit dynamical properties from the SM environment and the SM habitat. In general we can expect some of the most likely candidates for an SM coordination to originate in regularities of the SM habitat. However, not all regular patterns found in the SM habitat will necessarily be instances of SM coordination, as some of them may have no functional significance.

This qualification allows us to bring some clarity to O'Regan and Noë's example of the missile guidance system, of which they say that it is " 'tuned to' the SMCs that govern airplane tracking," while an out-of-order missile guidance system is said to only have "a kind of ineffectual mastery of its SMCs" (2001, p. 943). Within our proposed framework, we can avoid referring to seemingly paradoxical concepts such as "ineffectual mastery." A correctly functioning missile, which fulfills the goal of tracking airplanes, can simply be said to exhibit SM coordination. An out-of-order missile, in contrast, might still perform stable trajectories, but if these are not efficacious for the tracking of airplanes, we do not consider them cases of SM coordination. (We will return to other, deeper concerns with this example when we look into the issues of sensorimotor agency and subjectivity in Chapter 6.)

3.4.4 Sensorimotor schemes

Until now we have described different sensorimotor structures in terms of their dynamical properties and their functional contribution, but without reference to other normative or adaptive dimensions, for example, according to standards of efficiency or level of skill. The explanatory value of the sensorimotor approach is, however, often expressed in normative terms, such as "being attuned to SMCs," possessing "knowledge," or "skillful mastery" of the laws of SMCs, things that can happen in better or in worse ways, appropriately or inappropriately. Consider the discussion by O'Regan and Noë (2001, p. 942) about congenitally blind people who recover vision after undergoing a cataract operation. These patients must learn from scratch different visual attributes (e.g., the changing shapes of a disk that depend on the subtended angle, which in the case of normal vision are nevertheless perceived as always belonging to a circular disk). Presumably, the lawful covariation of shapes with respect to eye movements is the same in the case of these patients as in people with normal vision, in other words, they all share a similar SM environment. And after some basic training, they would share a similar structure for the SM habitat and progressively similar patterns of SM coordination as they learn visual tasks. Yet, for normally sighted people, visual attributes are deeply ingrained and their perception is already adjusted to a level of visual performance required by constraints of speed, efficiency, etc., while for people with recovered vision, the task of learning them to the point they must not think explicitly about them takes quite some effort and time. This difference motivates the proposal of a sensorimotor concept able to account for variations in the larger organization of SM coordination patterns according not yet to a level of mastery or lack thereof, but more neutrally, to norms in general. The need for this concept will become clearer in Chapters 4 and 5 as we begin to address questions of mastery and agency.

standard of skill / efficiency should not be defined in numeric

A normative framework involves reference to given criteria that distinguish or value some possible outcomes as preferable to others: a dexterous movement versus a clumsy disaster, achieved know-how versus lack of expertise, and so on. We are not going to discuss the origins of such norms at this point (this comes up later in the book), we will just assume that they exist and that one can provide some kind of normative gradation to behaviors in a given situation, such as efficiency, fitness, optimality, or even subjective criteria like hedonic value.

Performing a complex task requires various SM coordination patterns spread out in time, sometimes in sequential order, sometimes in parallel, and involving different action and perceptual systems. Picture again our museum visitor blindly attempting to find her way to the exit. After a few trials, she probably will be able to quickly distinguish the slow curvature of the wall; after almost bumping into something, she will also move with one arm stretched out forward to warn her of possible collisions, and so on—her patterns of SM coordination will become progressively better integrated.

Even if several combinations of SM coordination patterns may be efficacious, some may be more efficient than others. Within a normative framework, we can introduce the notion of a *sensorimotor scheme* (or sensorimotor organization or strategy). The idea describes an organization of SM coordination patterns that is regularly used by the agent because it has been evaluated as preferable (along some relevant normative framework) for achieving a particular goal. The development or acquisition of a SM scheme describes how an agent becomes attuned to a specific situation, by selecting and modulating SM coordination patterns in accordance with relevant norms. The notion corresponds to an even more agent-centered and history-dependent idea of SMCs and it seems close to some of the uses of the term that link SMCs and personal level phenomena.

A discussion of possible interpretations of terms such as "task," "goal," or of the origin of norms will have to wait until Chapter 5. For now, it is enough to say that it does not matter if one chooses to ground norms in cybernetic ideas about mechanisms for flexible and robust achievement of final conditions (Rosenblueth, Wiener, and Bigelow 1943), in evolutionary considerations of how a trait has contributed to a species' survival (Millikan 1989), or in enactive concepts of agency and sense-making (Di Paolo 2005; Di Paolo, Rohde, and De Jaeger 2010; Barandiaran and Egbert 2014). Our distinction of kinds of SMCs is independent of these considerations and can be applied once we have settled on a suitable understanding of these terms in the context of a particular case.

We summarize the main properties of the four kinds of SMCs in Table 3.1.

3.5 How to squeeze a sponge (and perceive its softness)

Let us consider how these notions can be applied to the example of perceiving the softness of the sponge. This is a very useful example, which was proposed by Erik Myin as a tool for thinking about the non-arbitrary relation between the sensorimotor engagement with the environment and perceptual experience. "In the sensorimotor account, the experience of softness comes about through a specific pattern encountered in a sensorimotor exploration, including such facts as that if one pushes on a soft object, it yields … there seems to be an intimate and intuitive link between this pattern of exploration and the experience of softness" (Myin 2003, p. 43).

Imagine you are holding a small spongy ball as in Figure 3.4. We can use this situation to see the four kinds of sensorimotor regularities at work. For simplicity, let us assume that the sponge is held between thumb and forefinger, and that the motion is constrained to pinching movements controlled by the activation of relevant muscles (**m**). The pertinent sensory variables (**s**) include the pressure felt on the surface of the skin, proprioception, and the sense of effort required to maintain a certain grip. The body variables (**p**) include the distance between fingertips, the fatigue state of the hand muscles, and so on.

The SM environment describes the sensory consequences of all possible motor commands in this situation. It expresses, for example, how tactile sensation in each finger is

Table 3.1 Summary of the four types of sensorimotor regularities (SMCs) indicating dependencies on different factors

	Environment	Body	Internal Activity	Task	Normative Framework
Sensorimotor Environment					
The set of sensory states as a function of motor variations independently of the agent's internal dynamics	✓	✓			
Sensorimotor Habitat					
The set of possible sensorimotor trajectories traveled by a closed-loop agent for a range of values of relevant parameters	✓	✓	✓		
Sensorimotor Coordination					
Individual trajectories within the SM habitat that occur reliably and contribute functionally to a goal	✓	✓	✓	✓	
Sensorimotor Scheme					
Efficient organizations of SM coordination patterns developed, acquired, or selected as a consequence of being normatively evaluated	✓	✓	✓	✓	✓

Adapted with permission from Buhrmann T., Di Paolo E. A., and Barandiaran X. (2013). 'A dynamical systems account of sensorimotor contingencies.' *Frontiers in Psychology*, Volume 4, doi: 10.3389/fpsyg.2013.00285. Copyright © 2013 Thomas Buhrmann, Ezequiel A. Di Paolo and Xabier E. Barandiaran.

related to pinching movements, and identifies the positive correlation between sensed pressure and the closing of the pinch. If motor commands decrease the distance between fingertips, the sensed pressure will increase. If the fingers are separated, the sensed pressure will drop. Contact sensation eventually vanishes altogether when the distance between the fingertips is larger than the maximum diameter of the sponge (in both fingers if the ball drops). Similarly, and depending on the material properties of the sponge (size, mass, and composition), the relation between squeezing and resistance will not be even (Figure 3.4 B, top). The more you squeeze, the greater resistance the sponge will offer—up to the point that you have more or less crushed it completely. There will also be similar correlations with other sensations apart from tactile pressure, such as proprioception and the sense of muscle effort. Because the sponge does not yield entirely (as if we were pinching water), some correlations in the SM environment will also be affected by the integrated amount of physical effort exerted. This may affect the accuracy of the tactile and proprioceptive sensation as a function of the material properties and the state of muscle fatigue.

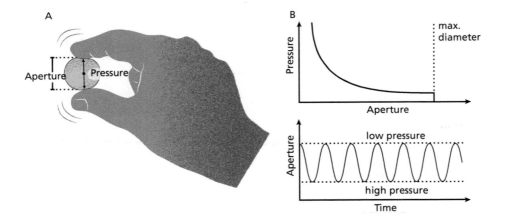

Figure 3.4 How to feel the softness of a spongy ball. **Panel A**: When the sponge is squeezed between the fingers, certain relationships will hold between motor-related variables, e.g., muscle activation patterns, opening of the grip (aperture), strength exerted, etc. and sensory variables, e.g., the sponge's resistance to squeezing felt as pressure at the fingertips, fatigue, and so on. **Panel B**: hypothetical sensorimotor contingencies. The upper plot shows a contingency in the SM environment: a static relationship between motor commands that realize a particular aperture of the grip and the pressure sensed as a result. The bottom plot shows a regularity in the SM habitat. As a result of a particular explorative strategy (here repeated squeezing with movement reversals at selected low- and high-pressure points) and the springy characteristics of the sponge, a regular oscillatory movement is enacted. The specific frequency and amplitude of the movement contribute to the perception of the sponge's softness and elasticity.

The SM habitat can be described only if we know how you will respond to the various sensations (i.e., if we close the loop and allow you to explore the sponge as you wish). We could assume, for example, that, like many others, you have a tendency to squeeze the sponge until a given level of resistance is met, at which point the direction of movement is reversed, and you release the squeeze, until a minimum pressure threshold is reached before fully losing tactile contact, and the squeezing restarts. In this way, and at least for a few cycles, a regular pattern of squeezing and releasing is enacted moving between high- and low-resistance points. This is just an assumption; of course, there are other ways of exploring a sponge. But assuming this is what you do, the SM habitat in this case exhibits regularities that reflect the resulting rhythmic pattern and the points of movement reversal. For instance, under the right conditions, the sponge will behave as a damped spring externally perturbed by oscillatory force (Figure 3.4. B, bottom). It is a well-known result from classical mechanics that such periodically forced systems can exhibit resonance (i.e., certain frequencies of forcing produce the largest spatial displacements). The resonance interval and optimal frequency of forcing depend on the springiness of the sponge (the relation between response force to a given displacement), on its mass, and on its internal friction or damping factors.

[handwritten: Difference betwee SM env and SM habitat?]

While in the SM environment we had a functional, point-by-point relation between finger movement and sensation, the corresponding relation in the SM habitat is expected to depend on the frequency with which you squeeze and release the sponge in a periodic movement. Such features cannot be deduced from the SM environment, and the reason is, as we have said, that the SM habitat includes not only sensorimotor variables, but also internal variables that play a role in the generation of the particular behavioral pattern with time-extended properties.

[handwritten: why are internal variables]

[handwritten: de/dm important?]

To speak of SM coordination, we need to link this sensorimotor activity to a task. Let us assume that you want to discriminate between soft and hard sponges. One solution to this task relies on using the rhythmic pattern just described. Proprioceptive feedback at the high-resistance turning point of the movement is indicative of how resistant the sponge is in term of spatial displacement. For a given amount of effort, small displacements between the high- and the low-resistance points mean the sponge is hard. And it is softer in the case of larger displacements (the more you can squeeze it with the same effort). Alternatively, a similar coordination could control the pressing movement to obtain always the same displacement. Here instead of keeping effort more or less constant, you squeeze only a given distance. In this case the pressure sensation would be indicative of the degree of softness. Both patterns constitute cases of SM coordination because they are efficacious for the distinction of soft and hard sponges. Other types of movements, such as moving the fingers along the surface of the sponge with minimal pressure, do not have this property.

[handwritten: why do we need a task of SM coordution.]

The SM coordination patterns described can be enacted in a variety of ways, all of which will solve the task. Repeated squeezing, for example, might be helpful in getting a more reliable estimate of sponginess. Depending on the material and other physical properties it might be possible to find the resonant frequency of forcing, and that may also provide you with quick clues as to the "kind" of softness of the sponge (whether it is rather "springy" or disposed to deformations). How to approach these choices among different instances of SM coordination will depend on a normative framework in which your activity is immersed (e.g., whether you care about speed, accuracy, efficiency, and so on). Imagine the choices you would face if you were a worker whose daily task is to sort out thousands of sponges into soft and hard. Putting the right SM coordination patterns together according to the relevant norms results in what we have called a SM scheme.

[handwritten: a Help hae to lo grd onese d.]

[handwritten: How to approch choosing different instances?]

3.6 **Tactile discrimination: a minimal model**

The sponge example suggests that the four notions of sensorimotor environment, habitat, coordination, and scheme, while intimately linked to one another, are indeed distinct and can be used for different purposes. But to propose a conceptual distinction that seems to hold and potentially be of use is not the same as demonstrating that the resulting notions can be profitable in a scientific context. For this, we still need to show that these

[handwritten: What is tactile discrimination.]

notions are not only operational, but that their application does not demand an unrealistic amount of presuppositions and idealization.

In this section we intend to briefly show the four notions at work in a simple model originally presented in Buhrmann, Di Paolo, and Barandiaran (2013) (where more details can be found). And in the next section, we discuss ways in which work in psychology, neuroscience, and robotics has already made use, in different ways, of the different SMC concepts.

Once again, let us imagine that circumstances require us to solve a problem using only tactile perception, for instance, finding the light switch on the wall. To make matters more interesting, we can imagine that there is more than one switch on the wall, one for the light, the other for opening a garage gate, or anything else. They are similar in shape but one is a bit bigger than the other. Put more abstractly, we can imagine our problem as identifying two similar shapes of different sizes as depicted in Figure 3.5 using only simple, one-point tactile perception, for instance, by touching the object only with the fingertip.

If the objects that need to be discriminated are of similar shape and not too different in size, then the instantaneous tactile sensation at the fingertip will not be very informative in itself. It will be rather ambiguous, and the task will require an active exploratory approach. We will examine a possible solution to this problem using a minimal agent-based model. But before we do, in case the task seems rather trivial, we suggest actually trying this out before continuing. Pick, for instance, three or four coins of different sizes but similar thickness. Ask someone to place them on a table in a random order without you looking. Then with your eyes closed, feel the coins with the tip of only one finger, try to order them by size. It may be harder than it seems. Eventually, you may get better after a few trials. If so, try to figure out exactly how you are doing it. What is your strategy? What if you tried with different objects, say pens of different sizes?

Let us approach this tactile discrimination problem with a minimal cognition model of active categorical perception (see Box 3.1). This will allow us to illustrate and analyze in detail the agent's sensorimotor dynamics. With this example, we will be able to illustrate the operational nature of the notions of sensorimotor environment, habitat, and coordination (we present illustrations of sensorimotor schemes in Chapters 4, 5, and 6, using scenarios that are more complex). We keep technical details to a minimum (see Buhrmann, Di Paolo, and Barandiaran [2013] for more information).

In our model, an agent can move along a one-dimensional environment that contains stimuli in the form of two bell-shaped objects of different widths (Figure 3.5). These objects need to be discriminated. The agent moves along a line and can sense the shapes via a single sensor. The activity of this sensor (s) increases proportionally to the proximity of the object directly in front of it.

The agent is controlled by a very simple neural system. It takes in the rate of change of the sensor signal s and feeds it to a small dynamical neural network that delivers continuous motor commands (m) controlling the agent's velocity. The network is minimal: it only

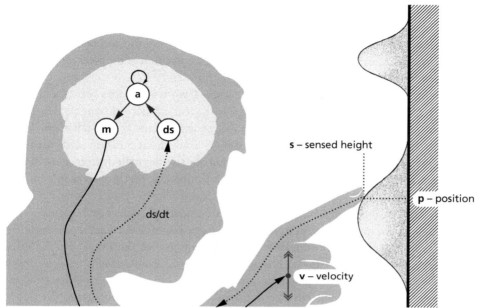

Figure 3.5 A simple sensorimotor task. Discriminating between two similar objects of different sizes using only the tip of one finger can be modeled using a minimal agent moving in a one-dimensional environment along the surface of different objects. The agent possesses a sensor (**s**) stimulated by the perceived height of objects in front of it (narrow or wide bell-shaped objects), which feeds its time derivative into a set of hidden neurons (**a**) of the agent's neural network. The hidden neurons in turn are recurrently connected to themselves and drive the motor neuron **m** controlling the agent's movement. This changes the agent's velocity **v** along the object's surface resulting in change of position **p** and corresponding change in the sensed height **s**.

has two neurons, and each neuron acts as a leaky integrator with its own time constant and other parameters.

The task to be solved by the minimal agent is the discrimination between wide and narrow shapes, requiring it to move away from the former, and approach the peak of the latter. No instantaneous sensory input can be used to discriminate one shape from the other (the height and horizontal positions of the two shapes are varied such that within each trial they can be distinguished only by their steepness and horizontal extent). Consequently, an active sensorimotor strategy is needed. We optimize the parameters of the neural network using a search algorithm. We test each candidate agent in different trials of the task and assign it a score. From trial to trial, the two shapes are presented in random positions with a minimum guaranteed separation. Agents are selected according to how well they approximate the ideal performance. Toward the end of our search, we obtain agents that are able to find their way toward the position of the peak of the narrow shape and move away from the wide shape.

The best agent thus found distinguishes narrow from wide shapes across a range of widths and heights (exhibiting generalization with clear categorical boundaries). In the case of narrow shapes, the agent approaches at constant velocity and starts oscillating once in contact with the shape. This oscillation is asymmetric and such that the agent slowly approaches the peak, close to which it ultimately settles with decreased oscillation amplitude. For wider shapes, the initial approach is identical. But instead of moving in a cyclical fashion toward the peak, the agent moves away from it.

To explain this behavior, we study the SM environment and habitat of the agent.

The SM environment, to repeat, captures properties of the external environment and the agent's sensorimotor interface, without taking into account the agent's internal dynamics. In Figure 3.6, we see the change in sensory values that results from the agent being placed at a given position relative to the center of a shape and having issued a certain motor command (moving at a certain speed). Because the agent senses the rate of change of the height of the shape at its current position and for a given motor command, a specific change in sensor value is produced for each position and velocity of the agent. The surfaces shown in Figure 3.6 (narrow shapes on the left; wide shapes on the right) represent the agent's SM environment. They capture the functional relation between **s** and **m** (i.e., the sensory consequence of performing an action [moving at velocity v] taking position **p** as a parameter).

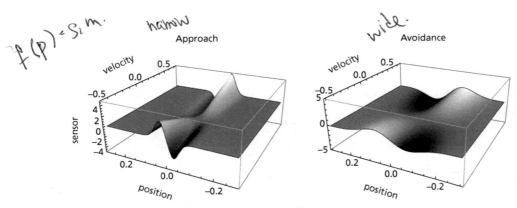

Figure 3.6 Mapping the SM environment. The figure shows the SM environment corresponding to the narrow (approach) and wide (avoidance) shapes. The surfaces show the sensory change that results from issuing a motor command that determines agent velocity at a given position.

Reproduced with permission from Buhrmann T., Di Paolo E. A., and Barandiaran X. (2013). 'A dynamical systems account of sensorimotor contingencies.' *Frontiers in Psychology*, Volume 4, doi: 10.3389/fpsyg.2013.00285.

Although the two SM environments look similar, the one corresponding to the wide shape (shallower gradient on the right of Figure 3.6) leads to lower absolute sensor values. What are the regularities of this surface (its "laws" of SMCs) and how do they contribute to explaining the agent's behavior? It is clear that there are two types of symmetries in the SM environment: those reflecting properties of the external environment (i.e., the bell-shaped figures) and those reflecting the agent's motor possibilities (i.e., the fact that the agent can move in one direction or its opposite). As a result, we notice that sensor values as a function of position and velocity observe the symmetry s(p, v) = s(-p, -v). That is, the sensor value at distance to the right of the peak of the shape and moving in a particular direction is the same as the sensor value at the same distance to the left of the peak of the shape and moving in the opposite direction. Specifically, the sensorimotor surfaces feature two symmetrically arranged peaks and troughs that correspond to the points of greatest slope of the bell shape. These peaks are found at either side of the bell curve (leading to the first symmetry) and change in sign if the agent's velocity changes direction (leading to the second symmetry). The surface also reflects other general properties of the sensorimotor coupling, such as the fact that sensor activity is continuous and smooth.

The structure of the SM environment, for example, its symmetry, can constrain but does not fully determine the possible behavioral strategies for solving the task. It does not provide us yet with the full explanation of the behavior of the agent. However, it is informative. For example, without any further knowledge of the agent's internal structure, the SM environment predicts—and this has been confirmed—that if the sign of the motor signal were reversed (effectively exchanging the directions of movement), then the same discrimination behavior would be observed, but with the difference that the agent would now "scan" the shapes on the opposite side of the peak. Thus, however the agent works, we know that an agent that responds with the exact same motor commands but with reversed direction will also be able to solve the task.

The SM habitat describes the relation between sensor activity and motor commands taking into account the internal dynamics of the agent. It is the set of all possible trajectories that the agent takes, given a range of boundary conditions and parameters. If we are dealing with a potentially complex, nonlinear dynamical agent–environment system, providing a full analytical description of this set may be infeasible. A typical approach is to adopt a quasi-static method. The idea here is to treat the variable that links two components of a complex dynamical system, such as the sensor variable in our model, as a fixed parameter. This allows us to study the tendencies of the remaining variables for this given value of the parameter. This removes temporal variation from the dynamical component of interest, and so we can calculate its qualitative behavior (limits sets, attractor basins, etc.) for the given, now fixed, value of the sensor state. Separately, as a next step, we can then study how the qualitative behavior changes as the parameter is varied (e.g., see how the tendencies change, how the attractors move, and so on; in other words, a bifurcation analysis). Together, these two analyses can approximately describe how the overall behavior of the component results from the change in its qualitative dynamics as the normally time-varying input changes.

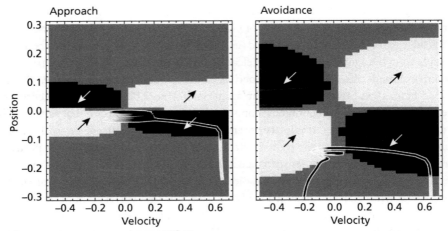

Figure 3.7 Mapping the SM habitat. This is a partial view of the SM habitat of the minimal agent for the case of the narrow (left) and wide (right) shapes. Regions are colored by calculating the asymptotic tendencies of the motor commands as sensor values are held fixed at the value given by the SM environment surfaces in Figure 3.6. The light and dark regions indicate opposing tendencies; gray regions have more than one tendency depending on the state of the agent. The arrows indicate the long-term attractors for changes in velocity and position, and the curve shows the trajectory of the behaving agent.

Reproduced with permission from Buhrmann T., Di Paolo E. A., and Barandiaran X. (2013). 'A dynamical systems account of sensorimotor contingencies.' *Frontiers in Psychology*, Volume 4, doi: 10.3389/fpsyg.2013.00285. Copyright © 2013 Thomas Buhrmann, Ezequiel A. Di Paolo and Xabier E. Barandiaran.

In Figure 3.7 we show the result of carrying out such analysis for our model agent (i.e., the qualitative behavior of the agent for different values of its sensory input). Here, for each fixed sensor state (which is given by the surface of the SM environment as a function of position and motor command), we have determined the steady state of all the agent's variables, that is, the state to which the agent would ultimately converge given enough time (its attractors). Of these states, we have plotted only the motor command, and have used different shades of gray to indicate the long-term tendencies at each position.

Imagine we are looking at the surfaces of the SM environment (Figure 3.6) from above. For each position, we read the sensor value given by the SM environment at that position and, keeping it fixed, we let the agent "run," measuring where it goes and at what speed. The result is a transformation of the four (smooth) peaks and troughs of the SM environment into four (discrete) regions of attraction. The light regions here correspond to positive velocity attractors, and the black ones to negative velocity attractors. Additionally, in some regions the system is bistable (gray). In these areas, which attractor the system tends toward depends on the internal state at the given time. The arrows in each region give an indication of the tendency of the change in velocity and position. This kind of plot indicating long-term tendencies for different fixed values of parameters is very useful and commonly seen in the literature on minimal cognition models.

What are the salient regularities of this attractor landscape? Firstly, it reflects the same symmetries as the SM environment, one owing to the environment and the other to the agent's motor capabilities. It should be noted that this is not by necessity. However, the requirements of the task here have led the agent (or rather the search algorithm) to preserve in its internal dynamics those features of the environment that are useful for achieving the task.

Secondly, the attractor landscape implements a binary choice. Depending on whether the sensory states are above or below certain value, the system will tend to move in either one or the other direction (indicated by the arrows). One prediction we can make based on this structure is that because there is no attractor that would lead to zero velocity, the only way the agent can stay close to a peak is by using changing sensory inputs to oscillate back and forth between different attractors. This is what we observe as we superimpose the actual trajectory of the agent on this plot (curves in Figure 3.7). Looking at the agent's trajectories within this attractor landscape, one can also explain the difference between the approach and avoidance behaviors. Even though the possible steady-state motor outputs are identical in both types of environment, the regions in which they can be found differ in size (left and right plots). This results in the initial approach of the agent toward the shapes being identical. But when faced with wider shapes, this trajectory does not carry the agent as far through the attractor region as it does for narrow shapes. Therefore, when the agent leaves the initial attractor region, it manages to enter into an asymmetric cycle of transitions between the two attractors in one case (approach), but it is fully captured by one of the attractors in the other (avoidance). The asymmetry of the oscillating approach pattern, in turn, can be explained because the absolute steady-state velocities are different in value for each attractor.

From a wider perspective, we can therefore say that the agent's attunement to the environment and task depends on its internal structure. The attractor landscape transforms the SM environment into movements that conform to the achievement of the task. The attractor landscape, itself modulated by sensory perturbations, also determines which areas of the sensorimotor environment the agent will visit and how. For every sensory perturbation, this landscape provides us with information about which action the agent will tend to take and, therefore, what the following sensory stimulation will be as a consequence.

The superimposed trajectories in Figure 3.7 already give an indication of the SM coordination used by the agent to stay near the narrow peak or escape from the wide one. In the case of approach (on the left of the figure), we see the trajectory converging to a low dimensional-stable oscillation touching the two opposing attractor regions. This contributes functionally to the performance of the task as the small oscillations occur near the peak of the narrow shape. In the case of avoidance, the oscillation does not converge to a stable pattern, and the agents moves away from the shape. The SM coordination takes the form of a reliable transient in this case.

In both cases the properties of the oscillations are a consequence of the mutual influence between sensor and motor activity. It is not a case of the sensors providing information to be processed internally concerning the geometrical properties of the shape and subsequently choosing an appropriate motor response. Moreover, the sensorimotor patterns do not correspond to, nor are they determined solely by, endogenous "brain" dynamics. We find a stable neural oscillatory pattern in the case of approach, but this cannot be trivially deduced from features in the agent's internal state space (limit sets) or changes therein (bifurcations). Rather, what happens here is that the agent's state is always "chasing" one attractor or the other, without ever being captured by either. Because the agent's sensory input changes as a result of its movements (on the same timescale as the behavior), so does the asymptotic steady state that the agent tends to. In other words, the agent is always on a transient toward an attractor that is regularly changing its position. Interestingly though, the overall pattern that emerges forms a stable oscillation, which corresponds to a limit cycle of the coupled agent–environment system. In other words, the environment plays as much a part in the creation of the observed SM coordination pattern as the agent's internal dynamics. This is what in Chapter 2 we have called a *world-involving* explanation.

In summary, the different kinds of SMCs we have introduced contribute to a full explanation of the observed behavior. SM coordination patterns determine a relation between lower dimensional reliable trajectories in sensorimotor dynamics and their contributions to task functionality. The SM habitat establishes the "rules" that explain the presence of SM coordination by examining the dynamics of the coupled internal and environmental variables. And the SM environment establishes the wider constraints to the SM habitat such as symmetries that depend solely on the environmental and embodiment structures.

In addition to verifying the fact that the proposed kinds of SMCs are operational, our minimal model also highlights some aspects of the sensorimotor structures that may not be immediately obvious. These include the observation that regularities in one sensorimotor structure can be preserved in another, if this helps solving the task, even though this need not generally be the case; the non-reducibility of SMCs to "brain" dynamics alone; the fact that SM coordination patterns serving different functions (approach and avoidance) can be implemented in a single, non-differentiated system (functional diversity without structural modularity); and the role of the environment as an equal player in the "selection" of which coordination to enact at a given time. The model brings home these subtle aspects of the definitions in a way that would have been difficult to arrive at otherwise.

3.7 Investigating SMCs

How are the different concepts of SMCs applied in various scientific contexts? We find that the four notions are regularly used whether in mathematical formulations or through approximate statistical measures, but that, as expected, they are not described using the terminological conventions we have introduced here.

Let us first look at an example from acoustic perception. Inspired by the work by Philipona, O'Regan, and Nadal (2003) on the mathematical reconstruction of spatial invariants through the active exploration of sensorimotor topological properties, Aytekin and colleagues (2008) have shown how to reconstruct the acoustic features through which it is possible to locate distant sound sources based on compensable movements of the listener's head. They demonstrate that with the application of infinitesimal changes in head orientation, it is possible to learn properties of the manifold of sound source locations. The idea of reconstructing spatial invariants based on learning correlations between sensor movement and sensor activity goes back to Poincaré (1902; see also Cassirer 1944; Vuillemin 1972), who wondered how it is possible to construct a perception of spatial relations based upon the (supposedly) space-less impressions that impinge on our sensors. The answer to this question came from noticing that some of the changes in sensations could be compensated or cancelled out by a self-generated response (a voluntary movement). This realization provides the foundation for distinguishing between different classes of sensory changes, those that can be compensated by movements and those that cannot. The properties of the set of motor responses that an agent needs to execute for cancelling compensable sensory changes can provide an agent with clues about the kind of space she inhabits. For instance, its dimensionality, whether it is smooth or not, whether it is homogeneous or warped, and so on.

Aytekin and colleagues apply their Poincaré-inspired framework to stored examples of head-related transfer functions in humans and bats. Their method measures the SM environment. They map the sensory input changes caused by a set of infinitesimal head movement in the presence of a sound source at a given location. These sensory changes are smooth for small variations of the source location and they correlate when the same head movements are applied for sound sources at different locations. The resulting topological maps represent smooth transformations of the hemisphere of angular orientations of head movements relative to source. Interestingly, these auditory spaces can differ significantly, not only between humans and bats but also among humans themselves, depending on the frequency-dependent acoustic filtering introduced by the head and the shape of the ears (pinnae), which change from one person to the next. In keeping with our definition, the SM environment depends crucially on features of the body.

We see in this example, and also in the work by Philipona and colleagues, as well as others (see Laflaquière et al. [2015] for an application to a simulated visually guided robot) how the SM environment, the most abstract of our SMC concepts, can be highly informative, its "laws" providing information regarding, for instance, the classes of compensable changes, dimensionality, smoothness, warping, and relative resolution.

Research into the structure of SM habitats necessitates relatively unconstrained agents who can move more or less in the way they would normally do in everyday situations. This can make it hard to design appropriate lab studies, but not impossible. Recent investigations of toddler visual experience by Linda Smith and her colleagues at Indiana University are a good demonstration of how sensorimotor habitats and sensorimotor coordination can be studied empirically in quasi-naturalistic settings. They investigate a situation involving a toddler (between 1 and 2 years old) sitting comfortably at a small table facing her mother. On the table there are a few different toys. Both mother and child are able to freely move their arms and heads, to direct attention where they wish, and to interact normally. The mother is instructed to encourage the child to play with the different toys, sometimes exchanging them. Both parent and child wear head-mounted cameras, and their interaction is recorded, as well, by a bird's eye view camera (Smith, Yu, and Pereira 2011).

The authors study the statistical properties of the video from the head-mounted cameras and in particular the proportion of the image frame occupied by the different objects. Without using the term, their observations yield information about the differences between the SM habitats in mother and child. While the mother keeps a broad view of the tabletop, the child, and the various toys, the child's view is highly selective, focusing on a single toy at a time, typically bringing it very close to the face so that it occupies most of the visual field. Smith and her colleagues discuss several possible reasons for these differences. They include the differences in body size, movement patterns, and general interest in the toys. Because toddlers have short arms, their activity is constrained to a smaller visuomotor space near the body (this is something that in our terminology we would describe as a structure of the SM environment because it involves general properties of the body in relation to the surroundings). Within these constraints, objects are likely to be held close and to occupy a large proportion of the child's vision, partially blocking other objects. In combination with highly synergistic movement patterns typical of toddlers, the result is a prolonged focused manipulation of the object very close to the face, inevitably implying a strong visual attention directed at the object (this is a structure of the SM habitat in our terminology as it arises from typical patterns of activity in the child). The authors derive some developmental implications from these observations; for instance, the extreme focus on one toy at a time may be helpful for learning about objects and their properties. Later work in similar settings examines the quality of social interaction between mother and child (Yu and Smith 2013), which is also affected by object visual dominance in the child, who in these circumstances pays less attention at the parent's face and gaze. Coordination seems to be induced via shared attention to the object itself (we come back to the interplay between interactive and sensorimotor development in Chapter 8).

Being task-related, sensorimotor coordination may be easier to study in the lab. An interesting area of research in active perception attempts to understand how visual flows coordinate with action-related clues in the body, such as proprioceptive or somatosensory signals or reafferent feedback, that indicate preparatory or actual activity of the muscles. For instance, some visual flows projected in front of an observer on a two-dimensional screen can have different three-dimensional interpretations: they may look as if a plane is rotating vertically or as if it is rotating horizontally and moving toward the observer. Typically, one interpretation is dominant if the observer does not move. If the observer moves toward the image but keeping optic flow intact, there is still an ambiguity between two three-dimensional interpretations, but now none of them is dominant (Cornilleau-Pérès and Droulez 1994). This result indicates that the active movement of the observer coordinates with the visual flow as sensed by the eyes and contributes to the production of a different perceptual experience (see Wexler and van Boxtel [2005] for a review of this and similar examples; see, also, Simons et al. [2002] on the effect of retinal and extraretinal coordination in active observers on the recognition of complex objects).

Another way to study SM coordination and schemes is to contrast how a person learns to use different sensory substitution devices for a same task. Bermejo et al. (2015) study the head and arm movements of participants who are sitting blindfolded and must recognize simple geometrical shapes in the space in front of them using either a head-mounted or hand-held echolocation device, or alternatively a head-mounted image-to-sound system (the vOICe device, Auvray, Hanneton, and O'Regan [2007]). The three substitution devices induce different kinds of task-related coordination as participants learn to recognize the shapes. Two of the devices allow for wide and narrow sweeping movements of the head or the arm, which often dominate the initial stages of recognition, whereas the vOICe system, because it transduces visual images by an automatic sweep to convert them into sound, does not allow this, and so must be used differently. Similarly, the three devices induce different exploratory coordination patterns relevant to the task.

We see in these examples, that even though we have formulated our notions of SMCs using dynamical systems concepts, the notions themselves can be studied without necessarily having to have a full dynamical description of the agent and her situation. Still, a criticism that could be leveled against our dynamical definitions is that their practical application will be severely limited by the fact that we seldom know all the relevant processes that govern a cognitive system to express them in the form of equations. This is indeed true if we were to demand a full dynamical account but as we have seen this is not always necessary.

Furthermore, although formulated in the language of dynamical systems theory, the four kinds of SMCs can be studied using a variety of methods, like information-theoretic measures and other probabilistic approaches. Let us mention a few examples.

Using a simple robotic platform, Maye and Engel (2011, 2013) have developed an analysis of SMCs as fundamentally probabilistic: for a given action, given sensory results are more or less likely. Thus, a probability distribution over action–outcome pairs is one way of capturing the current sensorimotor landscape in which the agent finds itself. They implement this approach in a wheeled robot using discrete-time Markov models. The robot generates action commands and records the sensory consequences. It turns out that for a real (or simulated) simple embodied robot, only a small percentage of the action–outcome space is actually encountered. Therefore, the model is soon able to estimate the probabilities of action–outcome pairs that occur in the robot, based on the frequencies with which these have been observed in the past. In order to generate actions appropriate to a given task, the robot also needs some form of evaluation function. In this case simple evaluation functions are hard-wired into the robot: for instance, the robot may be "punished" for falling over or for hitting anything with its bumper. Given this setup, it is possible to choose the next action that maximizes the expected reward. Applying this idea to the Robot Puppy (a puppy-shaped robot platform using four-legged locomotion) results in the robot learning which one of several possible gait patterns is the most stable while walking on different surface materials, and to adopt the gait (or pattern of gaits), which gives it the greatest overall stability (Hoffman et al. 2012). It should be noted that the parameter of interest (the kind of supporting surface) cannot be read off directly from any sensor, but is rather implicit in the patterns of sensorimotor interaction.

It is possible to build progressively a probabilistic model of sensorimotor mappings in real systems using these methods, and such a model will reveal structures of what we have called the SM habitat. In these cases, the model is used to drive learning in a robot, but similar methods could equally be applied to studying the behavior of an animal.

Another methodology for describing sensorimotor relationships makes use of information-theoretic measures. Schmidt et al. (2012) have studied the statistical information transfer relationships between the different sensors and motors of a compliant quadruped robot that is controlled by either random or coordinated (but open-loop) leg actuations (again the Robot Puppy). Instead of analyzing the particular sensory consequences of given actions, however, the approach here extracts more abstract information about the agent–environment coupling, namely the directional correlations between the various sensors and motors. The random controller, which does not exhibit the same kind of correlations in motor signals as the coordinated controllers, is shown to reflect mostly the effects of the environment (contact with the floor surface) and the body structure (hip actuators driving hip angle sensors in the same leg). The information extracted from this open-loop test corresponds to the SM environment in our account, where we eliminate the sensorimotor loop and study the sensory consequences of motor commands, in this case sampled randomly. The information obtained in this way provides a description of the constraints of the agent's body in a given environmental context.

Schmidt and colleagues have found that by introducing coordinated movement patterns in their quadruped robot, further informational dependencies are induced through the regular time-extended behavior of the robot's gait. For example, information transfer now also occurs among variables belonging to different legs, and certain dependencies reflect properties of the task, such as more information being transferred between some variables when turning rather than walking straight. Here, though the system is still an open loop with respect to externally imposed motor commands, the analysis takes into account the ongoing interaction between agent and environment during a "meaningful" task.

We can see that our definitions of SMCs, though derived from a dynamical perspective, are general enough to be investigated with different methodologies. We do not need to have a full dynamical description of a given situation in order to apply these concepts as tools for investigation. The example of the information-theoretic approach demonstrates that knowledge about dynamical sensorimotor structures can be obtained even without analyzing (or indeed being in possession of) a complete set of equations describing the agent and the environment. And, as we have seen in the research by Linda Smith and her colleagues, the investigation of sensorimotor structures is not restricted to simple cases only. The four SMCs concepts allow us to think differently about cognitive and perceptual problems of all kinds.

3.8 Emphasizing the perspective of the situated agent

Arguably, an interesting aspect of the sensorimotor approach to perception is its agent-centeredness (i.e., taking into consideration the agent's embodiment, situatedness, skills, and goals). As suggested, the four kinds of SMCs can be arranged along a dimension ranging from an external perspective of analysis to one increasingly focused on the agent. The SM environment requires for its definition the least amount of detail about the agent. It corresponds to all agents with similar sensors and effectors in a similar environment. The SM habitat is more agent-specific and it adds the agent's internal dynamics and closes the sensorimotor loop. The SM coordination patterns bring on a particular task-oriented dimension. Finally, SM schemes add a normative dimension with the inclusion of some idea of value (efficiency, degree of skill, etc.).

In a similar vein, time plays different roles in each of the four kinds of SMCs. All of the structures can be time-dependent in the sense that external dependencies on time can alter both the environment and the agent (e.g., seasonal rhythms or the effects of age and wear). Other than this, the SM environment is "atemporal": it describes all possible sensory consequences of freely introducing a motor change. The SM habitat involves the set of possible trajectories. As such, it provides dynamical information and introduces relevant ideas such as trends, attractor landscapes, oscillations, etc. SM coordination patterns entail a more "local" element of temporality than that of dynamical trajectories because they rely on the fine-grained exercise of specific agent–environment engagements with the added constraint of contribution to functionality. Elements of duration, rhythms, etc., become crucial for this contribution; for instance, oscillatory patterns different from those observed in our simulated agent around the narrow peak may not contribute to the task. Finally, SM schemes add to this latter aspect that of a temporal organization of SM coordination patterns themselves. Efficiency, resilience, and other normative evaluations will be affected by how patterns are coordinated in time, whether they run in parallel, or in sequence, whether there are hard deadlines, delays, and so on.

The four definitions also highlight the relevance of the determinants of action, a question that has been rather absent in the sensorimotor approach. Except for the SM environment, which describes a dependency of sensory activity on motor changes, the other kinds of SMCs are strictly speaking sensorimotor *co*-dependencies, as the loop is closed. Sensorimotor theory has been formulated in terms of SMCs, but these have been almost exclusively illustrated as one-way sensory dependencies on motor commands (recall the line and retina example). Action has been treated more or less as a free variable in many of these illustrations. By this we mean that the appropriate action in a perceptual context is brought into an explanation of perception as required and without constraint (a topic we shall return to in later chapters). The squeezing movement of the fingers constitutes in part the perceived softness of the sponge and the stroking movement of the hand constitutes the perceived smoothness of the table surface. But what calls forth these particular movements in each case? Why don't we stroke the sponge and squeeze the table? In each case what counts as appropriate action is in part also constituted by the perceptual context—action is perceptually constituted—and this aspect has been underdeveloped in sensorimotor theory even when the mutuality between motor and sensor activity is acknowledged in general terms. Except for SM environment, the other kinds of SMCs revert this situation at the most basic level by including closed-loop dynamics explicitly.

The four kinds of SMCs can clarify the similarities and differences between the sensorimotor approach and ecological psychology (Gibson 1979). According to the latter, a much-neglected constitutive factor of cognition is the structure of the environment: agents are thought to directly perceive the world by picking up invariants in the sensory array, that is, properties that remain constant across transformations produced by self-motion (see Mossio and Taraborelli 2008). This view thus conceives of a rich structure arising from the world and the properties of the body—in our terms, the SM environment—without which we could not explain the behavior of an agent. The sensorimotor approach acknowledges the importance of this structure. However, in the ecological approach, the origin of the particular motor patterns that bring about the invariant-revealing transformations is not always considered relevant; instead, what matters in many cases is simply the structure of movement-induced flows. This suggests that in such cases, the SM environment could be deemed exhaustive for the constitution of perception. We have seen, however, that it is not the case. Key regularities are found in the closed-loop scenario in which the agent's internal dynamics play a co-constitutive, irreducible role. These regularities are described by the SM habitat and, as we have shown, cannot be deduced from the SM environment alone. In this sense, the enactive interpretation of the sensorimotor approach afforded by the four SMCs definitions may provide a more complete picture of perception compared with the ecological approach, because it gives an account of motor-independent, open-loop sensory invariants as a special case, but explicitly acknowledges, in addition, the role of agent-specific, closed-loop sensorimotor invariants.

The difference between SMCs as dependence of sensory activity on motor changes, and SMCs as mutual co-dependence can be summarized in the following way:

$$M \rightarrow S : \Delta S = F(\Delta M), \text{ Sensory dependence,}$$

$$M \circlearrowleft S : SMC = G(\Delta S, \Delta M), \text{ Sensorimotor } co\text{-dependence,}$$

where S stands for sensory and M for motor states or changes and F and G correspond to the functional relations between sensor and motor changes in one case, and between sensorimotor contingencies and co-occurring sensor and motor changes, in the other. The former corresponds to what Dewey describes, in the quote at the opening of this chapter, as a defective notion of reflex arc, while the latter explicitly acknowledges his observation that both sensory stimulus and motor response gain their significance as such only in so far as they contribute to "maintaining or reconstituting the coordination" as a whole.

As we shall see in Chapter 4, this distinction is relevant for testing the implications of the approach for novel perceptual situations. We come to a novel perceptual context for the first time already equipped with our bodily know-how. We do not confront the new situation in a naive manner but use our existing skills instead. There is not only a dependence of perception on motor activity but also a dependence of bodily movement on proprioception and the emerging perceptual awareness. This situation can converge into a stable way of exploring the perceptual situation and the development of this stability is what we shall call the progressive *mastery* of SMCs.

ecological psychology

Why is the ecological approach insufficient?

↳ emphasis on invariant-revealing transformations.
↳ emphasis on structure-induced movement flow.

Why not?

Sh env. = executive enough for perception.

regularities in closed loop in scenario's which agent's internal dynamics play a co-constitutive role

→ why is the 4SMC account more complete?

sensory + motor required
stimuli

contribute to marking or directing in reconstituting the concrete as a whole.

Chapter 4

Mastery: learning to act and perceive

As a system of motor powers or perceptual powers, our body is not an object for an 'I think': it is a totality of lived significations that moves toward its equilibrium. Occasionally a new knot of significations is formed: our previous movements are integrated into a new motor entity, the first visual givens are integrated into a new sensorial entity, and our natural powers suddenly merge with a richer signification that was, up until that point, merely implied in our perceptual or practical field or that was merely anticipated in our experience through a certain lack, and whose advent suddenly reorganizes our equilibrium and fulfills our blind expectation.

—*Maurice Merleau-Ponty, (1945/2012, p. 155)*

4.1 **Masters of the sensorimotor universe**

As we have seen in Chapter 3, an everyday sequence of acts, such as those enacted when operating a door handle and opening a door, can involve different kinds of sensorimotor regularities. They range from those that are already given by the properties of our bodies and our surroundings, to those that we bring forth when we mobilize our bodies and engage the environment in particular ways.

But the success of an action depends on how well we bring these regularities together, how we combine several sensorimotor engagements into a coherent whole. In other words, to perceive and act successfully, we must demonstrate certain sensitivities and certain mastery of their circumstances. Our living bodies must pre-reflectively understand how they move in the world and how the world changes in response.

This is the same whether we happen to be focusing on action or perception. In fact, both are organically linked as moments of how embodied agents make sense of the world. This is the reason why we speak of action that is perceptually guided and of perception that is active. In both cases, we require sensitivity and mastery in bringing together the right sensorimotor engagements into an enactment.

To act and perceive in the world therefore involves a series of skills. And, as with all skills, depending on experience and circumstance, we may be more or less proficient at perceiving the world or doing things in it. We may learn new ways of seeing and we may face the possibility of having to readapt our perception and action to compensate for the effects of illness, injury, or systematic changes in our environment or lifestyle. The processes that underlie the organization and honing of these skills operate over a wide range of timescales, which include the enaction of a particular perceptual act at one end and the lifetime development of perception-action capabilities at the other. The goal of the present chapter is to propose an account of the principles according to which these transforming processes are organized.

We proceed along similar lines as before, offering dynamical systems interpretations of important concepts introduced by the sensorimotor approach to perception. In particular, we examine a crucial but neglected aspect of this approach: sensorimotor learning. This will be another step on the road to an enactive account of action and perception. Other missing pieces, such as the concept of agency, will be the topic of the later chapters. However, some notions that pertain to this concept, such as the idea of normativity, will already play a role in the current discussion. As before, these concepts are not presented in operational terms yet. This will happen in Chapters 5 and 6.

If action and perception depend on the mastery of the laws of sensorimotor contingencies (now understood as any of the four dynamical concepts introduced previously), then our approach will not be complete without an account of how such mastery is achieved and implicitly of what exactly constitutes it. One of the goals of this chapter is to arrive at a theoretically articulated concept of *mastery*.

Of course, the initial formulation of an enactive theory of sensorimotor learning will be provisional and as before, will initially assume a simplified situation involving a single agent in interaction with a rather generic environment. This is not how human beings actually learn to act and perceive. On the contrary, evidence shows that most of our perceptual capabilities are massively influenced by our history of social interactions and our embeddedness within our particular sociocultural milieu, from parental scaffolding during triadic engagements in the first year of life to the fact that very few things (if any) in our world can be said to be untouched by social production and cultural norms (Di Paolo 2016). We will come back to these issues in Chapter 8. As is often the case, enactive accounts are crammed with (one hopes virtuous!) circularities and complex relations between various systems, levels, and timescales, as expected from the three interrelated dimensions of embodiment mentioned in Chapter 1. Yet we must start somewhere and understand the principles of organization at a given level in order to be able then to make sense of how this level is influenced by others.

We mentioned in Chapter 3 that it is possible to formulate mathematical models to extract the invariant properties of a sensorimotor structure, such as the sensorimotor environment of an embodied agent. Philipona, O'Regan, and Nadal (2003) demonstrate how a series of articulated actuators and light sensors can be exploited to deduce mathematically spatial properties such as dimensionality, homogeneity, and so on. Aytekin, Moss, and Simon (2008), as we have seen, apply similar methods in order to extract the properties of auditory sensorimotor spaces. But does the acquisition of a perceptual skill *by the agent* amount to acquiring certain pieces of information like those extracted by mathematically trained researchers? Or is there something different involved?

Putting the accent a bit more on the perceptual side, to keep in dialog with the sensorimotor approach to perception, we propose to address these and other questions such as what counts as perceptual mastery and by what kind of processes such mastery is acquired. This should let us resolve an apparent paradox implicit in the claim that to perceive is to exercise certain pre-existing sensorimotor skills: How is it then possible to learn to perceive anything new? Moreover, human perception seems able to exceed any prescribed set of relevant species-ecological criteria by constantly opening up novel domains of significance (e.g., wine tasting). How is it possible that the new things we can learn to perceive seem to be part of an *open-ended* space of possibilities, i.e., that perceptual learning, without being arbitrary, does not seem to have an endpoint?

Fortunately, some of these questions have been studied before. For this and other reasons discussed below, we propose to use Jean Piaget's theory of equilibration as a suitable starting point. It will prove helpful in interpreting the claims of the sensorimotor approach as part of a complex, dynamic organization of sensorimotor and cognitive schemes. In order to demonstrate the compatibility and various complementarities between Piaget's theory and the enactive approach, we will reformulate Piaget's proposal in dynamical systems terms that render it compatible with the dynamical definitions of SMCs given in Chapter 3.

4.2 **What kind of theory do we need?**

As we have seen, the sensorimotor approach to perception emphasizes the claim that all perceiving involves some kind of understanding (Noë 2004; O'Regan and Noë 2001). However, this is not to say that abstract, disengaged conceptual knowledge is required in order to perceive. On the contrary, the claim is that practical, engaged, often non-conceptual, sensorimotor skills underlie all of our perceptual acts. This is the know-how that we display when we move about in our environment using our senses. Together with this know-how, the way we actually deal with our world is also a constitutive factor of our perceptual experience. When we speak about understanding, therefore, we mean it in the practical sense of the term.

While the idea that some kind of understanding underlies perception seems unproblematic, it leads to an apparent paradox. If understanding is required for perception, how can we learn to perceive something new, which we do not yet understand? The contrary claim is that the very existence of perceptual learning shows that some experience must exceed our current understanding (see Roskies [2008], who makes a similar point against the idea of perception being conceptual).

This is an old philosophical problem. It resembles the foundational problem of epistemology as expressed by Plato in the dialogue *Meno*. In trying to determine the essence of virtue, Socrates admits not knowing what it is but invites Meno to inquire into its nature together with him. Meno asks how will they manage to search for something of whose nature they know nothing at all; how will they even recognize it if they find it? Marjorie Grene (1966), following Maurice Merleau-Ponty's and Michael Polanyi's conceptions of practical, embodied knowledge, comments that the structure of Meno's problem is particularly puzzling, not to say unsolvable, only if we assume that knowledge must be fully explicit. However, Grene argues, the problem does not involve any paradox if we admit the possibility of knowledge or understanding having degrees of explicitation, from what we can verbalize, to the practical knowledge that we tacitly embody in our everyday skills. The solution to the problem of perceptual learning may be sought in between fully expressible understanding and total ignorance, along a "continuous" space of different degrees of skillful coping.

As a desideratum, to avoid paradoxes, we need a theory of perceptual learning that builds on a conception of understanding as something that is present but admits quantitative and qualitative variations. Also as something that admits different forms of manifestation, from performing the correct action at the right time, to giving an account of something we know to be the case.

A consequence of this desideratum is the acknowledgment that the starting point from which perception develops is always already a form of perception. Perception does not start from disjointed sensations, which are gradually organized. It starts always from an existing sensorimotor organization, some basic form of sense-making or understanding of the relation between agent and world, and it develops from there into novel forms, differentiated forms, forms that become extinct and replaced by others, and so on. But always as transformations of pre-existing forms of perception.

Consider Kohler's (1964) famous work involving long-term adaptation to vision-inverting prisms and lenses. Initially, on putting on such devices, the world stops making sense to Kohler's subjects: objects move around in completely unexpected ways; solid objects no longer even appear solid, but rather rubbery and distorted, and changeable in size and shape. Nevertheless, for subjects who actively engage with the world, after a long period of using such devices, the world slowly "rights" itself. Starting from existing sensorimotor schemes and then modifying them, actions become more and more adequate, and eventually perception itself becomes more and more "correct" (i.e., less uncanny). It is notable that (a) adaptations, both behavioral and perceptual, are partial, and (b) they are very situation dependent: subjects come to deal correctly only with those situations in which they have practice. Kohler's subjects often used explicit strategies to react in the initial stages of rehabituation, but these strategies eventually became implicit, and automatic; and concomitantly, the visual world itself came to look more and more normal. This and other cases of adaptation to inverting or distorting prisms are important verifications of sensorimotor claims about perception, especially given that adaptations only occur in the context of active, personal effort in remastering the visual world.

Tactile-visual sensory substitution (TVSS) provides us with another relevant example. Here, the image from a camera is fed to a two-dimensional array of vibrating touch actuators on the body. In early experiments (Bach-y-Rita 1967, 1972, 2004), this was a relatively large array, placed on the subject's back or on the belly. In more recent experiments (Sampaio, Maris, and Bach-y-Rita 2001), smaller arrays of electrodes are used, for instance, in contact with the tongue because of its higher tactile sensitivity. The electrodes give a tingling or bubbly impression reproducing spatially the rough structure of the video image. Strikingly, subjects who are passively "shown" the world via this system do not learn to perceive anything new. In contrast, those who are allowed to use the camera to explore the world actively can begin to get a sense that the system is providing visual-style access to the world within only a small number of days of training (Guarniero 1974). Hurley and Noë (2003) argue that there is a distinctively visual phenomenology to this new way of experiencing of the world—even though it is of course not nearly as detailed as normal vision. According to others, the quality of perception in these cases is neither strictly tactile nor strictly visual, but rather it indicates the possibility of novel or mixed perceptual modalities (see Auvray and Myin 2009; Farina 2013).

Because it relies on a practical, action-based concept of understanding, the enactive account of sensorimotor learning is what we have called a world-involving account, i.e., it involves a relation to the dynamics of the world beyond the mere supply of sensory inputs, but as enabling relevant sensorimotor engagements. Ours is not an empiricist account, one in which the world provides a series of data whose regularities are extracted by the agent's brain and stored in world models. Nor is it an intellectualist account, in which a form of consciousness imposes a certain organization on the information coming from the world according to pre-existing categories. It simply does not relate to the world in terms of inputs or outputs, but in terms of coupling and engagement.

For example, the passive subject using a TVSS device cannot make sense of the tactile sensations on the skin or the tongue if somebody else moves the video camera: this is not because the tactile patterns (the supposed "inputs") are unstructured. A machine-learning algorithm could very likely extract many regularities out of these patterns and build some predictive models of what state of the array is likely to follow a given history of states. The fault is not on the subject's side either or any inability to find meaning in a complex pattern of tactile stimulation. Within the existing forms of perceptual understanding in her possession, the passive subject could indeed choose to pay attention to the activity of the array in specific ways, considering rhythms, sweeping movements, whether the patterns move in this or that direction, and so on. These are not meaningless inputs; they are simply, for the passive subject, not meaningful in terms of "visual" perception of objects in space. Tactile attention, however, never moves beyond the existing perceptual understanding (tingling on the skin or the tongue), even if the person is conscious that something more "visual" should happen (as participants in these experiments often are).

So, it is not the lack of "input" structure, nor the lack of knowledge that prevents the passive subject to learn a novel perceptual relation with the world using the TVSS device. It is her passivity, i.e., the lack of a chance to engage the world and attune to a new form of coupling with it. This is the difference between a world-involving account and one that sees the world as an informational resource. A similar point can be made about Kohler's experiments.

Of course, the need for an active coupling does not mean that the traditional empiricist and intellectualist positions are fully defeated yet. They could argue that the active participant is now able to extract information from the world more efficiently by biasing the sampling statistics. Or they could argue that the sensations that obtain from bodily movement complement tactile inputs in a way that better fits pre-existing knowledge of spatial categories. As we shall see, the path we take in this chapter implies neither of these alternatives. For the moment, let this simply mean that these are not the only options, that treating the relation between agent and world in terms of inputs from which meaning must be extracted or perceived directly, as in empiricism or ecological psychology, or on which it must be imposed, as in traditional cognitivism, are not the only games in town. Later we will come back to review some of these issues and show what other reasons we have for preferring the enactive alternative for a theory of perceptual learning.

4.3 **Sensorimotor schemes**

Before we proceed, we must briefly say a few things about sensorimotor schemes. Of the four concepts of SMCs we have introduced earlier, we will make extensive use of sensorimotor schemes in the present and later chapters. As we have seen, this notion is the more situated way of looking at SMCs, involving not only aspects of the agent's embodiment and environment, but also the internal processes that regulate the agent's activity, and the task scenario or situation in which the agent is involved. In a way, other sensorimotor structures (sensorimotor environments, habitats, and coordination patterns) are implied and inform what is relevant when examining sensorimotor schemes.

The notion is useful because it provides us with a mesoscopic level of description. Schemes are organizations of sensorimotor coordination patterns that are often enacted together and in combination with other schemes. While they can be easily related to neural and musculoskeletal processes, the fact that these processes are brought together according to a certain organization and a certain normativity makes the level of sensorimotor schemes equally close to physiological and action/intentional phenomena (these relations, of course, will have to be worked out explicitly in specific cases).

But perhaps the most important reason to use the concept of sensorimotor schemes is the undeniable fact that as acting embodied individuals we are *made up* of a complex bundle of these schemes (we will formalize this idea in Chapter 6). We always come into any situation not with the task of adequately constructing from scratch every single muscle activity and the specific movement of each joint. On the contrary, we are equipped with a rich repertoire of ready-made, highly organized ways of engaging the world. Ours would be a poor theory of the embodied mind if we did not recognize the richness with which we confront the world, a richness that is literally embodied as a repertoire of potential sensorimotor schemes, some of which are rather widespread across the species, others are acquired as part of our sociocultural milieu, and some are idiosyncratic; all of them together reflect the history of each particular body.

The best way to think of sensorimotor schemes is as reusable, interlocking, organized sets of coordination patterns between body and environment. More often than not, they involve the synergistic activation of various muscle groups, in ways that are already pre-attuned to the environmental dynamics. The latter enable in turn the enactment of sensorimotor schemes. Walking along a path, we recurrently activate the schemes of stepping forward, maintaining balance dynamically, and compensating for different properties of the terrain such as ruggedness and slope, not to mention environmental processes that are more complex, such as walking around moving obstacles, and so on. Grasping a full coffee mug without spilling the coffee involves various visuo-tactile-proprioceptive patterns of sensorimotor coordination in fine temporal order. Dribbling a basketball demands a regulated transfer of force in magnitude and direction from hand to ball in ways that the whole process is rhythmically brought back to its starting point, while compensating for variations on the floor surface, wind, and so on. All of these are examples of sensorimotor schemes. Walking while bouncing a basketball is an example of two simultaneously enacted schemes. So is walking while grabbing a mug full of coffee. Schemes can be combined in parallel or in sequence with more or less difficulty. Grabbing the full coffee mug and bouncing the basketball is a particularly demanding combination of schemes.

It is worth noting that sensorimotor schemes are not strictly speaking something that the agent *possesses*, but something that the agent *enacts*. A sensorimotor scheme must always be enabled by the right environmental participation. Nevertheless, we sometimes refer to schemes as "belonging" to the agent, or to an agent acquiring the skills for a sensorimotor scheme, or as "having" conflicting schemes, and so on. Although one must try to be careful and avoid confusing uses of the term, it is sometimes practical to use such phrases. They should be understood as a shorthand way of referring to the agent-side of the constituents of a sensorimotor scheme, for instance, a particular neuromuscular organization that becomes active when coupled to the right environmental circumstances.

Another important aspect, as we have already mentioned, is that a sensorimotor scheme implies a certain normativity. Schemes may be involved in actions that either succeed or fail, or do so better or worse according to some relevant standard. Not spilling much or any coffee is one norm when I grab the coffee mug. But so is not taking so much time doing it that the coffee gets cold. As we have said, we will need to come back to discussing the origins of such norms. But we think the experiential fact is undeniable that sensorimotor schemes are subject to norms in terms of how adequate they are for a given situation, how well they are organized internally, whether they conflict or not with other schemes, and even how they fit within the overall organization of the embodied agent (e.g., in terms of behavioral style or personality).

As we shall see, these important properties of sensorimotor schemes are already part of Piaget's theory of sensorimotor development in the child to which we turn next.

4.4 **Piaget's theory of equilibration**

Jean Piaget has addressed many of the difficulties of the problem of perceptual learning (Piaget 1936, 1947, 1969, 1975; Chapman 1992; see, also, Boom 2009; the relevant literature is large). In this section, we briefly go over the central points of his theory of equilibration of cognitive and sensorimotor structures. A good part of his work is very relevant for the enactive approach (see Box 4.1). Piaget's research program can be seen as the quest to determine how abstract and explicit (e.g., conceptual, mathematical, formal, and rational) human capacities of understanding stem from early and less explicit forms of sensorimotor and cognitive organization. The set of developmental transitions that span this continuum are conceptualized under the general notion of adaptation, which

Box 4.1 Piaget, an enactivist?

In the mid-1980s, Francisco Varela drew a polar chart summarizing his view of the cognitive sciences at the time. It consisted of different radii corresponding to subdisciplines such as Neuroscience, Linguistics, Philosophy, etc., and three concentric circles, the inner one labelled Cognitivism, the middle one Emergence, and the outer one Enactive. Scattered on this chart—later appearing in (Varela et al. 1991, p. 7)—were the names of influential 20th century figures. Close to the radius labelled Cognitive Psychology, the name closest to the Enactive circle is Jean Piaget.

It comes as no surprise that the enactive perspective on action and perception should resonate with Piaget's work. Like enactivists, Piaget saw cognition in terms of the meaningful engagement of a whole, embodied organism with its world. He also stressed the continuity between life and mind, between *Biology and Knowledge* (the title of one of his major books; Piaget 1969), not only as an empirical matter but also as a conceptual and methodological pillar of his ideas. Often in his work on genetic epistemology, he adopted a complex systems' perspective, drawing inspiration from work by W. Ross Ashby and L. von Bertalanffy and making use of notions such as feedback and circular causality.

This is not to say that the enactive framework can easily assimilate the whole edifice of Piagetian psychology (although as this chapter shows, interpreting his ideas enactively can be enriching). The young Varela felt attracted to the ideas in *Biology and Knowledge*, though its "language and idiosyncrasies left [him] unsatisfied" (Varela 1996, p. 412). Later, Varela and colleagues (1991) expressed both their admiration and their qualms about the Piagetian progression from early sensorimotor intelligence to the adult understanding of the external world. On the one hand, the infant only knows the world by engaging with it as an active agent, a notion dear to enactivists. But on the other, for Piaget this context-grounded subjective engagement with the world is merely a stage to be ultimately transcended in favor of abstract, unsituated, logical structures with which the world can be seen objectively. Thelen and Smith (1994) also complain about Piaget's placing a rational adult as the end state of development. And if we look at the other extreme, the beginning of development, Piaget's intellectualist bias is manifested in the assumption of the infant as living in a disjointed world of meaningless sensations. Merleau-Ponty (2010) criticizes the phenomenological coherence of conceiving the infant's world as deficient and involving a kind of meaningless stage.

Yet, despite these worries, there is something inherently resonant for the enactive approach, not so much in Piaget's assumptions about either endpoint of development, but in the logic of process, activity, and transformation implied by his theory and his rich empirical observations.

itself is seen as a process of equilibration between processes of assimilation and accommodation. These notions present us with a suitable candidate solution for the problem of perceptual learning for, as we shall see, they assume the possibility of different degrees of explicitation and kinds of understanding.

The ongoing adaptation and transformation of sensorimotor schemes, according to Piaget, follows a logic of equilibration between two kinds of processes: assimilation and accommodation. These are the processes by which a challenged agent–environment coupling may be adaptively steered back into its normal or into a new way of functioning. Of course, these processes may fail and the enactment of a sensorimotor scheme may not be appropriate for the current situation, as we shall see.

Following the standard interpretation, by *assimilation* we refer to a process by which an environmental aspect (a perturbation, a new object, or a novel situation, etc.) is integrated, coupled, or absorbed into an existing physiological (metabolic, neuromuscular, etc.) or cognitive/behavioral (sensorimotor, perceptual, and reflective) supporting structure in the agent. This is one way of saying that the agent and environmental sides of a sensorimotor scheme are in agreement according to the relevant norm. In Piaget's famous example (1936, 1947) a baby assimilates the mother's nipple into a suckling scheme (a sensorimotor structure involving a complex patterns of muscular coordination, proprioceptive, tactile, temperature and taste sensory signals). The baby may have a propensity to suck but this does not mean she is immediately able to. The baby has to learn to "latch on" successfully, to become comfortable with the shape and feel of her mother's breast. That is, the baby has to learn new patterns of sensorimotor organization.

To understand this process better, Piaget subdivides a complex sensorimotor scheme into its component sensorimotor coordination patterns, and in the latter, he distinguishes between contributions on the part of the agent and those of the environment, both being necessary to complete a closed-loop coordination. On the agent's side, we will here call the set of variables involved in enabling a coordination the *coordination support structure*, and we denote this set with capital letters, A, B, C, …. In Piaget's notation, an agent's coordination structure A assimilates an environmental aspect or process A′. This is a *sensorimotor engagement* or *coordination* and is denoted as A × A′. The effect of this combination between an agent's coordination support structure or process and environmental dynamics leads in turn to another coordination structure B that assimilates B′ as an ensuing sensorimotor coordination, and so on. The sequence of pairing of assimilated moves in the agent and the environment forms a *sensorimotor scheme*. This is expressed as A × A′ → B; B × B′ → C; C × C′ →. … In terms of the example (see Figure 4.1), A can denote the suckling support structure and A′ the mother's breast or milk bottle, B the swallowing support structure and B′ the milk, C the breathing support structure and C′ air, etc. Note that for Piaget the environment is not a set of pre-existing stimulus conditions that affect the organism to produce a perceptual or cognitive effect. Only what can be assimilated in an already existing scheme or sensorimotor coordination pattern (i.e., an action or operation of the subject) can be engaged with, and thus perceived.

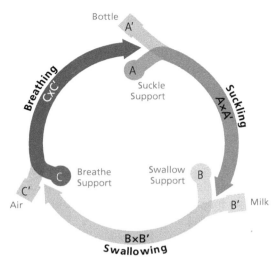

Figure 4.1 An example of a sensorimotor scheme. Bottle-feeding consists of three individual coordination events: suckling, swallowing, and breathing. Each sensorimotor coordination is produced as the result of a coordination support structure on the agent's side (A, B, C) assimilating an aspect of the environment (A', B', C').
Copyright © 2017 Ezequiel A. Di Paolo, Thomas Buhrmann and Xabier E. Barandiaran, with permission.

By *accommodation* Piaget refers to the process by which the structures in the agent that enable a sensorimotor scheme (A, B, C, etc.) are modulated or transformed to facilitate or encompass a not-yet-assimilated aspect of the environment. So, for instance, the suckling sensorimotor coordination of the baby is progressively attuned to the size, texture, and shape of the nipple. So, variations of A are progressively attuned to variations in A'.

Equilibration is the process by which a sensorimotor organization reaches a new form of stability as a result of a maturational process or in the presence of a changing environment or any internal sources of tension. Piaget denotes the stability of the organization as the closure of a cycle of sensorimotor engagements: $A \times A' \to B$; $B \times B' \to C$; $C \times C' \to$ ... $\to Z$; $Z \times Z' \to A$. We refer to this cycle as a circular sensorimotor scheme (similar to Baldwin's [1894] concept of circular reactions). Even though the simple cycle is a paradigmatic case (suckling, walking, breathing, bouncing the ball, etc.), Piaget allows for the possibility of additional complexity (short-circuits, intersections between cycles, etc.; see page 10, note 2 in Piaget 1975). He hints at the fact that ultimately a cycle should be understood as conservation of the conditions of viability of an organism or a cognitive system as a whole (Piaget 1975, p. 11). As we will discuss in Chapter 5, this is close to the enactive understanding of normativity (Barandiaran 2008; Di Paolo 2005, 2009; Di Paolo et al. 2010; Di Paolo and Thompson 2014; Thompson 2007a). Accordingly, the conservation of a precarious self-sustaining circular organization is proposed as grounding certain kind of vital norms. In the particular case of circular schemes used in most of this chapter, the evaluation of whether equilibration occurs (or not) corresponds to the cycle closing back on itself (or not).

A sensorimotor scheme that has undergone equilibration will potentially be affected by higher order internal tensions with other schemes and by new possibilities for action that have been brought about by adaptive changes during accommodation. For instance, if a baby had acquired a suckling skill and then is presented with a milk bottle for the first time, she faces the challenge and opportunity of accommodating a new object with new properties (texture, shape, etc.). These differences induce instabilities in the suckling scheme. As a result, a new accommodation is triggered, and repeated exposure to these conditions can lead by the interplay between assimilation and accommodation to a new equilibration of the suckling organization eventually splitting it into two broad categories: $A_1 \times A_1' \to B_1; B_1 \times B_1' \to C_1; C_1 \times C_1'$, corresponding to breastfeeding and $A_2 \times A_2' \to B_2; B_2 \times B_2' \to C_2; C_2 \times C_2'$, corresponding to feeding from the milk bottle. The new equilibration may open new possibilities for action: the milk bottle affords grasping with both hands, new feeding positions are available, etc.

A more radical equilibration process can result from changes in the baby's muscular and sensorimotor capacities. When the baby starts to grasp objects and bring them to her mouth, the suckling coordination might sometimes be triggered in an assimilation attempt. But subsequent sensorimotor coordination may be severely challenged if the object, say a puppet, is not fully compatible with the enacted coordination. The inability to assimilate this new object, and to force an impossible accommodation upon the suckling scheme may then lead to a higher order equilibration where environmental objects are now split into suckable and non-suckable (yet graspable, movable, chewable, etc.) objects. Further equilibration may result from a habit of bringing the thumb or a dummy to the mouth to assimilate it into the suckling scheme as breastfeeding becomes less frequent. In turn, this new assimilation $A_3 \times A_3'$ might now lead to a modification of the swallowing scheme transforming the original sequence $A \times A' \to B; B \times B' \to C; C \times C'$ into $A_3 \times A_3' \to B_3; B_3 \times B_3' \to C_3; C_3 \times C_3'$ that might later be further transformed into a repetitive biting pattern when teeth start to grow, resulting in $A_4 \times A_4' \to B_4; B_4 \times B_4' \to C_4; C_4 \times C_4'$. These processes of differentiation of sensorimotor schemes, their grouping, branching, and sequential ordering, lead to higher order equilibrations that are richer in diversity and combinatorial potential than what the previous sensorimotor organization made possible.

Piaget describes two kinds of perturbations that may be encountered by an established sensorimotor scheme: *obstacles* (contradictions or disturbances) and *lacunae* (gaps in the current organization) (Piaget 1975; Boom 2009). Both types are manifested in the concrete encounter between agent and environment regardless of whether they originate from external changes or from internal contradictions. This is important because Piaget's theory implies, but perhaps does not sufficiently emphasize, that failed encounters with the world (i.e., obstacles and lacunae) drive equilibration, and there is always a possibility that the world may "guide" part of the equilibration process (and as we shall see in Chapter 8, this guidance is often carried out by the social world of the infant and child). The theory might otherwise be interpreted as too internalistic, relying only on the reorganization and adjustment of existing patterns of sensorimotor coordination. It might also be seen as too deterministic, as if aiming toward a predetermined adult stage with standard cognitive

[handwritten marginal note: What kinds of perturbations might an agent, an SMC, encounter?]

capabilities. In fact, systematic socio-environmental biases and chance events have an important influence in the history of equilibration processes for a given agent.

A class of similar situations will present variations between concrete instances. Every time we grab it, the coffee mug is slightly different, the mug is more or less full, our hands more or less sweaty, and so on. When the equilibration process is such that the sensorimotor organization readily assimilates most of the potential obstacles and lacunae for a given class of situations, we say that the agent has achieved *maximal equilibration*. In other words, maximal equilibration is that (sometimes unattainable) state where the enactment of sensorimotor schemes requires no further accommodation to a class of situations.

As already suggested, equilibration processes are not limited to agent–environment dynamics, but can occur between the schemes themselves. This needs some clarification, since we have defined equilibration as a relation between patterns in the agent–environment coupling. What does it mean for such sensorimotor patterns to equilibrate between themselves? This is a shorthand expression for a variety of relations between sensorimotor schemes. For instance, some schemes are usually followed by other schemes in time (turning the door handle down, pushing the door open, and entering the room), and so equilibration between them means relating movements in time, coordinating the activity and preparatory stages in various muscle groups, etc. Other patterns may be enacted in parallel (walking and having a conversation), and equilibration between them can mean minimizing interference between the two schemes. Engagements that are more complex may involve selecting an appropriate scheme out of several that a situation affords. Being ready to inhibit those schemes that will not be used at the time the chosen one is enacted implies also a form of equilibration between schemes, only one that involves not only individual schemes but also aspects of the totality of the agent's sensorimotor organization. In other situations, accommodation of one scheme requires adaptations in others for the repertoire corresponding to a whole activity to remain internally consistent. Consider, for example, the case of a continental European driver driving for the first time in the United Kingdom, and in a car manufactured to be driven on the left side of the road (i.e., having the driver seat on the "wrong" side from the continental perspective). In this case, accommodating the sensorimotor schemes involved in shifting gears now to operate with the left hand may also require adapting the schemes involved in steering, such that the right hand may now become the dominant hand on the steering wheel. Only when all schemes involved in driving are equilibrated with the controls of the car, as well as with each other, is the skill of driving completely restored.

We have then three broad categories of equilibration:

1. Forms of equilibration that involve sensorimotor engagements between agent and environment and result from the disequilibrium between coordination support structures in the agent A, B, C, … and environmental aspects A′, B′, C′.

2. Forms of equilibration owing to the reciprocal accommodation and assimilation between sensorimotor schemes themselves, so tensions originating in the inability to assimilate or accommodate relations between schemes of the form (A × A′ → B; B × B′

$\rightarrow ...) \leftrightarrow (X \times X' \rightarrow Y; Y \times Y' \rightarrow ...)$ lead to new forms $(A_1 \times A_1' \rightarrow B_1; B_1 \times B_1' \rightarrow ...) \leftrightarrow (X_1 \times X_1' \rightarrow Y_1; Y_1 \times Y_1' \rightarrow ...)$ either by the transformations in the schemes involved or in aspects of their relation, or both.

3. Forms of equilibration that result from tensions between a particular scheme and the system's totality. This is, according to Piaget, a new form of equilibration, since it involves a hierarchical dimension of relationships among schemes.

Piaget's equilibration framework provides a progressive microgenetic conception of how an agent moves from one kind of sensorimotor understanding to another. This is important because, as we have said earlier, conceiving of understanding as something that can only be either present or absent leads us back to the Platonic conundrum: I can only perceive what lies in front of me if I understand it with the categories and skills I already possess; yet new categorizations are required to perceive something new, and there seems to be no source of categories other than those I had before. In contrast, a microgenetic approach works on a graded conception of understanding, and so it allows us to specify the processes involved in the emergence of new perceptual categories, habits, and sensorimotor schemes out of those that existed previously.

4.5 **A dynamical approach to equilibration**

In the above formulation, Piaget's theory lacks a detailed dynamical formalization that can make justice to the microgenetic processes he conceptualized. Let us redescribe some of his ideas in dynamical systems terms and relate them to our definitions of SMCs in Chapter 3. Because we cannot offer a full dynamical systems account of Piaget's theory of equilibration, we will simply try to demonstrate how a dynamical systems interpretation of this theory, even if only partial, can act as a common language between Piaget and the enactive approach. Nevertheless, this section will be a bit technical.

Earlier we distinguished between four concepts of SMCs: sensorimotor environment, sensorimotor habitat, sensorimotor coordination, and sensorimotor scheme or strategy. These concepts are formalized as functional mappings involving variables such as the activity of sensors and motors, internal (neural) activity, relative positioning and configuration of the body, and so on.

In Piaget's terminology, instances of sensorimotor coordination in the SMC sense correspond rather straightforwardly to the agent–environment engagements $A \times A'$, $B \times B'$, $C \times C'$, etc. The variables on the agent's side involved in supporting a sensorimotor coordination (i.e., its support structure) correspond to Piaget's labels A, B, C. The parallel environmental variables A', B', C', again in Piagetian terms, are also taken into account in our dynamical formulation through the equations describing the environmental intrinsic and responsive dynamics. The engagement of the agent and environmental sides constitute the total sensorimotor coordination.

As an example of an agent-side support structure A and an environmental aspect A' coming together to form a sensorimotor coordination $A \times A'$, consider the dynamical description in the following simplified case. A bipedal agent's neural dynamics may feature a pattern that consists of two stable attractors: one corresponding to a stepping motion that brings the foot forward during the swing phase and another for the motion that propels the body forward during the stance phase. Each of these attractor states involve the formation of dynamic linkages, or synergies, in the agent that coordinate the activity of the muscle groups involved. The presence of the attractors may further be dependent on sensory signals that correspond to certain events in the environment. Now, when this agent is placed on a supportive surface, such as an inclined plane, receptors in the foot that are sensitive to pressure may trigger the presence of the swing phase attractor in the dynamical landscape, bringing the leg forward. Moments later, another sensory signal— for example, one indicating maximal forward leg extension—may trigger the second attractor and the corresponding stance phase motion. The result of the coupled agent–environment system may in this case correspond to that of a limit cycle, and the observed behavior to that of walking. Note that the limit cycle coordination does not exist in either the agent or the environment alone, but is the result of the proper contributions of the two (see, e.g., Izquierdo and Buhrmann 2008; Vaughan, Di Paolo, and Harvey 2014).

At the next level, an organized set of coordination patterns—Piaget's sensorimotor scheme—corresponds to the notion of sensorimotor scheme we introduced in Chapter 3. However, although they are directly comparable with Piaget's cyclic organizations, sensorimotor schemes in the SMCs sense also refer to aspects that are more detailed and remain implicit in the theory of equilibration. For example, sensorimotor schemes as defined in Chapter 3 need not present a circular organization at the level of coordination patterns. This is because their normativity can be grounded elsewhere, either in the self-constitution of the organism (Di Paolo 2005; Di Paolo 2010; Thompson 2007a) or in norms originating externally that the organism incorporates (efficiency in labor time, craftsmanship, etc.). For these reasons, sensorimotor schemes in general can be more complex than cycles, but it is nevertheless possible to apply the concepts of assimilation and accommodation by adopting a criterion of equilibration that follows the organismic or the externally imposed norm.

Barring these differences, that might later be exploited to inform the Piagetian approach, it seems so far that our dynamical approach to SMCs promises to establish a compatibility between equilibration and enactive theories. We summarize the terminology in Table 4.1. What remains to be seen is how we interpret the concepts of assimilation, accommodation, and equilibration in these terms. First, we look again at equilibration, describing it as simply as possible in dynamical terms.

The agent and the environment are two coupled systems, which means that some parameters in each of these systems are affected by the state of variables in the other. To simplify things, throughout the following analysis, we will focus on two lower dimensional projections of the full coupled system: one projection looking at the state of the agent's sensorimotor variables (s and m) and another projection looking at the subset of environmental variables that have a direct effect on the agent (e) (this is the same notation used in the Chapter 3). By *projection* we simply mean that we focus only on what happens with these sets of variables, although we know that many other variables are involved in

Table 4.1 Summary of Piagetian and dynamical systems concepts for a theory of equilibration (SM = "sensorimotor")

Piagetian Concept	Dynamical System Definition	Notation	Example
SM coordination support structure	Class of agent variables that support SM coordination patterns	A, B, C ...	The class of movements and sensations that belong to the subject's experience of pushing objects toward the ground, absorbing impacts with the hands, etc.
Environmental response structure	Environmental variables directly affecting the agent's variables in A, B, C; i.e., the projection of the whole dynamical system, when engaged in SM coordination instances onto relevant environmental variables	A', B', C'...	Gravity, sound of the ball hitting the floor, height of the ball above ground, force exerted by the ball on the hand, etc.
SM trajectory or pattern	Instance of SM coordination support structure belonging to class A, B... ; i.e., a trajectory in SM space that belongs to the respective SM coordination support class	$a(t)$, $b(t)$, ...	A particular instance of pushing the ball toward the ground
Environmental response	Instance of environmental response of class A', B', ...	$a'(t)$, $b'(t)$, ...	The trajectory of the ball, the sound of the impact for this particular bounce, etc.
SM coordination instance	Simultaneous occurrence of SM pattern $a(t) \in A$ and corresponding environmental trajectory $a'(t) \in A'$ in the coupled system	$<a, a'>$	A successful instance of pushing the ball downward
SM coordination	The set of all tuples $<a, a'>$	$A \times A'$	
SM scheme	An organization of SM coordination patterns	O: $A \times A' \rightarrow$ $B \times B' \rightarrow ... \rightarrow$ $A \times A'$	Ball bouncing sequence of coordination patterns that includes pushing the ball toward the ground, hearing the impact, waiting for its return, preparing muscles for new contact, absorbing the impact and pushing it down again
Assimilation of A' by A in O	Fulfilment of stability condition and transition condition in the SM coordination $A \times A'$		Continuous, stable ball bouncing despite small variations in motor pattern or wind speeds

(continued)

Table 4.1 Continued

Piagetian Concept	Dynamical System Definition	Notation	Example
Accommodation of A to Z′ in O	Plastic changes that re-establish a scheme O when confronted with new environmental pattern Z′ by modifying the coordination structure A such that A now assimilates Z′		Learning to bounce a ball on a slope
Lacuna	Violation of the transition condition; something is manifestly "unknown" about the world because the presumed "right" handling of the situation (A × A′) does not lead to the next stage in the cycle (B × B′)		Bouncing a ball on a slope for the first time; the ball does not return to the expected position
Obstacle	Violation of the stability condition; something in the sensorimotor coordination has failed, where in the past it used to work	-	Attempting to bounce a new ball that is significantly heavier than the one that had been accommodated; bouncing demands more strength
Equilibration	A potentially never-ending series of parametric changes of the totality of SM organization, aimed at maximizing the stability of each scheme against violations of the transition and stability conditions resulting from environmental perturbations or internal tensions		The process of learning to bounce the ball under a variety of conditions (size and weight of the ball, slope, and friction of the floor, etc.)

Adapted with permission from Di Paolo E. A., Barandiaran X.E., Beaton M. and Buhrmann T. (2014). 'Learning to perceive in the sensorimotor approach: Piaget's theory of equilibration interpreted dynamically.' *Frontiers in Human Neuroscience*, Volume 8, p. 551. Copyright © 2014 Ezequiel A. Di Paolo, Xabier E. Barandiaran, Michael J. S. Beaton and Thomas Buhrmann.

the agent and the environment. The variables in the sensorimotor projection are affected by other variables belonging to the agent as well as by variables in the environmental projection. And similarly for the environment. In other words, these projections do not describe the whole agent–environment system. The agent variables (say, hormonal concentrations or neural activity) that are not directly coupled to the environment, and similarly the environmental variables that are not directly coupled to the agent, do influence the equilibration process but we do not express them explicitly. If necessary, the analysis below could be reformulated to include these variables as well.

Consider an equilibrated sensorimotor scheme, which we will denote as $O = A \times A' \rightarrow B \times B' \rightarrow C \times C' \rightarrow A \times A'$, where the notation $A \times A'$ indicates a sensorimotor engagement or coordination (i.e., a combined state involving the agent's sensorimotor variables in A and the corresponding co-occurring environmental variables in A′). Dynamically speaking, each class of sensorimotor support structure (A, B, C) involves establishing patterns of motor and sensor covariation in a task-related context (what in other contexts is referred to as *coordinative structures* or *synergies*; Kelso 2009; Latash 2008). In the current analysis, while the classes A, B, C, etc., involve sensorimotor *and* other variables in the agent, we restrict our analysis to the sensorimotor projection. In this space, we denote a trajectory over time of the sensorimotor variables as $a(t)$. These can take the form of a reliable transient (one that will likely occur in the right conditions) or a metastable set of states that fulfils the condition of belonging to the same sensorimotor class A. We denote this relation as $a(t) \in A$. Think of A, for example, as a set of neuromuscular processes typically involved in moving the arm and the hand so as to grab the coffee mug effectively. Think of $a(t)$ as one particular instantiation of these processes. The class A is defined as those sensorimotor trajectories that assimilate those aspects of the environment that contribute to generating environmental trajectories that belong to A′ (e.g., gravity, the presence and physical properties of the coffee mug, lighting conditions, and so on). We now clarify what this means in dynamical terms.

Given a sensorimotor scheme, we shall say that the class of sensorimotor structures A assimilates the class of environmental features and processes A′ when the following two conditions apply (see Figure 4.2, left):

1. *Stability condition*: the occurrence of a sensorimotor pattern $a_1 = a_1(t)$, $a_1 \in A$ enables the occurrence of an environmental pattern $a_1' = a_1'(t)$, $a_1' \in A'$, without disrupting a_1, and conversely the occurrence of an environmental pattern $a_2' = a_2'(t)$, $a_2' \in A'$ enables the occurrence of a sensorimotor pattern $a_2 = a_2(t)$, $a_2 \in A$, without disrupting a_2'. This is what it means to say that the sets A and A′ are mutually stabilized.

2. *Transition condition*: if a combination of mutually enabled trajectories a and a' in the coupled system is produced such that $a \in A$ and $a' \in A'$, then this leads in time to the production of sensorimotor pattern $b = b(t)$, $b \in B$ in the agent and the production of pattern $b' = b'(t)$, $b' \in B'$ in the environment, where B × B′ follows A × A′ as the next stage in the cycle that defines the sensorimotor scheme.

These conditions are then applicable to other links in the cycle that defines the scheme, so that as the agent approaches a pattern in B and the environment approaches some pattern in B′, these patterns tend to stabilize each other (condition 1) and lead the coupled system to the next stage (condition 2).

In the example of the baby suckling at the mother's breast, we mentioned various coordination patterns (suckling the breast, swallowing milk, and breathing air), which correspond to the different engagements in the cycle: A × A′, B × B′, C × C′. But each particular instance of suckling, swallowing, or breathing can be slightly different from other instances. These instances are made up by the engagement of specific sensorimotor—$a(t)$, $b(t)$, $c(t)$—and environmental—$a'(t)$, $b'(t)$, $c'(t)$—patterns in the relevant variables. What makes $b(t) \times b'(t)$ a successful instance of *swallowing*? The fact that $b(t)$ and $b'(t)$ belong to the classes B and B′, respectively. And how are these classes defined? By the stability and transition conditions, that is, by the fact that when $b(t)$ and $b'(t)$ are combined, both remain within their respective classes (i.e., neither is pushed outside their class) and both lead eventually to the next stage in the cycle. In other words, there is a dialectical relation between parts and whole, between the sensorimotor coordination that make up the sensorimotor scheme and the scheme that defines whether a coordination successfully counts as one or not.[1]

Graphically, the sensorimotor scheme **O** can be represented as a set of closed loops when the coupled system is projected onto the space of sensorimotor coordinates, each loop corresponding to a particular enactment of the scheme (Figure 4.2, right). The loops are not necessarily identical. As the scheme is enacted, the agent's sensorimotor activity will keep on cycling through the equivalent sensorimotor states $a \in A$, $b \in B$, and $c \in C$ in equilibrated coupling with environmental states $a' \in A'$, $b' \in B'$ and $c' \in C'$, respectively (Figure 4.2, left). We take the bundle of all these possible trajectories as the graphical description of the cycle **O**. The gray areas then represent the different sets A, B, C (and A', B', C' in Figure 4.2 left). In the dynamical formulation, the distinction between stages must somehow be pregiven and related to meaningfully distinct (at least from the agent's or the observer's perspective) sensorimotor engagements (suckling, swallowing, breathing, etc.). The separation into different coordination stages may involve factors such as, for instance, how similar coordination patterns are used in different schemes. In other words, the different labeling between sensorimotor engagements A × A' and B × B' is defined externally to what is shown in this figure. The shape and extent of the gray areas, however, is defined by the combined stability and transition conditions in a way that a closed cycle is formed. We draw the gray areas as smooth and continuous for illustration purposes; in general, the sets A, B, … and A', B', … need not have such properties. We assume that equilibration has been maximized if all cycles for a given set of conditions occur within the gray band.

This dynamical description describes the conditions under which we can say that the sensorimotor coordination that makes up a scheme is equilibrated.

Let us now consider the process accommodation in these dynamical terms. Imagine a perturbation to the sensorimotor scheme **O** that disequilibrates it. Something has

---

[1] Notice that the dynamical systems description is formulated without assuming a clear delimitation into well-defined agent or environmental stages (A × A' → B × B' → …). This notation, occasionally used by Piaget himself (e.g., Piaget 1967, p. 172), is more general than the more typical Piagetian notation we have used earlier (A × A' → B; B × B' → C) (e.g., Piaget 1975, p. 10). The difference lies in that the latter expression implies that a sensorimotor engagement at a given stage gives rise first to the agent-side pattern at the next stage, which later combines with the environmental dynamics, as if it were a staged action-response sequence. This need not be the case. The notation we use includes this possibility as a special case. In other words, since the dynamical interpretation assumes a necessary coupling between agent and environment, neither behavioral (as suggested by Piaget's notation) nor environmental consequences alone are taken to be solely responsible for transitioning to the next sensorimotor coordination within a given scheme.

occurred that, at least in some point in the cycle, does not fulfil either the stability or the transition condition. This means that the agent no longer assimilates the environmental dynamics.

This can happen, for instance, when the stability condition fails: during the $A \times A'$ engagement, and for reasons that can originate in internal or environmental processes, the sensorimotor trajectory $a \in A$ and the environmental trajectory $a' \in A'$, or both, become unstable (i.e., they cease to enable each other and at least one of them falls outside the coordinated sets). This leads either the agent or the environment, or both, to new trajectories d and/or d' with $d \notin A$ and $d' \notin A'$. The agent experiences this as an *obstacle*; something in the relation between environmental variables and the enacted sensorimotor coordination has failed where in the past it used to work. In the situation when we are opening a door, an example of an obstacle would be if the door handle were stuck midway as we turn it down.

Alternatively, the perturbation may provoke the transition to the next stage in the cycle to fail. Even if $a \in A$ and $a' \in A'$ are both within their respective sets A and A', the conditions of the coupling change such that instead of leading to $b \in B$, they lead to $e \notin B$ in the agent and/or instead of leading to $b' \in B'$, they lead to $e' \notin B'$ in the environment. This is the case of a *lacuna*, i.e., something is manifestly unknown about the world because the presumed "right" sensorimotor engagement ($A \times A'$) does not lead "as expected" to the next stage in the cycle. An example of a lacuna would be if we were confronted with a type of door that does not open by turning the door handle down, but through a different movement. If we are able to turn the handle down successfully, we find that the door still cannot be moved.

A particular failure in conditions 1 or 2, in principle, can originate from either internal or environmental proximal causes. The origin of a perturbation is invisible to the agent; only its effect is manifested as a disruption of the sensorimotor scheme: the loss of "control" over a previously stable sensorimotor coupling or the failure of an effectively achieved coupling to lead to its usual result (we shall return to this point when we discuss the sense of agency in Chapter 7). The terms obstacle and lacuna are used here for their effects on action and perception, not to indicate their proximal causes, which are not immediately perceivable.

If we assume that the gray zones in Figure 4.2 represent trajectories belonging to the condition of having achieved equilibration and that these trajectories define the subsets A, B, and C (and the same condition of equilibration defines the sets A', B', and C' on the environmental side), then any new perturbation as defined above (either an obstacle or lacuna) will make sensorimotor and environmental trajectories escape from the equilibrated sets (the gray zone). And, all other things remaining equal, it cannot be expected to return to the gray zone, except fortuitously, for instance, through an independent environmental change. Anything that at the personal level could be described as an attempt to deal with an obstacle or lacuna (i.e., an attempt to bring the unexpected situation back into the sensorimotor organization **O**) will imply at the dynamical, subpersonal level that things do not remain equal (i.e., that some form of plastic change must occur).

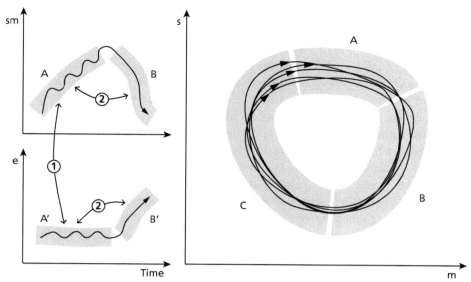

Figure 4.2 Left: Illustration of the two conditions describing assimilation. Condition 1: *stability*. A trajectory *a(t)* in the projection of sensorimotor space (**sm**) belongs to a set A (sets are represented by gray bands), which is mutually stabilized in coupling with a trajectory in the relevant projection (**e**) of environmental variables *a'(t)* that belongs to region A'. In other words, the sensorimotor trajectories in the upper panel and the environmental trajectories in the lower panel are both projections of the whole, coupled system onto the respective subspaces during sensorimotor engagements of type A and B. Condition 2: *transition*. Trajectory *a(t)* in coupling with *a'(t)* lead respectively to *b(t)* ∈ B and *b'(t)* ∈ B', the next stage in the sensorimotor scheme O.
Right: Projection of O onto sensory (s) and motor (m) coordinates.

Let us examine the concept of plasticity. Both the agent and the environment, as dynamical systems, are described by a set of variables and a set of constraints and parameters. We call this set of constraints and parameters R for the case of the agent and R' for the environment. A constraint or parameter is anything that has some influence on the variables of interest but that for particular reasons, the observer has decided not to consider as a variable. Say for instance, the ambient temperature may have an effect on how well and for how long I am able to bounce a basketball. This would be an environmental parameter that can affect the sensorimotor engagement. Whether I am very thirsty or tired, or the fact that certain neural pathways have recently been reinforced through training in a related activity may also be parameters that affect the way I play with the basketball. A coupling between systems, as we have said, implies that at least some of the parameters in one system vary with the state of the variables in the other. But other processes apart from the coupling may also drive parametric changes. In this context, we refer to these as processes of explicit plasticity.

For accommodation to occur in the conditions we have described, some form of explicit plasticity is needed. Contrary to what is traditionally assumed, it is possible for a system to exhibit adaptation, learning, and other history-dependent behaviors without explicit plasticity (e.g., Izquierdo, Harvey, and Beer 2008). Those forms of adaptive behavior rely on the rich dynamical possibilities of systems with sufficiently high dimensionality. Such systems learn by selecting different regions of their dynamical landscape in a history-dependent manner, a form of implicit plasticity. Thus, agents controlled by dynamical neural networks can perform some forms of learning and history-dependent categorization without any changes to the parameters of these networks (e.g., to the connection weights). In our description, if we started from the assumption of equilibration for a typical situation, such systems would, by themselves, without the need for parametric change, already assimilate what may look like a perturbation at the local, short-term timescale. Externally, and with respect to the timescale of behavior the agent is seen as "perturbed" and then adapting to this "perturbation." But on a sufficiently long timescale, in the absence of explicit plastic changes, a reliable (not fortuitous) return back into the sensorimotor scheme **O** implies that the original "perturbation" was not such, and had been assimilated all along, only that this did not seem to be the case at the timescale of observation. (This is a formal point; in practice, other normative considerations would enter into play as well, such as the time it takes the agent to enact the sensorimotor scheme.) Since we assume that accommodation can fail, this implies that the process will require some form of explicit plasticity. A change in the parametric conditions is needed precisely when a return to equilibrium is not guaranteed.

Let us describe what happens in dynamical terms when accommodation works. We denote as $<z, z'>$ the simultaneous occurrence of sensorimotor trajectory z and environmental trajectory z' in the coupled system. Thus, by the notation $<z, z'> \in Z \times Z'$ we simply mean that in addition to occurring together, $z \in Z$ and $z' \in Z'$. With this notation, we do not assume that the simultaneous occurrence of z and z' means that the conditions of assimilation are necessarily met. This could be the case or the trajectories in the agent and the environment could correspond also to a moment of breakdown, when the coordination fails. Consider, for instance, the case of a lacuna, a failure in the transition $A \times A' \rightarrow B \times B'$. This means that, after producing the trajectories $<a, a'> \in A \times A'$ (two coordinated sets), a new combined state $<e, e'>$ is produced where at least one of the following conditions is true: $e \notin B$ or $e' \notin B'$, and, in addition, the combined state is uncoordinated (unstable) and does not lead to the following stage in the cycle ($C \times C'$). This breaks down the cycle. Let us suppose that, on attempting the same transition again, a plastic change has occurred in the system (for the moment it does not matter where). This change has the effect of producing a different sensorimotor transition $<a, a'> \rightarrow <b_1, b_1'>$. That is, a new environmental trajectory b_1' is now produced instead of e' and a new state of the agent b_1 is produced instead of e. We assume that like e, $b_1 \notin B$ or that like e', $b_1' \notin B'$, i.e. either of these new patterns or both are still outside the previously assimilated sets. By itself this change is not sufficient for accommodation to occur yet. However, we assume that unlike the combination $<e, e'>$, now the combination $<b_1, b_1'>$ is stable, and that it leads back to $<c, c'> \in C \times C'$. Then the altered factors that lead to the new trajectory have been accommodated. If the accommodation does not disturb the already assimilated sets (which may or may not be the case), then the new combination can be added to the previously coordinated sets. The set $B_1' = B' \cup \{b_1'\}$ defines the newly assimilated environmental conditions, and $B_1 = B \cup \{b_1\}$ the accommodating class of sensorimotor coordination. The sensorimotor scheme has been transformed. Its new organization is now: $O_1 = A \times A' \rightarrow B_1 \times B_1' \rightarrow C \times C' \rightarrow A \times A'$. In longer sensorimotor schemes, a return to the cycle may occur at a later point in which case the sets describing the intermediate links need to be redefined accordingly. In the case of the strange door mentioned before, turning the handle down does not lead to opening the door, but after trying a few things, we may discover than turning the handle in the opposite direction does, at which point we are back into a modified door-opening scheme.

The case of an obstacle can be treated similarly (plasticity would be involved in transforming the new situation $<e, e'>$ into $<b_1, b_1'>$, such that b_1 and b_1' reliably stabilize each other within the accommodated new set $B_1 \times B_1'$).

In Piagetian descriptions, accommodation seems always to imply plastic changes in the agent and not in the environment. This is fitting because Piaget was interested in human development starting from its biological roots. However, in principle, plasticity may occur in the agent or in the environment (in either R or R'). We often equilibrate our sensorimotor operations by modifying the environment (or indeed other agents contribute to our equilibration). Like most species, we are active constructors of our environments, purposefully or not. If it is too cold to perform our activities outdoors, we accommodate this obstacle by wearing warmer clothes or building a shelter, not by growing fur. There may be a range of less obvious cases where the agent's activity is not directly aimed at transforming the environment and such transformations occur nonetheless (e.g., the formation of trails on grass or the emergent spatial ordering of workspaces; Agre 1997; Kirsh 1995, 1996). In the rest of this section, we stay on the agent's side to keep things simple. However, the analysis permits equally well the consideration of cases in which accommodation occurs through environmental plasticity such that the sensorimotor patterns are modified without requiring any physical alteration to the organization of the agent itself.

How should changes in the set of parameters R occur? What triggers them? These important questions largely depend on the case at hand. Ashby's (1960) general formulation for a dynamical theory of adaptation postulates that these changes could happen at random as soon as they are triggered by the mismatch between the current situation and the acceptable (equilibrated) set of possible agent–environment states. In such a scheme, random changes in the parameters R governing the agent's sensorimotor coupling would lead to exploration of the space of possible alternative sensorimotor coordination, and this process would terminate if re-equilibration were achieved. It is clear that natural adaptive behavior involves strategies that are more sophisticated, about which we can say little in general terms here. But we can affirm one implication that arises from our formal description: open-ended accommodation (accommodation to new and unknown breakdowns), if it occurs, always involves an element of randomness. If this were not the case, if a sure, deterministic accommodating strategy existed for the agent to deploy, this would mean that the "perturbation" had been assimilated all along as the closure of the cycle was guaranteed (although by external standards it may look as if the agent was struggling to accommodate a new environmental feature). Therefore, open-ended accommodation implies some degree of random search in how internal parameters are affected and/or randomness in how the environment responds to these parametric changes.

If accommodation has only additive effects (i.e., they add to the set of assimilated states without subtracting previously assimilated conditions), then maximal equilibration is conserved. If not, maximal equilibration may be re-attained through a sequence of further accommodation steps (learning the new but also re-learning the old).

In practice, in many cases we witness a tendency toward increasing equilibration as learning progresses. As we mentioned in Chapter 2, this tendency can be measured by dynamical signatures, for instance, by studying long-term correlations and variability across repeated trials (e.g., Dotov, Nie, and Chemero 2010; Van Orden, Kloos, and Wallot 2011; Wallot and Van Orden 2012 Wijnants et al. 2009; Wijnants et al. 2012). Such measures are indicative of the degree of fluency in sensorimotor engagements and other important aspects, for instance, whether the action is more driven by the agent or by the environment.

Without radically altering the present analysis, we can also account for Piaget's second type of equilibration (between sensorimotor schemes), simply by noting that plasticity in the agent may occur not only in parameters that regulate the sensorimotor schemes themselves, but also their relation with other schemes. In fact, it seems unlikely that in complex adaptive systems, a parametric change will affect only one sensorimotor coordination without affecting other coordination patterns and other schemes. The condition of equilibration in such cases would not necessarily be a return to some later segment of the original cycle, but a mutual accommodation of the various elements of a sensorimotor scheme with respect to each other, a transformation of the scheme as a whole and of its relations to other, also possibly changing, schemes.

The purpose of the foregoing technical discussion is to demonstrate that equilibration can be described in dynamical terms using concepts of SMCs we developed in Section 4.4. The gist of this analysis can be illustrated in Figure 4.3. Panel A shows a maximally equilibrated scheme **O** and an instance of a trajectory within the gray zone that defines it, corresponding to a case of successful assimilation. Owing to a perturbation in a second enactment of the scheme (dashed line), the coupled system moves away from the maximally equilibrated area. A series of plastic changes are induced (Panel B), with the result that the scheme now also encompasses the new situation. Re-equilibration is not yet achieved maximally, as indicated by the jaggedness of the modified scheme (although this is merely a graphical convention; in some cases, equilibration could indeed look jagged in the plot and still be near maximal). But as further accommodation events occur, the scheme may increasingly approach this condition (Panel C). Here we also show an additional possibility. It may happen that new metastable regions can be discovered by the plastic exploration of sensorimotor couplings while the system is in the process of accommodating the original perturbation. This will result in the creation of a new sensorimotor

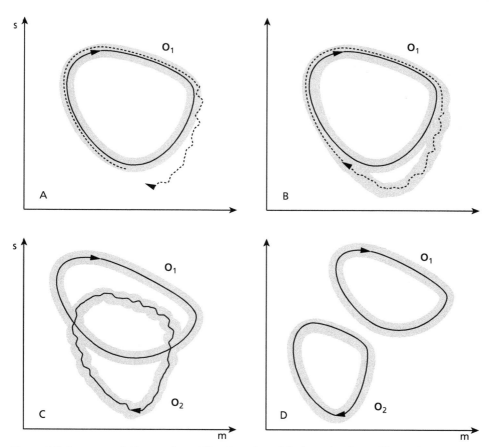

Figure 4.3 Accommodation and equilibration. **Panel A:** A maximally equilibrated sensorimotor scheme O_1. A trajectory in sensorimotor space is shown traveling within the gray band defined by O_1 (solid line), indicating successful assimilation. Owing to a perturbation, a second trajectory (dashed line) leaves the gray band. **Panel B:** The perturbation is accommodated thanks to plastic changes and the scheme re-equilibrated (though not yet maximally, as indicated by the jagged trajectory). **Panel C:** Through repeated perturbations and adaptations, the scheme O_1 has been reshaped, and another scheme O_2 has been equilibrated. They correspond now to two different schemes. The new scheme O_2 is not yet maximally equilibrated. **Panel D:** O_2 attains maximal equilibration (represented by a smoother trajectory), and the two schemes become further differentiated.

scheme O_2 without the disappearance of the original one. As before, this scheme may not be maximally equilibrated, but in the right circumstances may increasingly become so (Panel D), and in the process, also progressively differentiated from the original scheme. This would be like learning that there are two kinds of doors, those that open turning the handle down and those that open turning the handle up.

In terms of our earlier example, O_1 could correspond to the sensorimotor breastfeeding cycle (Figure 4.3, Panel A). A perturbation occurs (dashed line), for instance, the first attempt at feeding from a milk bottle. The breastfeeding cycle serves here as the departure point for accommodating the new situation. Through plastic changes and "reuse" or adjustment of sensorimotor elements in the breastfeeding cycle (e.g., the baby changing the relative intensity, duration, and timing of suckling, swallowing, and breathing), the milk bottle is accommodated for the first time (Panel B). Through further disequilibrium and subsequent accommodation, the milk bottle cycle begins to be maximally accommodated. A separate gray area begins to be defined (Panel C), which corresponds to a new milk bottle-feeding sensorimotor scheme O_2. The breastfeeding scheme remains, though possibly modified in shape. There may be some "interaction" between the two schemes because both involve similar support structures in the baby (muscle groups, neural paths, etc.), that is, higher order equilibrations. O_1 is likely to change as O_2 becomes maximally equilibrated (Panel D).[2]

We may also come back to the example of the visual inversions studied by Kohler (1964), and try to interpret his behavioral and phenomenological reports broadly in terms of processes of equilibration. Kohler's main finding was that re-adaptation of the visual field always occurs only partially and stepwise. The first successful adaptations are always those involving co-occurring non-visual (e.g., tactile) sensations and/ or overt bodily interactions. Thus walking, avoiding obstacles, reaching, etc. become increasingly better and eventually nearly maximally equilibrated, while the visual experience itself is still judged non-veridical. After such skills are recovered, this is usually followed by recognition of objects that can be "brought into a new behavioural relationship with one's body" (Kohler 1964, p. 158), such as one's face seen in the mirror or one's shadow projected on the floor. Adaptation of visual experience is fragmented: objects might appear in the correct place, yet remain mirrored in orientation (the position and direction of movement of a car is perceived correctly, while the letters on its license plate remain reversed). This fragmentation of visual skills led Kohler to interpret his experiments as a means to probe the transformation of "the structure and mutual interweaving of [perceptual] habits" (1964, p. 139), and the process of adaptation as one of rehabituation.

Identifying "perceptual habits" (Kohler 1964, p. 140) with sensorimotor schemes and rehabituation with equilibration allows us to find further similarities. Firstly, new attempts at equilibration always start from what is already known (i.e., from existing patterns of coordination). During rehabituation this results in a failure to grasp objects or move in the intended direction because there is now a mismatch (a lacuna or an obstacle) between vision, proprioception, and intended movement. Interestingly, conscious strategies for

---

2 For an idea of what the differentiation of sensorimotor schemes may involve at the neural level, see, for instance, the splitting of neural sequences that lead to the differentiation of syllables in songbird vocal development (Okubu et al. 2015); or the differentiation of goal location schemata in the rat hippocampus (McKenzie et al. 2013).

controlling (compensating) fast grasping movements are not successful or stable in the long run. Kohler reports that only repeated and essentially random reach attempts eventually lead to the gradual adjustment of movements and the co-occurring visual experience, supporting our comments about the need for some randomness in the exploration of new coordination.

A second observation is that adaptations to wearing reversion goggles always seem to be task specific, that is, progressive trials are guided toward achieving closure of some kind (e.g., reaching the object). Equilibration also occurs gradually. The first cases of success are often unstable until equilibration is maximized and new metastable sensorimotor schemes are fully formed. There also exists, according to Kohler, a "particular splitting in the difficulty or ease of correct mastery" (1964, p.144) of different visually guided actions: equilibration in one task does not transfer to others. The kind of left-right perceptual schemes involved in walking or cycling are different from those used in reading letters, for example. Each perceptual scheme is mastered by undergoing an equilibration process of its own, although each equilibration process is likely to have some influence on other schemes. Only when all tensions between individual sensorimotor schemes are resolved, in other words, after appropriate re-equilibration (or equilibration of the second and third category), is the world as a whole perceived as coherent, that is, adequate visual perception re-established (although this stage may never be fully achieved in practice).

Another useful example is learning to bounce a ball under different conditions. This example and a summary of the different dynamical concepts introduced appear in Table 4.1.

4.6 **World-involving principles for open-ended equilibration**

Before we turn to discussing implications of the dynamical reinterpretation of Piaget's equilibration theory, we take a detour to discuss some general principles and requirements that we believe apply to any approach to learning sensorimotor categories and skills. These principles and requirements are not necessarily all novel, but they derive neatly from the dynamical approach to equilibration. Open-endedness is not necessarily entailed whenever these principles and requirements are present; rather, they make open-endedness a possibility.

Why is the possibility of open-ended learning of interest in this context? At least, in the case of human beings and possibly other animals, it seems that sensorimotor learning, while constrained by history and biology, is not restricted to a set of species-typical adaptations. In fact, the learning and refinement of action and perception skills in some cases, if not unbounded, at least seems to have no obvious predictable bounds. If an enactive approach to sensorimotor learning did not allow for the possibility of open-endedness, something would be missing. And we shall see that certain open-endedness will be required in order to speak of some aspects of agency at the sensorimotor level (in Chapter 6).

Our first observation is that learning agents need to be able to respond quickly to varying environmental situations. This requires that they never reach strictly stable equilibrium. Sensorimotor coordination patterns, therefore, must only be temporarily "stable," or *metastable*. They must retain a residue of dynamic criticality without which they would simply be unchangeable automatisms. This is accounted for in the requirement that each sensorimotor coordination within a scheme must not remain forever stable and should naturally lead to another as part of the transition condition.

Although we have not discussed it at this abstract level, it should be clear that in order to respond appropriately to different environmental contexts, *mechanisms of selection* are needed to choose which sensorimotor scheme to enact in a given situation. Random selection might be necessary in situations never encountered before, and might underlie the exploration of new patterns of sensorimotor coordination during equilibration. But more directed strategies can be expected. They need not be based on abstract deliberation or complex decision making. Dynamic mechanisms that allow environmental conditions to "call for," or resonate with, certain sensorimotor schemes are one possibility (Buhrmann and Di Paolo 2014b). Also, in the same way that certain patterns of sensorimotor coordination follow each other within a single scheme, there might be propensities for some schemes to be followed by certain others, and such interdependencies may be established during development. For instance, a sucking reflex might usually be preceded by a scheme that guides the baby's attention in the necessary direction, like the rooting reflex during the first four months, which in turn might be preceded by an organization that seeks proximity to the mother, and so on. Other mechanisms of selection might involve the similarity of sensorimotor schemes (and/or environmental conditions), such that similar but different schemes can be tried in similar contexts. This would also work through hierarchical organization of sensorimotor schemes without requiring a quantitative similarity measure. For example, if through accommodation a new bottle-sucking scheme is established that in turn derives from a previous breastfeeding scheme, then both schemes together could form a higher-level class of related schemes. We will come back to discussing the dynamics of scheme selection and the interrelation between schemes in Chapter 6.

Another important lesson we derive is that accommodation involves *normative evaluation* of sensorimotor schemes. Firstly, this is required for assessing when an assimilation attempt has failed and whether adaptive processes need to be triggered. Secondly, in the process of accommodation, new engagements need to be evaluated as better or worse than the failed one in order to determine which adaptations to fixate in the system. There may be multiple sources of normative evaluation, including external ones. But, as a principle, it is necessary to have a least some internal sources of normativity (e.g., closure of sensorimotor scheme and/or viability of the organism) in order for the agent to evaluate situations in which externally imposed norms (e.g., an external utility function or value system) break down, cease to apply, or endanger the organism. An *exclusive* reliance on external norms and value systems in models of learning by definition imposes limitations on the universe of learnable behaviors (i.e., they are restricted to those possibilities considered by the externally imposed value system) and therefore are an impediment to open-ended learning.

Accommodation also requires *adaptive mechanisms* for modifying existing sensorimotor coordination patterns. An important principle is that accommodation always starts from some pre-existing sensorimotor scheme, which undergoes adaptation if it cannot assimilate a given environmental feature. To achieve this goal, random "mutations" (i.e., generation of new potential sensorimotor schemes) are in principle required for true open-ended learning (in addition to more guided forms of learning). Otherwise, a system cannot transcend, even potentially, its current "laws" of operation. The randomness in question need not involve complete, but only partial, statistical independence from previous states. "Motor babbling" is one example of randomness creating new interactions with the environment. But while randomness is required for true open-ended learning, it will typically not be the most efficient route to learning in those cases that are recognizably similar to what has been encountered before. In such cases, an agent can also learn from *the way* she fails. Directed learning could rely, for instance, on gradients in the normative evaluation of SMCs or on details of perturbations encountered in failed assimilation attempts.

Finally, in a complex system involving many different sensorimotor schemes, for instance, in the case of hierarchically organized sensorimotor structures, the accommodation of one scheme may interfere with others established previously, as we have mentioned before—this is analogous to the stability-plasticity dilemma in traditional learning theory. Not only must accommodation produce valid sensorimotor schemes, but these are also subject to a *global coherence* constraint. In other words, equilibration not only involves adaptation of individual sensorimotor schemes but also the re-equilibration of the sensorimotor repertoire as a whole.

In sum, the building blocks of open-ended learning are meaningful patterns of sensorimotor coordination, and the schemes they are part of, rather than, say, individual sensorimotor states or trajectories. We learn to perceive and interact with something never before encountered through equilibration of an organization of such sensorimotor coordination patterns. Through a process of incremental, adaptive differentiation of existing coordination patterns—bootstrapped by simple sensorimotor loops either already present at the earliest stages of development or discovered through self-organizing processes—previously established know-how is adapted to a new context or new patterns of interaction are generated and integrated with an already existing set of sensorimotor schemes.

Several of these principles are already present in other approaches. For example, with his concept of ultrastability, Ashby (1960) formulated perhaps the first mechanistic account of open-ended learning, namely the random exploration of a large space of sensorimotor loops with the aim of achieving homeostatic equilibrium.[3] Parallels with reinforcement learning (Sutton and Barto 2009) and related sensorimotor approaches (e.g., Duff, Fibla,

---

[3] For recent investigations and applications of Ashby's notion of ultrastability see (Di Paolo 2000b, 2010; Di Paolo and Iizuka 2008; Harvey et al. 2005; Iizuka and Di Paolo 2007; 2008; Izquierdo, Aguilera, and Beer 2013; Manicka and Di Paolo 2009).

and Verschure 2011; Maye and Engel 2011, 2013) can be drawn as well. For instance, the exploration-exploitation trade-off characteristic of such approaches is related to the assimilation-accommodation dialectics in equilibration; and the global equilibrium toward which these systems tend is one of maximum expected reward, in analogy with the state of maximum equilibration.

Another parallel can be drawn between equilibration as the incremental differentiation of patterns of sensorimotor coordination, and evolutionary adaptation through natural selection. Both processes of adaptation rely on a pool of units (in our case, patterns of sensorimotor coordination), which produce copies similar to themselves with at least some random variation (through accommodation), and which are exposed to selective pressure (normative evaluation determining whether new sensorimotor schemes become fixated). The idea that selection might play a role in learning, and specifically, that it may operate in the brain, has a long history. Neural models, such as those based on the idea that (functional groups of) synapses are differentially stabilized by reward (Edelman 1987; see, also, Dehaene, Kerszberg, and Changeux 1998; Izhikevich 2007; Seung 2003), as well as those proposing proper Darwinian replication in the brain (Fernando, Szathmáry, and Husbands 2012), may well constitute one direction for a theory of how equilibration can be implemented.

However, equilibration differs from most of these approaches in crucial ways. Ashby's (1960) ultrastability, for example, fails to account for types of adaptation more directed and efficient than randomness, and, at least in its original formulation, is at odds with the incremental nature of equilibration, according to which learning always starts from where you are now and tends to conserve previously learned behavior for small accommodations. And selectionist models often fail to include intrinsic sources of normativity and rely on the relative stability of units of selection, a condition that may not always be true in the case of sensorimotor schemes.

In contrast with most other approaches, equilibration does not assume a "functional" source of normativity guiding adaptive change, for example, in the form of a central value system decoupled in function from the changing behavior of the agent. What drives sensorimotor transformation is rather the stability of individual schemes, along with their holistic coherence in the sensorimotor repertoire, that is, ultimately the viability of a particular "form of life" (in the case of humans, for example, as a sportsperson, as an academic, as a student, and so on). These are norms inherent in the processes of equilibration themselves, not functionally decoupled from them. As such, an equilibrating agent is not limited to learning what a pregiven value system knows a priori how to evaluate. The agent can also adapt to situations that involve norms that are intrinsically behavior dependent.

Piaget's account also distinguishes itself from other approaches in that learning is not seen as the discovery of existing structure in a pregiven space of perception-action states, as in most reinforcement learning architectures. It is rather the combinatorial construction of new patterns of sensorimotor coordination in a potentially ever-growing space of possibilities. Equilibration is thus akin to the evolutionary radiation of species that leads to the ever-branching phylogenetic tree. It can also be compared to the adaptation dynamics in the immune system, which enables production of new antibodies for virtually every possible antigen. As Stuart Kauffman has noted (2002), the space of solutions created by such mechanisms is "unprestatable," in the sense that one cannot determine ahead of time the set of all possible kinds of sensorimotor coordination a typical person might produce in the course of her life.

The open-endedness of Piagetian equilibration is in part due to the fact that most sensorimotor engagements only become available in a history-dependent manner, when other sensorimotor engagements that they depend on have been discovered (like pre-adaptations in evolution). In the human case, such a developmental system will in all likelihood be non-ergodic (see also Kauffman 2002) and path dependent. It will only ever visit such a small part of its "state space" (i.e., produce only a small number of sensorimotor coordination patterns out of all the possible ones) that it will move along a unique developmental trajectory (although commonalities in biology, embodiment, and culture will constrain the possibilities). This is what allows human to "specialize," in other words, to refine skills and sensitivities in directions that stand out of the ordinary in areas of craftsmanship, connoisseurship, artistic and social skills, etc. and do so with different styles and idiosyncrasies.

Additionally, equilibration is open-ended because the world itself provides an open-ended set of possible "behavioral niches." There is no predictable end to the variety of social and material couplings offered by the world. This is a consequence of proposing a world-involving account of equilibration, which implies that the world is a constitutive part of any instance of sensorimotor coordination. It is not the agent's learning architecture that is open-ended per se, but open-endedness is possible only in virtue of a coupling to an open material world, whose influence goes beyond simply that of being a source of information (in which case the space of possibilities would be limited by definition).

This point can be seen more clearly in a dynamical analysis of SMCs, rather than in an account based on manipulation of internal representations. In the dynamical perspective, the world plays a role in learning that is different from that of providing inputs to internal processing. Nothing in the formalism prevents aspects of the dynamics of the world forming constitutive parts of the learnt sensorimotor schemes, and indeed this is exactly what we see in dynamical models. Aguilera et al. (2013) provide a good example of the difference between dynamical world coupling and world as input. This constitutive involvement of the world further distinguishes equilibration from other proposed frameworks for adaptation. In selectionist brain theories (e.g., Edelman 1987), for example, the units of selection considered are usually groups of neurons (or synapses), that is, entities entirely in the head.

It should be noted that at this stage the dynamical systems approach to sensorimotor equilibration is not a fully developed theory. It outlines the essential elements that such a theory will eventually have to contain, but several details, for example, regarding its possible implementations, have yet to be filled in. Progress in this area will need to involve further work on the nature of open-ended learning: for instance, additional examination of the processes assumed to be open-ended in nature (such as biological evolution and the dynamics of immune networks) and their relation to processes that could be operating in the brain (see, e.g., Fernando et al. 2012; Watson and Szathmáry 2016) and in the non-neural body.

Future work should also be aimed at identifying empirical evidence supporting the dynamical formulation of sensorimotor equilibration. In this regard, the non-ergodic, history-dependent nature of this process suggests that it is necessary to study individual subjects' learning trajectories as a function of their pre-existing behavioral repertoire. A good example is the study by Kostrubiec et al. (2012), in which dynamical systems analysis is used to describe different subjects' strategies in learning a new sensorimotor skill. The authors find that the routes of learning, i.e., the dynamic adaptations involved, depend on the relevant sensorimotor coordination repertoire each individual learner brings to the learning task. The observed adaptations are either small incremental modifications of an existing sensorimotor coordination pattern, if it is similar enough to the coordination that is to be learned or otherwise abrupt bifurcations that qualitatively change the underlying sensorimotor repertoire and create novel forms of sensorimotor coordination. In further support of the equilibration approach, the authors also show that the stability of the desired coordination, rather than detected errors in performance, can serve to guide sensorimotor learning. In general, new methods of investigation will be needed to study the development of non-ergodic systems whose qualitative properties change over time in path-dependent ways (see Medaglia et al. 2011; Molenaar and Campbell 2009).

4.7 **The meanings of mastery**

We started this chapter with the aim of elaborating a theoretically rich conception of mastery from a world-involving perspective. How close to this goal has the dynamical equilibration framework taken us?

Hutto (2005) warns us about how easy it is to fall into a representationalist interpretation of the sensorimotor approach to perception. And one of the main reasons is precisely the formulation of mastery as a kind of knowledge about sensorimotor regularities that mediates the exercise of what we have here called sensorimotor schemes. The endgame of such formulations in terms of knowledge and mediation is a representationalist reduction of the theory, a replacement of its world-involving properties in favor of an in-the-head interpretation. Such is the case, for example, of Seth's (2014) attempt to formulate a notion of mastery in terms of predictive generative models, which we carry in the head (see Di Paolo 2014).

In this chapter, we have stuck to our promise of developing a world-involving theory. For this reason, we have presented a non-representational interpretation of the concept of mastery as involving the ongoing equilibration of sensorimotor schemes. This concept fits the necessary requirements for a theory of perceptual learning and transformations. At no point in our account are agents required to model the world or the body or to process informational inputs to update generative predictive models or any representational construct of any sort. Granted, not all of the key aspects of a full non-representational account have yet been provided (the same can be said for those representational accounts that have been offered until now). We still lack a theory of agency and a naturalistic account of how agents can act in relation to norms (internal or external). And an understanding of how our proposal would account for more complex aspects of cognition, even situated cognition, such as planning ahead, changing attitudes toward our current situation, and so on. Some of these issues are dealt with in the rest of the book.

But at the moment, we have produced a concept of mastery based on equilibration that has the following properties. It is grounded in dynamical systems conceptions and hence it is operational. It can be measured and indeed it has been (e.g., Dotov, Nie, and Chemero 2010; Wallot and Van Orden 2012; Wijnants et al. 2009). And it connects the right levels of description, it is not something subpersonal, like a mental representation or a neural pattern, but something that belongs to the sensorimotor and cognitive organization of the agent. In fact, we will soon see that for this reason, our concept of mastery is deeply connected with the sense of agency and coherent with the phenomenology of this sense.

Mastery, in a static sense, refers to an equilibrated, coherent sensorimotor organization involving many schemes. In this sense, an agent masters the schemes she is capable of enacting. This is true at any stage in which we can say that schemes can be enacted successfully. But in a diachronic sense, mastery also refers to the equilibration process itself, not just to its achievements. In this sense, mastery is the ongoing process by which the agent continuously adapts to the challenges of a changing world.

In both of these senses, mastery is a world-involving concept since it relies on dynamic engagements with the world, enacted or potential. Mastery avoids the extreme conceptions of empiricism (the structure of the world must be learned by extracting patterns and regularities from it) and intellectualism (the agent imposes its notions on the world to order it following some pre-existing categories).

In our framework, perception involves embodied understanding. But we steer clear of the paradox of perceptual learning, according to which we cannot learn what we do not yet understand. This is precisely because we have avoided seeing mastery as the accumulation of internal representations, whose relevance and applicability would escape the agent in novel, unlearned contexts. Instead, mastery is a regulated openness to be coupled to the world and to be guided by it starting from what has worked in the past. Mastery involves as much the agent as the world as sources both of metastability and of novelty.

Ours is a deeply embodied, deeply embedded conception of mastery. It is a property of how sensorimotor schemes can attain equilibration and relate coherently to a network of other schemes. In each enactment of a scheme, there is a fingerprint left by all the "neighboring" schemes that we master, those that link with the current enactment in sequence or in parallel and those alternatives that we inhibit. The mutual accommodation between schemes means that my skillful grabbing of tea mugs is colored by my command at pouring hot liquids from a kettle, as well as my elegance at sipping hot beverages.

At no point does any of this involve representations of any sort whatsoever. This is non-representational cognitive science at work.

Chapter 5

The missing theory of agency

In order to alter matter, living form must have matter at its disposal, which it finds outside itself in the alien "world." Life is thereby turned toward the world in a special relationship of dependence and capability. Its need reaches out to where the means of satisfying it lie. Its self-concern, active in acquiring new matter, is essentially openness for encountering external reality. In its need dependent on the world, it is turned toward the world; in its relationship, it is ready for encounter; in its readiness for encounter, it is capable of experience; in its active self-concern, primarily apparent in its active acquisition of matter, it continually brings about encounter and actualizes the possibility of experience; experiencing, it "has a world." Thus, "world" is there from the very beginning: a horizon opened up by the transcendence of need, which breaks the isolation of inner identity to embrace a circumference of vital relationship. In other words, life's self-transcendence consists in having a world in which it must reach beyond itself and expand its being within a horizon. This self-transcendence is rooted in an organic need for matter, and this need is based in turn on its formal freedom from matter. In its ability to sustain a relationship with the world, i.e. in its behavior, this freedom takes control of its own necessity.

—*Hans Jonas (1996, pp. 68–9)*

5.1 **Where the action is**

Grazing on steep slopes of wet volcanic rock, such as those a mountaineer can encounter on a trip to North Wales can be a risky way of life. But sheep are often found doing precisely this. The rock surfaces are smooth and slippery, only a few bits of grass here and there that could break the long fall should you miss a step. On this occasion, a sheep and a lamb have moved far along the smooth rock face. Turning back is difficult. They have very little room for maneuvering safely. Their hold is uncertain. The heavier sheep misses a step but quickly recovers. A few small rocks roll downhill and bounce on the smooth surface, gaining momentum. The small lamb follows the sheep cautiously, a few steps behind. The sheep moves her head toward a bit of grass barely out of reach. Maybe if she stretches out a bit more ... At which point she knows she is in trouble. Her front legs begin to slip. She looks up, tries quickly to spot a safer place, and ventures to jump upward. But she does not find a good hold to push herself up; and so, she begins to slide down the slope. What happens to the sheep now is not unlike what happened to the falling rocks a few seconds ago. She is unable to regain control. Her front legs are in the wrong position. When she finally manages to put them forward, she has gained too much speed to break the fall. At this point she bounces very much like the rocks did earlier. The lamb looks on anxiously until the sheep is lost to view. He hesitates for a few seconds and decides to follow the sheep down. But he does it by carefully running down first for a short stretch, then another. He tries not to run too fast. As the momentum becomes too difficult to control, he aims for a patch of grass that helps him break the movement. And there he stops. He assesses the situation. The slope is less steep here, and he can resume the search for the lost sheep a bit more safely and, with some luck, find her.

We can find similarities between the bouncing rocks, the poor sheep that could not control her fall, and the lighter lamb that runs down the slope while breaking the descent as much as possible. In all three cases, the physical forces at play in propelling the movement are the same. And the trajectories are complex but similar. Yet we know there is a difference between the downhill running lamb and the falling rocks. We see this in how the lamb controls the descent even if the same force drives it in both cases. We tend to think that this is precisely what agents do: engage in interactions with their surroundings that to some extent they can influence and control. And what about the hapless sheep? It would seem that in this case she was very much like the falling rocks: unable to *do* anything except submitting to forces beyond her control. If we said that the sheep and the lamb, unlike the rocks, are agents of their actions, it would appear that the sheep was not behaving as one during the fall. Yet, this would not be entirely right, because she was aware of something going wrong and tried to do something about it. She tried to reorient her falling body to the right angle, her legs outstretched. If the surface had been less slippery or if she had been moving a bit more slowly, she might have succeeded in breaking the fall. It just did not happen this way. Was she still behaving like an agent?

This and other questions regarding the nature of agency are the topic of this chapter. As we can already appreciate, a scientific approach to this subject is bound to be difficult. We might be tempted to look for the constitutive elements of agency in the exercise of a certain power by a system on its environment but we would run into problems because sometimes agents, like the lamb in our story, behave by influencing external flows of energy and not by themselves doing much physical work. We might consider that some complexity in the objective movements or some kind of goal-directedness is the mark of agency, but then again, sometimes agents perform very simple movements and things we do not attribute agency to, like the falling rocks, move in complicated ways in apparent pursuit of a goal. We may then notice that perhaps agency is not a property that belongs exclusively to a system but is a property of a *relation* between that system and its surroundings. And this relation is variable. A sheep that falls without control does not seem to be behaving as an agent at this point, even if she has done so on other occasions. Yet we may even question our intuitions in such cases too: should the mark of agency be successful action? And if not, how are we to tell the difference between a non-agent and an agent who is overwhelmed by a situation outside her control, like the sheep in our tale?

Up to this point, we have relied on an intuitive notion of agency in our accounts of sensorimotor contingencies and mastery. Similarly, we have used an unexamined concept of normativity when we needed it. If we cannot somehow propose a way of grounding these notions in a naturalistic manner (i.e., without recourse to explanatory elements outside the realm of science), the enactive project will be incomplete. This is our goal in this chapter. The story is complex and involves going over some of the more interesting but also more controversial developments of the enactive approach in its attempt to connect the organizing principles of life and mind. We will introduce what we consider are the three requirements that a system must meet in order to be called an agent. This will lead to a concept of agency that will help us address some of the questions raised by the above example. But this concept of agency is rather general. It will be the focus of Chapter 6, to discuss a more concrete derivation of this concept.

5.2 Requirements for agency

The aim of any scientific approach to agency is to provide a naturalized account of what it means for a system to act on its own behalf. This phrase needs to be unpacked. How do we distinguish a "system" from the background network of processes in which it is embedded? What constitutes an "act," as distinct from other physical exchanges undergone by a system? What does it mean for a system to act in its own interest, to have a concern or a stake in a situation? If we assume as in the previous chapters that we can apply the concepts and techniques of dynamical systems theory to approach these questions, can we justify, beyond matters of convention or convenience, the claim that a specific set of variables or processes constitutes "the agent" and another set "the environment?" Where do we find "actions" or "concerns" in a list of dynamical equations?

In this section we overview how the enactive approach has attempted to answer these and related questions.

To say that an agent is a system acting on its own behalf is to imply that we can clearly identify the system as an individual, that this individual is able to do something by itself, and that it does so according to some norms. Following lines initially explored in Barandiaran, Di Paolo, and Rohde (2009) we start by examining three requirements—self-individuation, interactional asymmetry, and normativity—that capture what are commonly understood to be the necessary and (jointly) sufficient conditions for a system to be considered an agent in the context of the sciences of mind.[1] We introduce these conditions as independent of each other, but come back to discussing their relation after we have proposed an operational definition of agency. We examine the living cell as an example of a minimal system that meets the three requirements. In Chapter 6 we will propose that the same ideas can be used to describe other kinds of agency, particularly at the sensorimotor level.

5.2.1 Self-individuation

In many everyday uses of the word *agent*, for example, when referring to persons or institutions, what distinguishes the entity we are referring to from its surroundings is rarely

---

[1] These requirements should be understood as attempting to capture a general conception of agency, one that includes animals and potentially artificially created agents. Discussions concerning the nature of *human* agency, in contrast to the agency of other animals, are normally concerned with aspects that may not be captured by these requirements (but which would entail them nonetheless). These may include forms of self-evaluation, commitment, reflectiveness, forethought, and responsibility (e.g., Bandura 2006; Jonas 1984; Korsgaard 2009; Taylor 1977).

questioned. Yet the distinction between agent and environment is critical for any theory of cognition. On reflection it becomes clear that even everyday uses of the term are far from straightforward. Do we include in the description of a human individual the approximately 10^{14} bacteria co-inhabiting her body, or the tools she uses to complete a particular task? Similar questions can be asked for chemical reaction networks in the prebiotic soup or for robots and their environments. Often what distinguishes a system from everything else that is not the system is a decision made by observers, using particular criteria such as the practical or conceptual usefulness of the distinction. One could therefore conjecture that being an agent is no more than a status attributed to systems by the observer as a matter of convention or convenience.

The enactive approach—while recognizing that any observation, even any scientific description, requires an explicit perspective—proposes that once the domain of observation has been agreed upon (whether we decide to study atoms, markets, or ocean currents), the question of whether a particular system in this domain is or is not an agent need not rely on conventions or mere convenience. The enactive approach suggests that agents are systems that *actively define themselves as individuals*, and may be identified as such without arbitrariness. Only systems that manage to sustain themselves and distinguish themselves from their surroundings, and in so doing define an environment in which their activity is carried out, are considered as candidate agents in this approach.

The enactive view seeks to explain such seemingly mysterious processes of self-individuation by reference to the properties of certain systems that actively conserve their own organization (see, also, Jonas 1968). A fundamental concept in understanding the material and energetic conditions under which initially undifferentiated processes organize and through their operation lead to the individuation of a well-defined unit is that of *operational closure*. The term originates in the mathematical concept of closure, which defines a set as being closed under certain operations if the application of those operations to elements of the set only produces other elements of the same set (the integers, e.g., are closed under the operations of addition and subtraction, but not division). Closure in this formal sense is sometimes called *organizational*. The enactive concept of operational closure is similar but the term "operational" highlights that closure is achieved through the actual work and transformations done by processes in time (i.e., we are not merely talking about closure in a formal, mathematical sense). The concept then refers to a network of processes whose *activity* produces and sustains the very elements that constitute the network.

The concept of operational closure may sound suspicious, as if some strange kind of backward causation was implied. This is not the case. Consider as one example of such systems an autocatalytic set of chemical reactions (Figure 5.1). This is a network of molecules and reactions in which each molecule is produced by a reaction that is itself catalyzed by other molecules produced by the network, such that as a whole the set is able to catalyze its own production (see, e.g. Kauffman 1986; Hordijk, Steel, and Kauffman 2012, 2013; Hordijk and Steel 2015).

In Figure 5.1, for example, we have a network of three reactions, r_1, r_2, and r_3. Each of these reactions is catalyzed—that is, *enabled* in the sense of being accelerated to sufficiently

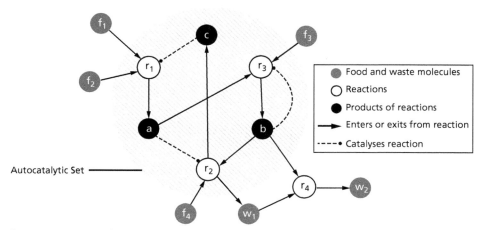

Figure 5.1 Autocatalytic closure. A network of chemical reactions is organized such that each reaction is catalyzed by products of other reactions in the network. The resulting system maintains itself as a whole as long as certain environmental conditions hold (e.g., sufficient concentration of high-energy food molecules or other energy sources—not shown here for simplicity).
Copyright © 2017 Ezequiel A. Di Paolo, Thomas Buhrmann and Xabier E. Barandiaran, with permission.

fast rates—by the molecules of type *a*, *b* and *c*, which are themselves the products of the same reactions. Reaction r_1, for instance, produces molecule *a* from two precursor, or food, molecules f_1 and f_2. The reaction is catalyzed by molecule *c*, which is the product of another reaction r_2, which itself is catalyzed by the product of the first reaction. One of the reactants in the second reaction is molecule *b*, which is produced in turn by reaction r_3, which, as a special case, is autocatalyzing. In other words, the product *b* of r_3 acts as a catalyzer for the reaction r_3 itself.

When embedded in a chemical soup, this set may also engage in other reactions that do not contribute to catalyzing its own production. For example, some molecules provide energy to keep the cycle running (the f_i molecules in Figure 5.1). They are "consumed" as "food," while others are released into the environment as "waste products" (the w_i molecules), where they may enter into yet other reactions (i.e., r_4, which is not catalyzed by any molecule in the autocatalytic set). While it may seem that thereby the boundary between the autocatalytic set and its environment is open, at any moment there exists a well-defined network of reactions that "catalyze" their own production.

In general, and without reference to any particular instantiation, an *operationally closed system* is defined as a network of precarious processes in which each process enables at least one other process in the system and is, in turn, enabled by at least one other process in the system (Di Paolo 2009; Di Paolo and Thompson 2014). Varela (1979) calls such systems simply *autonomous*. Through the active self-maintenance of a particular form of organization, the operationally closed network thus defines its own identity without

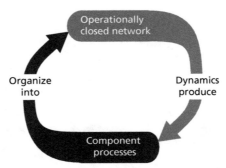

Figure 5.2 Operational closure. A network of interacting processes is organized such that the continued operation of each process depends on the activity of at least one other process in the network and every process contributes to enable at least one other.
Copyright © 2017 Ezequiel A. Di Paolo, Thomas Buhrmann and Xabier E. Barandiaran, with permission.

requiring the drawing of arbitrary boundaries. The diagram in Figure 5.2 captures the circular enabling relation at the heart of this self-production and self-distinction.

A system's closure in the domain of its constitutive processes does not mean, however, that it is in any way *independent* of its environment. In fact, any physical system of this kind can only maintain itself by constantly reasserting its organization despite opposing thermodynamic tendencies. It is a far-from-equilibrium dynamical system (Nicolis and Prigogine 1977) that channels energy and matter to produce the work required to maintain itself. Such a system is thus at the same time operationally closed and thermodynamically open.

There are different senses in which an operationally closed network of processes can be said to distinguish itself from its environment. As illustrated in the example of the autocatalytic reaction network, operational closure already implies a set of enabling conditions and constraints under which the system can remain operational. Certain food molecules, for example, though not a constitutive part of the network itself, are nevertheless required for its continued self-maintenance. They act as constraints, that is, as material configurations that influence the dynamics of the system without being themselves a product of the system. Importantly though, these constraints are *implied* by the organization of the system itself, rather than drawn to the convenience of an external observer. In this sense, the operation of the system determines what counts as its own enabling conditions. So, there is, on the one hand, an implicit, formal sense of self-distinction, which is given by the fact that the organization of the system maintains itself. It simply tells us that any given process either belongs or does not belong to the self-sustaining organization. On the other hand, there is an active sense of self-distinction in that the determination of what counts as enabling conditions is the result of the actual operation of physical processes, and not merely a formal question.

There is also a stronger sense of self-distinction, which is different from the formal, organizational sense and the simple active sense. In many cases the thermodynamic constraints may require the system to further separate itself *functionally* and *spatially* from environmental influences, for example, by "shielding" the operation of some processes from direct external perturbations or introducing "filters" between some "inner" processes and the environment. This can happen through the creation of spatial boundaries that encapsulate and protect internal activity, although this is not the only possibility. This is a stronger, functional sense of self-distinction, one that is active, as it emanates from the activity of the processes involved in the system, and functional, since it results in properties that contribute to keeping parts of the system partially isolated from the environment in the context of organizational requirements. Some processes in the network are thus involved in establishing a minimal spatiality, a difference between an inside and an outside space.

While we do not find this stronger sense of self-distinction in the autocatalytic cycle, the living cell is a prototypical example of such a self-distinct (and self-producing) entity. It is also a complex system situated far from thermodynamic equilibrium. At its core, a cell is a metabolic network of chemical reactions that produces and repairs itself. These reactions generate a membrane that encapsulates the reaction network while actively regulating (pre-formed) matter and energy exchanges with the environment. Exchanges through this semipermeable membrane are necessary to maintain a sufficient level of precursor chemicals required to fuel the cell's metabolism, as well as to expel its waste products (see e.g. Ruiz-Mirazo and Mavelli 2008).

Figure 5.3 summarizes self-individuating systems of this kind as the fundamental operational closure between network activity and component processes, the creation of

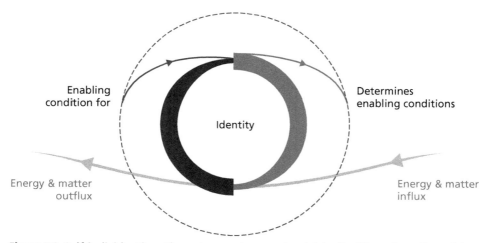

Figure 5.3 Self-individuation. The system produces and maintains itself in so far as the activity of the system as a whole is required to maintain its component processes (the circle at the center of the figure symbolizes the principle shown in Figure 5.2). Through its re-entrant activity, the system defines its own enabling conditions, which may or may not involve an actual physical boundary (such as a semipermeable cell-membrane). Though operationally closed in the domain of its constitutive processes, the system remains open to, and in fact requires, energetic and material exchange with its environment.

enabling conditions, and thermodynamic openness. Such systems are individuals beyond any question of convention or convenience. Ongoing *individuation* of this kind, involving active self-production and active self-distinction at some level of realization, is our first requirement for agency.

An important concept to avoid including "trivial" cases of self-maintenance among those we would like to identify as self-individuating is that of *precariousness*. For example, it might seem reasonable to include crystals in the kind of systems that produce themselves. After all, given the right conditions (presence of a seed in a supersaturated medium), chemical interactions lead to the spontaneous growth of a clearly identifiable entity, which thereafter is maintained over time. But note that once the crystallized lattice has reached equilibrium, it does not need to "do" anything at all to maintain its existence, even if there may still be ongoing individuation occurring at the crystal surface. Precariously self-individuating systems are different. Individuation is never finished as such, as in the formed crystal lattice; it is ongoing and this continuity is what defines it as an autonomous system (cf. Simondon 2005). Here, each process depends on the activity of the others for its own sustained operation. If isolated from the network that provides the conditions for its existence, the process would stop running. At the same time, its removal from the network would lead to a cascade of other processes failing, and ultimately to the disappearance of the system as a whole (or, in some conditions, to its transformation into a different kind of self-individuating system).

5.2.2 **Interactional asymmetry**

As we have said, any autonomous system that has carved itself out from a set of material processes will nevertheless be subject to exchanges of matter and energy with its surroundings. Note, however, that in ordinary language, when we say that an agent *performs an act* or *engages in an action*, we do not normally mean that the agent is merely "undergoing" an intercourse within its environment. Instead, we imply that the agent somehow is *the source of*, precisely the *agent of*, such exchanges. Rather than implying an equal relation between agent and environment, we mean that the agent is, at least on occasions and with some regularity, the source of certain activity, not always just a passive sufferer of external forces.

What could this condition mean in dynamical terms? It is clear that the processes that make up a system—whether it is self-individuated as an autonomous system or defined by convention—both influence and are influenced by processes that are external to it. At a meso-level, we say, therefore, that the system and its environment are *coupled* (i.e., some variables in the environment affect the parameters and conditions of realization of processes in the system, and vice versa; e.g., Ashby 1960). This notion of coupling is a symmetrical one. Of course, this does not mean that the specific effects flowing in one direction need to be identical to those flowing in the other; simply that it is a symmetrical relation of co-dependence between systems. In the particular case of self-individuating systems, such as living or autopoietic systems, the notion of *structural coupling* is also symmetrical in this sense (Maturana and Varela 1980, 1987).

This idea simply adds one requirement to the dynamical concept of coupling: that the autonomous system undergoing interactions with the environment remains viable. It does not say anything about the autonomous system acting or behaving in its environment.

By themselves, the dynamical and the autopoietic notions of coupling are insufficient to capture the idea of a system acting or behaving. Apart from being normative (something we will return to), acts have the property of being asymmetrical in terms of the relation between agent and environment. Because the idea of coupling is symmetrical, one way to introduce an asymmetry in it is to go one level up and propose that an agent is sometimes able to modulate its coupling with the environment (i.e., to modify the way its own processes and those of the environments *relate*). This condition of *interactional asymmetry* is the second requirement for agency.

It is far from trivial, however, to give an operational definition of what exactly we mean by interactional asymmetry. Intuitively it would seem right to equate it with the idea of agents being the "cause" of certain events. In complex systems (e.g., our example autocatalytic set) causation itself, however, is a concept with a variety of different meanings and interpretations. Two possible approaches can be considered.

Starting from energetic considerations, we could try to express interactional asymmetry in terms of the capacity of a system to constrain energy flows to sustain coordinated processes, which are in turn reused by the system in a circular manner (Kauffman 2000). Cells, for example, by coupling endergonic and exergonic reactions and channeling energy flows, produce work and can thereby maintain themselves far from thermodynamic equilibrium. This perspective matches an intuitive notion of agency: the system is the energetic drive of an otherwise neutral or spontaneous coupling with its environment (actively pumping ions or performing chemotaxis, as opposed to passively suffering an osmotic burst or being moved by currents in a pond). Though intuitive, one can imagine cases where explanations of this kind are difficult to uphold, say, for example, in the case of a gliding condor that is being carried by the wind and updrafts without having to exert much energy, or the lamb that controls a downhill descent in our earlier example. One may want to argue that gliding is certainly an action that a condor can engage in (with the purpose of traveling or hunting while conserving strength), yet on grounds of energetic considerations alone, this would seem difficult to defend. There is a sense in which actions can have an element of passivity, of letting energy flows external to the agent carry a good part of the enactment.

Alternatively, one may try to use statistical considerations to assess the extent to which one system (a candidate agent) influences another (its environment). For example, one may identify a system as being the agent of an interaction when changes in its own processes precede environmental changes in time in a statistically significant manner. Apart from certain mathematical complexities involved (e.g., the typical non-stationarity and nonlinearity of the data in most relevant cases), we again find cases that seem to be difficult to capture in this way. For instance, to a first approximation, it would seem that the event of somebody falling off a cliff, and the action of somebody taking a dive into the ocean, are difficult to distinguish on the basis of correlations between the moving body and its environment alone.

A different way to look at asymmetry is not to try to find asymmetric elements *within* the agent-environment relation (in terms of energy or statistical correlations), but, as we have suggested, to think that the agent is capable (not necessarily all the time) of altering the parameters and conditions of this relation. A dynamical systems perspective may be used to capture the concept of interactional asymmetry. Similar to our definitions in Chapter 3, consider a system (represented now by a vector of variables $\mathbf{x}$) and its environment (represented by vector $\mathbf{e}$) as two coupled dynamical systems of the form:

$$\frac{d\mathbf{x}}{dt} = S\left(\mathbf{x}, \mathbf{p}_Q\left(\mathbf{e}\right)\right)$$

$$\frac{d\mathbf{e}}{dt} = E\left(\mathbf{e}, \mathbf{p}_R\left(\mathbf{x}\right)\right)$$

$$\{\mathbf{r}\} \subset Q \cup R$$

$$\Delta\mathbf{r} = H_T\left(\mathbf{X}\right)$$

where the coupling is represented by the functions S and E, each with their own set of parameters $\mathbf{p}_Q$ and $\mathbf{p}_R$. These parameters belong to sets of conditions and constraints, Q and R respectively, corresponding to the agent and the environment. Asymmetric modulation of this coupling is described by the system's influence on a subset $\{\mathbf{r}\}$ of these constraints (some of which could be constraints on the system, others on the environment), which the system modifies according to the function H_T that depends on its own states $\mathbf{x}$, and which is active during a particular time interval T. It is important to highlight that the asymmetric condition in the coupling between the two systems obtains, strictly speaking, only during a certain period; things may be different outside this period.

Consider the cliff diver. Before his jump, the diver is interacting with the ground underneath his feet. A sequence of muscle movements (changes in $\mathbf{x}$) results in a dramatic change in the constraints that modulate the coupling with the environment ($\Delta\mathbf{r}$) leading the system to engage in free-fall dynamics. Here we must notice that if the diver had not been poised at the edge of a cliff, the same sequence of muscle movements would produce a very different effect. This highlights the fact that actions are always both contextual (on Q and R) and temporally extended (over T).[2]

---

2 To what extent can the normal unfolding of a system's behavior be empirically distinguished from its asymmetric modulation? How can we identify signatures of such modulations happening in specific instances? We have mentioned above that energetic and statistical considerations seem insufficient on their own to distinguish proper acts from other physical events. This is not to say, however, that they cannot provide us with useful empirical clues. Moreover, our definition of agency, expressed in dynamical terms, in principle, could be formulated also using different mathematical formalisms, and different methodologies can be used to identify asymmetric modulations in a given system. From a statistical point of view, for example, the normal goings-on might be associated with a stationary situation, while modulation introduces an element of non-stationarity that typically bifurcates or reshapes

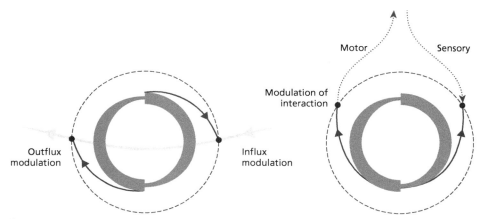

Figure 5.4 Interactional asymmetry. Left: The self-constituting system exerts control over its structural coupling to the environment, modulating both inflow and outflow of energy and material. This does not need to include a physical boundary, but may rely on the modulation of non-spatial enabling conditions. **Right:** A motile system may regulate not (only) its metabolic exchanges, but also its sensorimotor interactions with the environment.
Copyright © 2017 Ezequiel A. Di Paolo, Thomas Buhrmann and Xabier E. Barandiaran, with permission.

It is well known that it is often possible to redescribe dynamical systems by considering strongly influential parameters as variables or slow-changing variables as parameters. Does this possibility affect the definition of asymmetry we have just provided? The answer is no. Once we have settled on what we consider a variable and what a parameter and justified our decision, the above definition will describe an asymmetrical relation between a system and its environment. In some cases, our justification for the choice of variables and parameters may be given by simple convention or convenience. In other cases, as we put our requirements for agency together, the choice will correspond to whether what we describe as the system is an operationally closed network of processes. In both cases, the choice of the level of description depends on the observer, but only in the first case is it a matter of convention or convenience. In the second case, the self-distinction of the individuating system tells the observer that at this level of description, there is a potential agent, without regard to whether the system defines itself in accordance to convention or convenience.

With the above dynamical interpretation of asymmetry in mind, we propose that apart from self-individuation, an additional requirement for being an agent is that of being also a system that systematically and repeatedly modulates its coupling with the environment (see Figure 5.4). Of course, agents are not necessarily able to modulate

the dynamical landscape. Energetic considerations may uncover that the system needs to perform work in order to achieve modulation. Viewed dynamically modulation may correspond to a parametric change in the system, with the resulting qualitative change in attractor dynamics. And of course, a modulatory subsystem may be functionally or structurally distinguishable from the primary self-maintaining system, as is the case in most living organisms that develop sensors and effectors.

all conditions of their coupling all the time. Nor are agents the unique source of asymmetrical changes in their relation with the environment, which may also contain other agents or other processes that occasionally induce a modulation of the coupling. The concept of asymmetry only requires that a system be capable of engaging in some modulations of its coupling and that it does so at least in certain cases (the question of which cases is addressed when we consider the following requirement). As such, interactional asymmetry is not a unique capacity of agents. As we have said, the environment can also sometimes be a source of asymmetric modulations. But it is not asymmetry alone that marks a system's activity as that of an agent. It is rather the joint conditions of self-individuation, asymmetry, and normativity, which we discuss next.

5.2.3 **Normativity**

Something is still missing in our account of agency. Actions performed by an agent are commonly understood to be different from movements, or even modulations of the environmental coupling, in the sense that they are performed according to the agent's goals and norms. The spasms of someone suffering from Parkinson's disease, for example, are not usually considered actions, even though the person in question is undoubtedly a distinct individual capable of modulating his environmental coupling in other cases (i.e., an agent). It is not enough that a self-individuating individual be itself the active source of modulation of its coupling with the environment. For such a system to engage in proper actions, this modulation also has to be carried out in relation to goals or desires.[3] When systems actively modulate their interactions with respect to norms, we speak of *regulations* of their coupling. Such regulations introduce a normative dimension. They can succeed or fail. This *normative condition* is the third requirement for agency.

We use the term *norm* following traditional usage in psychology without involving at this stage a concept of social normativity. For instance, for Merleau-Ponty, norms are what distinguish the physical from the living order (see, also, the discussion in Thompson 2007a):

---

[3] We shall not enter here into the detail of the extensive debates about the nature of purposive explanations of behavior and whether these can be reconciled with scientific explanations. Lucid statements about this topic have been available since the early periods of cognitivism (see e.g. Boden 1972; Taylor 1964) and before. We only notice that the main tension provoked by notions such as goals, purposes, and norms is not with scientific discourse as such, but specifically with mechanistic and reductionist approaches. As we have mentioned in Chapter 2, dynamical and emergentist perspectives can offer interesting and consistent proposals for the naturalistic understanding of these terms (e.g., Juarrero 1999). In this respect, we subscribe to the words of Hans Jonas on the matter:

> How this finalism tallies, in the same world, with mechanical causality whose reality cannot be denied either is a problem not to be 'solved' by sacrificing an evidence (purposiveness) to a theorem (exclusiveness of *causa efficiens*) which was derived by generalization from another evidence; but, if solvable at all, only by treating it as the profoundly challenging and as yet completely unsettled problem it is.

Thus each organism, in the presence of a given milieu, has its optimal conditions of activity and its proper manner of realizing equilibrium; and the internal determinants of this equilibrium are not given by a plurality of vectors, but by a general attitude toward the world. This is the reason why inorganic structures can be expressed by a law while organic structures are understood only by a norm, by a certain type of transitive action which characterizes the individual.

(Merleau-Ponty 1942/1963, p. 143)

By accepting the fact that the organism itself modifies its milieu according to the internal norms of its activity, we have made it an individual in a sense which is not that of even modern physics.

(Merleau-Ponty 1942/1963, p.154)

We find similar concepts of norms, or organic norms, in Kurt Goldstein's (1934/1995) idea of the self-realizing organism, and in Georges Canguilhem's (1966/1991) work on the normal and the pathological (both Canguilhem and Merleau-Ponty were deeply influenced by Goldstein's work).

What is the origin of these norms? If we are not speaking of a self-individuating system, but one defined by convention, the relevant norms are also given externally to the system, unlike in the quotes above. Such would be the case of a machine designed to perform a particular purpose. What a machine "does" is thus evaluated normatively in accordance to what the designer or the user expects of it.

But it is possible also to conceive of a concept of intrinsic norms more along the lines of Merleau-Ponty, a concept not tied to the observer's conventions and convenience. Intrinsic normativity cannot be the result of observers making judgments on behalf of the agent about the "adequacy" of its behavior in relation to their own norms, standards, or goals. Rather, we need to justify this normativity based on the agent's own nature. For example, with the self-individuation requirement in mind, certain asymmetrical modulations performed by the system support the processes that distinguish it from its environment. Other modulations, in contrast, may interfere with these processes and threaten to break the system down. At least one source of intrinsic norms thus originates in the very organization of the system that maintains itself as a self-distinct, self-producing entity. In this sense an agent's actions and environmental events can be good or bad for its continued existence (this does not mean that an organism will always be sensitive to this

He goes on to reflect on this notion of "evidence" given to us by embodied experience:

the teleological structure and behavior of [the] organism is not just an alternative choice of description: it is, on the evidence of each one's own organic awareness, the external manifestation of the inwardness of substance. To add the implications: there is no organism without teleology; there is no teleology without inwardness; and: life can be known only by life (Jonas 1966, pp. 90–1).

Accordingly, Jonas argues against reducing purposefulness to the effect of feedback mechanisms as proposed by Rosenblueth, Wiener, and Bigelow (1943). This solution bypasses the question of how regulation toward a goal in a control circuit relies on this goal being externally imposed by a designer. It is for this reason that in this chapter we attempt to explain normativity as grounded intrinsically in the organization of self-individuating systems.

difference or, if it is, that it will behave by following only this one source of normativity as we discuss in Chapter 6).

Note that as described so far, the kinds of regulations we want to posit as essential for agency seem to result in a rather trivial if not tautological definition: actions are good as long as the agent is viable. It would appear that agents can be more or less robust to environmental perturbations, but at any point, the system would be either self-maintaining or dead, i.e., either within its so-called domain of viability or outside it. This either-or alternative per se does not leave any room for agents to be sensitive to their current conditions and to act so as to improve their prospects of remaining viable. This in turn would imply a rather poor notion of normativity: an encounter would be good if it does not instantaneously endanger an agent's self-maintenance, and otherwise bad and leading immediately to the system's demise. Clearly, agents need to be capable of more nuanced evaluations to remain viable, for example, by being sensitive to the *risk* of disintegration, or to *gradients* and *directions* in their viability conditions. A concept of regulation that captures these more subtle changes in normative conditions is the concept of *adaptivity*, defined as:

> A system's capacity, in some circumstances, to regulate its states and its relation to the environment with the result that, if the states are sufficiently close to the boundary of viability,
>
> 1. tendencies are distinguished and acted upon depending on whether the states will approach or recede from the boundary and, as a consequence,
> 2. tendencies of the first kind are moved closer to or transformed into tendencies of the second and so future states are prevented from reaching the boundary with an outward velocity.
>
> (Di Paolo 2005, p. 438)

Being adaptive (i.e., actively monitoring and regulating processes with respect to intrinsic norms) entails the capacity to improve living conditions or avoid or address threats to viability by assessing a situation relative to the norms given by self-individuation, and acting on it in a graded and directed manner. This capacity can be more or less efficacious, more or less general. It allows adaptive agents to make fine-grained distinctions between some situations that otherwise would be identical in terms of immediate implications for self-individuation (e.g., different levels of sugar concentration surrounding a bacterium that, for the moment, are equally sufficient to maintain its metabolism).[4]

---

4 In a perceptive criticism of the enactive account of intrinsic normativity, Nathaniel Barrett (2015) notices that the concept appears to be entirely proscriptive. As defined in terms of self-individuation and adaptivity, intrinsic norms seem to be always about what the agent should not do, but not about what it should do. Barrett also notices that any kind of normativity that is tied to the persistence of the identity of a "self" could never be relevant in regulating growth or adaptation aimed at transformations of the self. These are valid and insightful criticisms. Addressing them implies taking into consideration some of the phenomena we will introduce in Chapter 6, where we discuss multiple forms of individuation and their interrelation. Norms may initially be proscriptive and leave a "neutral" space of options as long as the system is viable and not under risk. This neutral normative space may not suffice to guide action in such circumstances. The problem, however, may never be faced by real organisms, which are restlessly driven toward the boundaries of viability by their own precariousness. Since negativity, and

Because *adaptivity* is defined as a capacity, an adaptive system is one that can show this capacity at play, but this does not mean that such a system is always able to make use of this capacity. An adaptive system may sometimes fail to generate adequate behavior. It is interesting to note that even in such cases the behavior produced still follows norms in the sense that the outcome of an action that fails can only be distinguished from a neutral movement in the context of the norm implied by the adaptive processes involved. These adaptive processes may provoke a change in the right direction, which is nevertheless too slow or whose effects are counteracted by the circumstances (see Barandiaran and Egbert 2014).

Adaptivity is a precondition for *sense-making* (Di Paolo 2005; Thompson 2007). This is one of the key concepts of the enactive approach. The idea was introduced by Francisco Varela in his later work to describe how a living system establishes its own space of meaning by differentially evaluating encounters with the environment according to their consequences for its self-individuation (Varela 1991; 1997; Weber and Varela 2002). A sense-maker is involved in interactions with its environment that are regulated with respect to their virtual consequences for the viability of the sense-maker's form of life. This is, for enactivists, the hallmark that distinguishes all forms of mindful activity, no matter how complex, from non-mindful activity. Bacteria swimming up chemical gradients and scientists trying to understand them are both cases of sense-making, albeit very different ones.

Sense-making does not imply sophisticated kinds of cognition, but it is implied in them. It is what is common to basic minds (Hutto and Myin 2013) and human minds. To be clear, by "sense-making," then, we refer to the notion that objects or events become meaningful for an agent if they are involved in the normatively guided regulation of the agent's activity (e.g., by triggering or mediating it). This mode of relating to the world is literally the active making sense of a situation and the orientation of the agent toward a course of action that is adequate to it. An approaching waiter, for instance, becomes

not neutrality, is a primary condition for life, its resistance may be construed positively ("life is the totality of those functions which resist death" as Xavier Bichat wrote at the start of the 19th century). Nor is it inconceivable for prescriptive (positive) norms to emerge out of the dialectical resolution of conflicting proscriptive (negative) norms. As we discuss later in this chapter, self-individuation itself implies both self-production and self-distinction, two requirements that tend in opposing directions in terms of how an organism should ideally relate to its environment. The resolution of this contradiction results in a new kind of normativity involving both negative and positive elements (reject certain flows of energy and matter; seek others). The "neutral" normative space may be quickly filled in with norms that are the consequence of interactions between norms at different levels of agency. And conflicts emerging between these levels can drive the transformation of the self at either one of them, although these transformations would also imply new emerging norms. Finally, we should remember that these questions must not be considered in the human case without bringing in the full social dimension of human existence. Social agency is subject to norms that can guide the transformation of individual agency (Cuffari, Di Paolo, and De Jaegher 2015). We do not treat these complex issues in full in this book but we return to some of these questions in Chapter 8.

meaningful in relation to the opportunity for asking for a glass of water. The glass of water, in turn, is meaningful in relation to one's intention to drink, and so on. The meaning that a particular event entails is thus enacted, brought forth, by the regulative response it induces. Two events that (in the same context) lead the agent to performing the same action are not meaningfully distinguishable from the point of view of the agent. The severity of an event tending to destabilize an agent's self-individuation corresponds to the amount of regulative resources required to compensate for it (Di Paolo 2005). In short, adaptive regulation is a precondition for agents having a subjective outlook on the world. Without it, the only meaningful events an agent would ever encounter would be those leading to its immediate demise. Without sense-making, an agent would not be able to appreciate its current standing relative to its own viability conditions while still alive. Equally, it would not be able to avoid risky situations or seek opportunities (we come back to this in Chapter 8). Sense-making inherently brings together aspects of the mind that are treated separately. It involves sensitivities and efficacies as co-defined pairs in adaptive regulation. It also involves affectivity at its core (see Colombetti 2014) in a way that is not distinguishable from "cognitive" aspects of adaptive regulation.

5.3 **Examples of the different requirements**

The three requirements of self-individuation, interactional asymmetry, and normativity capture most senses of the concept of agency in common use. Each requirement is necessary for agency. None of them is sufficient on its own, but the three together are jointly sufficient for agency. In other words, we cannot think of any examples of systems that satisfy all three conditions without considering them agents, nor can we think of any empirical agents that fail to satisfy any one of these conditions.

We can represent the set of entities fulfilling each of the three requirements as in Figure 5.5 and name some examples corresponding to the different regions in the figure. At the intersection of the three requirements, we have cases of agency (e.g., the case of metabolic-dependent chemotaxis in bacteria we discuss in the following section), but not in any of the other regions.

A patient with Parkinson's disease—who is naturally a self-individuating organism—undergoing involuntary spasms may at some points be inducing an asymmetrical modulation (the movements can provoke changes in the relation with the environment that are not sought for, such as dropping an object). Yet, exactly because they are involuntary, in other words, not performed to satisfy a norm or fulfil a particular goal, these particular movements cannot be considered proper actions.

Conversely, it is possible that some intrinsic norm in a self-individuating system is satisfied without the system's active intervention, that is, without asymmetric regulation. For instance, when a living cell is undergoing osmosis, water molecules are transported across its semipermeable membrane passively, driven only by the difference in concentration

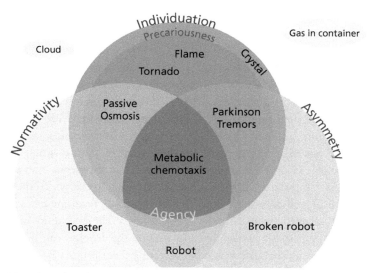

Figure 5.5 **Mapping the elements of agency.** Examples of systems satisfying none, some, or all of the requirements for agency.
Copyright © 2017 Ezequiel A. Di Paolo, Thomas Buhrmann and Xabier E. Barandiaran, with permission.

between the inside and outside of the cell. When this has the effect of maintaining a healthy equilibrated level of pressure, the process of osmosis is beneficial for the cell. But as long as it is not actively regulated, the cell is not the agent of the osmotic process (though it may be the agent of other *active* forms of transport across its membrane, such as the ATP-fueled activity of sodium-potassium pumps maintaining a normal electric membrane potential).

If a system does not meet the requirement of self-individuation, it is a system defined by convention. Such is the case of a typical robot designed to fulfil a task, for instance, pick up garbage. Because it is not a self-individuating system, it is not an agent in our sense. Its operations are subject to an externally imposed normativity: for example, how efficiently it performs its task, how much time it can function without maintenance, etc. Unlike self-individuating systems, this normativity is not intrinsic (Di Paolo 2003; Jonas 1966). The robot's "actions" can involve interactional asymmetry, though. To fulfill its task efficiently and reliably, the robot, for example, may modulate its usual garbage-seeking behavior, taking into account the energy level of its battery (perhaps operating more slowly or returning to a charging station periodically). In this case, the asymmetrical regulation follows the norm of remaining operational for as long as possible. But the robot may also exhibit interactional asymmetry divorced from any apparent norms. If, because of a loose wire, the robot's gripper spontaneously rises and drops such that its visual sensors are intermittently blocked and unblocked, for instance, then the robot in this situation would engage in asymmetrical modulations of its coupling, but it would not follow the requirement of doing this according to some norm. Similarly, we could also think of arti-facts (say toasters) that are designed according to norms but are not self-individuating nor do they engage in asymmetrical modulation of their coupling with the environment.[5]

In some cases it may be difficult to determine whether a candidate system meets the three requirements. Such is the case of mobile droplets, which are studied as models of protocells (Hanczyc 2014; Hanczyc et al. 2007). These seemingly simple oil-in-water or water-in-oil droplet systems are capable of producing surprisingly life-like phenomena, including self-motion, gradient climbing, and complex interactions. A droplet of an immiscible liquid placed in another will not dissolve; it remains distinct. The addition of surfactants, which in far-from-equilibrium situations concentrate non-uniformly at the droplet-medium interface, can produce internal convective flows. In the presence of chemical or physical gradients, these flows can self-organize, resulting in a self-propelled and life-like motion of the whole droplet (Hanczyc et al 2007; Sumino and Yoshikawa 2014). If the medium is spatially complex, like a maze, the droplets can show interesting "behaviors," like finding the shortest path or correcting the trajectory if they take a wrong turn (Lagzi et al. 2010; Čejková et al. 2014). It would seem that these systems would be good candidates for a sort of minimal prebiotic agency. There is an individual, the drop-let, in which asymmetries are induced at the interface with the environment, resulting in internal convection flows that organize self-motion. Gradient climbing indicates then that some kind of interactional asymmetry is in place in these droplets. What about normativ-ity? In principle, this requirement does not seem to be met if we mean intrinsic norma-tivity. If we define norms externally, such as climbing the gradient efficiently or avoiding wrong directions or correcting the path, this is indeed what these droplets can do. But the norms are not given by the conditions of viability of the system itself. The movement is not oriented toward changing the conditions of the droplet according to how viable the droplet is. And what about self-individuation? The droplet is clearly distinct from the external solution, so self-distinction would seem to be the case. But is the droplet also self-*producing*? The answer in the typical case is negative. And so the self-individuation requirement would not be fulfilled by these systems. This does not preclude the possibility of imagining a system based on these droplets, which, for instance, could maintain itself far-from-equilibrium by "recharging" energy and catalysts to keep self-motion from run-ning down. And if, in addition, this recharging occurred as a consequence of following chemical gradients, we would have to admit that all of the requirements for agency would be met in such a case.

Related to such borderline cases, there exists a class of complex dissipative and pre-carious structures, including tornadoes, candle flames, etc., that seem to self-individuate, at least once they get started. Whether and how exactly their self-production differs

---

5 Though the robots and artifacts we mention here serve as convenient examples of systems that are not agents, we do not mean to imply that artificial systems, in principle, cannot be agents also. In fact, in Chapter 6 we will propose that some systems may satisfy the three conditions for agency at the sensori-motor level, and this, in theory, may apply to systems that are not self-producing in the molecular sense.

from other cases, such as those of biological agents, is an interesting open question (e.g., McGregor and Virgo 2011; Mossio and Bich 2014; Mossio and Moreno 2010). But even if we decided to attribute to such systems the property of self-individuation and accepted that they can undergo processes that are good or bad for their continued existence (e.g., a tornado dissipating as a result of making landfall), nothing so far indicates that these systems have the capacity to adaptively regulate their interactions with the environment. If they did, and they fulfilled the three requirements, then we would have to consider them as agents.

5.4 **A definition of *agency***

We can now begin to formalize the idea expressed by Hans Jonas in the epigraph to this chapter. A living organism—through its dependence on interactions with the environment for its own self-individuation—has a world of significance to which it is sensitive and in which it acts. In order to sustain its precarious hold on itself, this entity must turn outward and engage this world. These engagements or behaviors are part of what constitutes the agent as a whole, not just something they do apart from their being agents (we come back to this issue in Section 5.5). In the words of Christine Korsgaard, "Being a giraffe is *doing* something: a giraffe is, quite essentially, an entity that is always *making* herself into a giraffe … this is what a giraffe's life consists in" (2009, p. 36).

Let us proceed then to a formal definition of agency that follows these intuitions in the form of an operational description of systems satisfying the three requirements (Barandiaran, Di Paolo, and Rohde 2009). In short, an *agent* is defined as an autonomous system capable of adaptively regulating its coupling with the environment according to the norms established by its own viability conditions. More formally put:

> A system S is an agent engaged in a coupling C with an environment E if and only if:
>
> 1. S is an *autonomous* system, meaning that:
>
>   a. S is an operationally closed network of precarious processes whereby every process belonging to the network is enabled by at least another process of the network and enables at least one other process in it, so that isolated from the network any component process would tend to run down or extinguish;
>
>   b. S actively and functionally distinguishes itself as a unity and the set of processes (not belonging to S) that can affect S and are affected by S defines S's environment (E); and
>
> 2. S sometimes exercises a capacity to *modulate* the coupling C in an *adaptive* manner:
>
>   a. where *modulation* indicates an alteration (dependent on the state of S) in the set of parameters and conditions that affect the coupling between S and E;
>
>   b. and *adaptive* means that modulations in the coupling C contribute to keeping S as a viable system.

This definition has the advantage of being at the same time generative—in the sense that the three requirements for agency follow from it—as well as non-circular and operational because it does not rely on terms presupposing the concept of agency. All of the concepts used in this definition have been defined in dynamical terms in the previous sections.

Condition 1 captures the requirement for self-individuation. The system is self-producing and actively and functionally self-distinguishing (that is, self-distinction is not merely formal or just active as in the case of a simple autocatalytic network, but also functional, i.e., establishing a differentiation between "inner" and "boundary" processes). An important aspect of this condition is the precariousness of the system. Not only does the organization define the system, but it is also thanks to it that the system continues to self-individuate over time; without the organization the component processes would run down.

Condition 2 captures the requirements of interactional asymmetry and normativity. There is a specific sense in which the system can be a source of actions, for not only can it modulate the coupling, it can do so in relation to intrinsic norms. Normativity in turn emerges from the autonomous organization, that is, from the manner in which specific interactions can either support or threaten to break down the system's self-individuation. In other words, the organization of the system determines what counts as adaptive regulation of the coupling.

By saying that the system exercises the capacity to regulate its coupling with the environment, we want to stress two things. First, the system is not constantly engaged in regulating the coupling, but may engage in this regulation only at some points and for certain durations. Second, regulation may not always succeed; it may fail by not inducing the necessary changes in the coupling or by inducing changes that do not meet the normative constraints. But as long as the capacity for regulation exists and it is sometimes exercised, the second condition of the definition is fulfilled.

It is important to note that the definition of agency provided here is associated with a certain timescale and spatial granularity. For example, what counts as variables depends on how the system under study is defined, and the requirement is that the changes in the variables sufficiently describe the states of the system. For this, we need to provide the equations describing these changes over time. As we have seen, anything else that is included in these equations and is not a variable is called a parameter or constraint. Typically, but not exclusively, the choice of parameters is associated with the timescales of observation. An autocatalytic set, for example, achieves its functional closure at the level of molecules and their transformative relations only, while no such closure is found at the quantum level in the same molecular interactions. It is thus perfectly possible to miss the presence of agency altogether when considering different levels of description. But once a level is chosen, there will be signs of agency in action or not, and an observer can then empirically test for the presence of agency at the chosen level of observation using our definition.[6]

---

6 This is not to say that testing for the presence of agency will not be difficult in specific cases given current tools and methodologies. But once a given level of observation is chosen and the relevant processes and relations identified, then, in principle, the three criteria for agency can be tested empirically.

Let us consider simple life forms to illustrate cases that satisfy all three requirements for agency (see also Box 5.1 for description of relevant models). This is useful to understand the various abstract elements involved in our definition. We discuss forms of agency that are more complex in Chapter 6.

Box 5.1 Models of agency

Models in science fulfil different purposes, from accurate data-driven prediction and simulation, to abstract and conceptual computer-aided thought experiments (Di Paolo, Noble, and Bullock 2000). The latter kind is concerned with exploring the coherence or plausibility of an idea. Several models of agency belong to this type, which means they explore some relations between the three requirements at the expense of simplifying others. This is fine and often useful despite (or because of) the fact that these are models, *not* instantiations of agency. In this sense a large part of the work in autonomous robotics since Grey Walter's (1950) *Tortoises* through Braitenberg's (1986) *Vehicles* on to behavior-based and evolutionary robotics (Brooks 1991; Harvey et al. 2005) can be said to contribute to the understanding of agency.

Most work in robotics, however, assumes an already-individuated system—the robot—with an externally given set of norms—the task it must perform, its efficiency, robustness, etc. A few models look into how processes of individuation interact with behavior and explore the emergence of norms out of this interaction. Egbert and Di Paolo (2009) study chemo-ethology in a simulated protocell able to select contextually between alternative resources. In a follow-up study, Egbert et al. (2010) present a simulation model of metabolism-dependent bacterial chemotaxis. Metabolism is modeled as a very simple autocatalytic reaction. Its product concentration affects the probability of running or tumbling. This straightforward link suffices to demonstrate a series of empirically observable phenomena, which are hard to explain if the chemotactic signaling pathways worked always independently of metabolism (as is sometimes assumed). These phenomena include chemotaxis to metabolizable compounds, inhibition of chemotaxis toward relatively scarce resources in case of local abundance, cessation of chemotaxis toward an attractant if that attractant stops being metabolizable, and chemotaxis away from metabolic inhibitors. None of these behaviors is programmed into the model; they simply emerge as norms that simulated bacteria follow by keeping their metabolism ongoing. The emergence of norms out of simulated operational closure in this kind of model is further explored in (Barandiaran and Egbert 2014).

Other models put emphasis on how processes of self-individuation entail an asymmetric relation with the environment, for instance, through following gradients or regulating exchanges of matter and energy through a membrane (Ruiz-Mirazo and Mavelli

2008). Work by Nathaniel Virgo explores reaction-diffusion patterns in simulated excitable two-dimensional media, which self-organize into distinct precarious, individuated dissipative structures or spots. These spots move by following chemical gradients in the environment. They become self-propelled by inducing "an asymmetrical distribution in the domain of individual-environment relationships" (Froese, Virgo, and Ikegami 2014, p. 66), essentially moving away from the waste concentrations produced by their own metabolism. A more sophisticated version shows the emergence of self-motion through the introduction of two different autocatalytic cycles, leading to two kinds of spots that spontaneously assemble into a spatially asymmetric symbiotic "spot-tail" system. These move spontaneously even if the environment is homogeneous. The relative simplicity of these models suggests that self-motion and other regulations of the relation to the environment may have co-arisen in history together with self-individuation. What emerged at the very origins of life may already have been an agent.

A living cell is an agent according to our definition. An example of its agency at work is the regulation of osmotic pressure within the cell, in contrast to passive osmosis. The natural tendency for a cell interior composed of macromolecules and surrounded by a semipermeable membrane is to create a solute gradient with respect to the exterior and thereby drive water through the membrane into the cellular compartment. To prevent bursting, the cell regulates properties of its membrane (e.g., the activity of ion channels) and stabilizes the osmotic pressure. Ruiz-Mirazo and Mavelli (2008) have shown a similar mechanism at work in a protocell model that achieves control of internal pressure resulting from the build-up of waste products. Here the same metabolic network responsible for creating and maintaining the cell produces peptides that are incorporated into the cell membrane, and they increase its elasticity and/or permeability. There is no clear demarcation in this case between processes of production and processes of regulation, although functionally the asymmetric regulation of the environmental coupling is evident. The protocell thus not only defines but also regulates its own enabling conditions.

A case where the separation between regulated and regulating subsystems can be more easily appreciated is that of the *lac* operon mechanism in *E. coli* bacteria (Jacob and Monod 1961). Under normal conditions, *E. coli* metabolizes glucose. But when the availability of this sugar is low, while that of lactose is abundant, certain normally inactive genes are expressed that open up a new metabolic pathway for the processing of the new sugar. In effect, the bacterium detects in the current virtual tendencies of the environmental conditions a change that puts its continued viability at risk and reacts by modifying the internal processes underlying its self-production. This induces a change in how the environmental coupling is regulated. It is easy here to distinguish the adaptive change in behavior from its normal condition, as the normally dormant genes are activated contingently on specific environmental situations.

The adaptive regulation required for agency does not only affect internal processes (e.g., the metabolism in the previous example). The system's mode of interaction with the environment can be directly regulated as well (see Figure 5.4, right). Consider, for example, a simple motile unicellular organism exposed to toxic chemicals. An internal adaptive response for dealing with this situation could involve internal processes of rendering the dangerous chemicals harmless, or expelling them at sufficient rates. Another option is for the cell simply to move away from the dangerous area. Motility thereby allows for regulation with respect to variables that the agent has no direct control over (unlike, say, the build-up of internal waste products).

One example of this is the chemotactic behavior exemplified by *E. coli*. Movement is regulated by changes in the rotation of the flagella that result in simple forward motion ("running"), interspersed with random changes of direction ("tumbling"). When the organism detects a positive gradient of attractants (chemicals that under normal circumstances it can metabolize), it alternates running and tumbling motions and moves as a result toward higher concentrations of the attractant. If it detects a negative gradient, the relation between running and tumbling frequencies is altered in search for more profitable locations. While the ability to sense gradients itself depends on adaptive processes (Alon et al. 1999), what is perhaps more interesting is the *regulation* of the chemotactic behavior. In principle, chemotactic behavior could be carried out independently of the organism's metabolism; i.e. sensed changes in the level of attractants are coupled directly to the motion of the flagella responsible for the bacterium's motion. However, a number of chemical pathways seem to exist in *E. coli* that link this basic sensorimotor loop with processes of metabolic regulation (Alexandre and Zhulin 2001). It has been shown, for example, that sensitivity to attractants is increased after starvation. It is also known that mutants that cannot metabolize a specific attractant do not show chemotaxis toward that particular sugar, even though the mutation has no effect at all on the function of the sensory receptors or other elements of the pathway underlying the sensorimotor loop. In addition, the presence of a metabolizable chemical prevents chemotaxis to all attractants. Finally, it has been shown that inhibiting the metabolism of one particular chemical stops chemotaxis toward that attractant (and only to it). Alexandre and Zhulin (2001) have demonstrated that these metabolism-dependent modulations of chemotaxis rely on the bacterium's ability to be sensitive to the state of its own metabolism (e.g., by sensing metabolites in the internal milieu directly, or the flow through the electron transport system). As a result, chemical substrates can be "dynamically interpreted" as attractants (or repellents), as a function of whether they can or cannot be metabolized at a given moment.

The case of bacterial chemotaxis also serves to highlight another aspect of the definition of agency. What happens if for some reason the bacterium performs chemotaxis toward toxins rather than metabolizable attractants: for example, because of a mutation or experimental manipulation? Does this still count as an action on the part of the bacterium? What about the more general case of self-destructive behavior? Clearly, the bacterium's modulations involved in this chemotactic behavior are not adaptive, and therefore, according to our definition, not proper acts. The observed activity is like the tremors of the Parkinson's sufferer. It is caused by the system, of course, but it is not a proper regulation following the norms of its self-maintenance. In fact, if the activity undermines the agent's survival, we can conclude that it is a malfunctioning on the part of the system (which can nevertheless be the agent of other behaviors). Even if the effects on viability are neutral these cases may also be classified as a malfunction of an adaptive system, but only given the context of how such mechanisms should work in general. In the absence of such a context (say, we witness for the first time the behavior of an extraterrestrial form of life), it is not possible to classify neutral activity as cases of failed actions. Note that this is different from the case of human self-destructive behavior. In systems that are more complex , as we will discuss in Chapter 6, a maladaptive behavior can still be an act if it satisfies norms that originate not in the subject's organismic survival, but in the self-maintenance of a particular "way of life" (specifically, in one of the other two dimensions of embodiment: sensorimotor or intersubjective).

Similarly, in some cases norms may be satisfied as a side effect of some other actions not regulated by them. For example, a bacterium may be following a gradient of chemicals that it metabolizes. It may be the case that the same gradient also leads to an environment that provides more sustainable temperature conditions. Yet, according to our definition, the move to a more viable temperature region in this case is incidental and not per se an act performed by the bacterium. The situation is not unlike that of an externally induced favorable change in local temperature without the bacterium performing any behavior.

To summarize, the view that emerges in the above examples of minimal motile adaptive agents is that of two interacting circular processes. Firstly, there is a cycle of self-individuation underlying the agent's basic autonomy, which itself may sometimes be able to induce regulations in the agent's relation with the environment. Secondly, we have the emergence of somewhat independent sensorimotor cycles, like signaling pathways in bacterial chemotaxis. Coupling of the two, as exemplified in *E. coli*'s metabolism-dependent chemotaxis, implies that behaviors can be regulated with respect to the norms originating in the agent's self-maintenance. These regulations represent, therefore, the most basic form of proper actions.

5.5 **Identity in interaction**

We have presented the three requirements for agency as independent conditions. And indeed we can see that they remain conceptually distinct by thinking of examples that fulfil one of the requirements but not the other two, or two of them simultaneously, but not the third (Figure 5.5). However, our proposed definition of agency not only results in systems that meet the three requirements, it also affects the conceptual relation between the three conditions.

An agent, as defined, is not just a system located at the intersection of three separate regions describing systems that achieve some kind of self-individuation *and* also asymmetrically alter their exchanges with the environment *and* do so by following norms, as if it were a matter of ticking three independent boxes. The requirements can help us understand the steps toward a useful definition of agency. Once we have arrived at this definition, our task is to look at the requirements again and examine their relations from the new vantage point. What this exercise reveals is that in the case of concrete agents as defined, the three requirements relate as mutually implied consequences of the definition. In other words, their relation is not one of independent aggregation, but one of internal entailments.

This is already something we can suspect if we recall Hans Jonas characterization of the relation between a living organism and its surroundings that we used to open this chapter. For him, the metabolizing entity sustains a relation of "needful freedom" with its environment (Jonas 1966). Because the organism moves through one flow of matter constituting its organic body to the next, it is able to avoid the entropic fate of any of the material configurations it temporarily "inhabits," hence its freedom or autonomy, as we have called this property. But for this same reason, it requires the availability of energy and matter rich in potentialities to allow it to surf across material configurations, hence the needful or dependent character of this freedom.

We could describe this as a dialectical relation or primordial tension between the organism and its environment. And if we examine the conditions we have postulated for describing self-individuation, we can find this primordial tension at play (Di Paolo 2017). Let us consider again the two conditions for self-individuation—self-production and self-distinction—and examine what each of them implies in terms of the organism-environment relation.

The self-production condition specifies that the closed network of component processes realizes the relations that give rise to the production (or regeneration) of the same processes. To realize such relations, in the material world, implies establishing the conditions by which the flows of matter and energy present in the environment can be successfully used in the regeneration of the self-producing network. If there were no other condition, what would be the ideal situation in which a self-producing system could realize these relations? This situation would correspond to the idealized condition in which every possible encounter of the organism with the external world produced a positive contribution to self-production. In other words, if we take self-production on its own, the ideal conditions would be one of *total openness* to the environment, such that every possible flow of matter and energy is taken advantage of. No relation with the environment would facilitate self-production more than this one, if it were possible.

Consider now the other condition: self-distinction: the autonomous system actively constitutes itself as a well-delimited unity with specific topological relations (e.g., inside versus outside). What would be the relation with the environment that would most ideally realize this condition? It would be one of total robustness to any environmental influence, in other words, a *perfectly shielded boundary* protecting the autonomous system. If this case were possible, no interaction with the environment could possibly put at risk the condition of being a distinct unity, simply because no event in the environment would have *any* effect on the organism.

There is a primordial tension in the concept of self-individuation in so far as the system-environment relations that best satisfy each of the two conditions tend in opposite directions. The self-individuated system must tend to be self-enclosed to assert its distinctiveness as an individual, but it must also tend to be open to sustain its self-production as a far-from-equilibrium system.

In previous relevant literature, for example, in the theory of autopoiesis (Maturana and Varela 1980), while the two conditions are described explicitly, the primordial tension they generate is not explored. It is resolved by a clarification: operational closure does not imply material or energetic closure. But how can we expect matter and energy to always flow in or out of a system in the "abstract" (i.e., with zero influence to organizational/structural relations)? This separation between organization and material flows is a hylomorphic abstraction, that is, it relies on the conceptual separation between form and matter. This is typically aimed for in existing human technology. Fuel is supposed to provide pure energy for the car engine and not alter its function. But we know that even in a system specifically designed to approximate this condition as much as possible, this is only an idealization. In biology especially, most of the matter and energy flows are pre-formed (high-energy compounds, proteins, plasmids, etc.). So it is problematic to say that matter and energy may flow freely across the boundary of the organism, because if this happened, it would soon become a violation of the self-distinction requirement owing to uncontrolled transformative effects. On the contrary, pre-formed matter and energy can only flow *conditionally* across the organismic boundary.

Let us clarify how. The ideal organism-environment relation for each requirement in the definition of autopoiesis negates the other requirement: total openness negates self-distinction, and total enclosure impedes self-production. Given this pull of opposites for the organism-environment relation, there is one solution, which is the dialectical overcoming of this tension. A material autonomous system able to overcome this tension would need to be adaptive, in other words, open to selected environmental flows (those that contribute to the condition of self-production) and closed to others (those that act against the condition of self-distinction). These options are presented schematically in Figure 5.6.

The overcoming of the primordial tension of self-individuation is resolved along a temporal dimension and the solution is no other than adaptive regulation of the coupling with the environment. It is not a question of finding a golden mean between the two extreme ideal conditions. Thus, a self-producing material system, which is able to sustain itself in time and navigate its own conflicting requirements, *must* relate to its environment as an agent. Moreover, it is through regulated interactions with the environment—both in terms of what is accepted but equally important in terms of what is rejected by the system—that its identity is sustained. Identity does not precede the possibility of interaction, as we would be tempted to postulate; they are co-dependent pairs.

In its particular relation to matter, the self-individuation of life entails a normative and interactional relation to the world, as Jonas suggests at the start of this chapter. These two aspects of biological autonomy are often treated as separate, for instance, in terms of the difference between so-called constitutive and interactional forms of autonomy

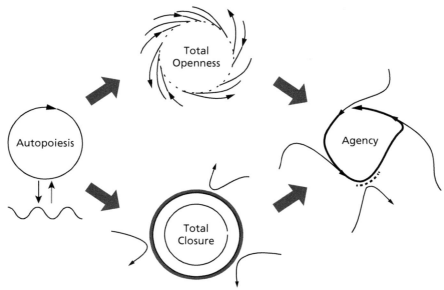

Figure 5.6 The primordial tension of self-individuation. On the left, a self-enclosing circle represents the circular condition of operational closure depicted in Figure 5.2. The ideal realization of the condition of self-production is shown at the top center, the arrows representing material/energetic flows in the environment. In this ideal case, all of the environmental flows contribute to producing the system. The ideal conditions for self-distinction demand the opposite situation (bottom center), which would be satisfied by building an impenetrable barrier, preventing any environmental flow from affecting the system. In both these cases, actual self-individuation is impossible (this is depicted by the open circles). The tension between the two requirements is overcome by managing their divergences over time, through adaptive, asymmetrical regulation of the coupling with the environment, accepting certain environmental flows and rejecting others (right). A system able to manage these inherent tensions in material self-individuation is an agent according to our definition.

(e.g., Moreno and Mossio 2015). But we have just shown that material living systems must also be agents, that the giraffe is a world-oriented entity whose defining feature is that it is constantly engaged in making itself into a giraffe. So the conceptual separation between constitutive and interactive domains is always ambiguous. In addition, the requirements of self-individuation, interactional asymmetry, and normativity are not simply cumulative; they co-emerge out of the dialectics of self-production and self-distinction under precarious material conditions.

5.6 **What is a sensor?**

As with other concepts we have used, so far we have spoken of sensors and effectors relying on the accepted meaning of these terms in biology and cognitive science. Using the concept of agency, we may now be in a condition to examine the meaning of these terms a bit further.

From a traditional information processing perspective, a sensor is rather unproblematic. It is simply a point of entry for the information to be processed by the cognitive mechanisms of the agent. An effector or actuator is also rather unproblematic from this perspective, being whatever allows the agent to induce changes in the relation between itself and the environment based on a plan that is the outcome of information processing.

But if we adopt a world-involving perspective, we cannot simply assume that the coupling between agent and environment carries relevant information. It may do so under particular conditions but not in general, and yet the coupling, informational or not, still is a constitutive part of ongoing action and perception. The material coupling between agent and environment is always already fundamentally richer than any informational account of what goes on in it, which inevitably is a reduction of its complexity to a few dimensions and sets of interest.

It might be tempting to speak of sensors and effectors in terms of the interface between agent and environment. Because a self-individuating agent is also self-distinct, this should be an easy matter. Accordingly, any parameter in the agent that is affected by environmental variables would count as a sensor, and any variable in the agent that affects the environment would count as an effector. While this option is operational, it is not sufficiently specific. After all, the self-individuation of an agent can be affected by environmental processes for which we have no adaptive sensitivity (e.g., ionizing radiation). Similarly, the body of a living agent in most cases will give off heat and scents into the environment, thus affecting it, but not in terms we would normally describe as an act.

The definition of agency involves an autonomous system able to regulate its relation with the environment adaptively. The concept of adaptivity in this context may give us a clue. The capacity for adaptive responses entails two co-defined moments. One is the moment of discrimination or differentiation of the virtual tendencies in the current situation that may have an effect on the viability of the system. The other moment involves the system inducing some change such that these virtual tendencies are modified toward sustaining viability over time. Now, these two moments mutually imply each other (Di Paolo 2005). There is no monitoring that is not linked to an adaptive response, whether the latter is actualized or not.

Successful adaptive changes modify the relation between the agent and the environment, by definition. But not all of these adaptations are necessarily manifested as regulations of the coupling between these two systems (i.e., as acts). Some regulations could conceivably be limited to purely internal changes triggered by a change in the environmental conditions, such that a pressure on the viability of the system is counteracted through some internal adjustment. But in many cases, as with the example of the *lac* operon in *E. coli*, even an internal regulation will have effects on how the agent and the environment are coupled.

We could then propose that when an adaptive response specifically involves an asymmetric regulation of the coupling between agent and environment, the *sensory* aspects of this process are related to the monitoring of current tendencies in viability conditions, and the *effector* aspects to the attempt to modify these tendencies adaptively. This functional distinction is not problematic in itself, but it does not quite guarantee that it will reliably point to dedicated structures and processes that fulfill these distinct functions. In fact, we have already seen some examples in which sensory and effector processes defined in this way are deeply entangled even in cases where dedicated structures are involved. For example, motile cilia form the basis of mechanoreception in protists like the *Paramecium* and the movement of flagella in the protozoan *Heteromastix* can be used to feel for obstacles or seizing them (Bloodgood 2010). There is little anatomical differentiation of the sensory and effector function in these structures.

Differentiation into dedicated sensory or effector function does occur during evolution, presumably in relation to the fact that adaptive sensitivities and effector responses can specialize along various dimensions of the material coupling they respond to or modulate. Thus mechanoreceptors, photoreceptors, gravity-sensing organelles, magnetic receptors, and so on, become available not to one single matching effector process in each case, but to many different ones, which conversely may be activated by different kinds of sensitivities, in a many-to-many mapping of possible adaptive responses.

Remarkably few models exist that address this sort of question, viz., how does a physical coupling become a sensorimotor one.[7] It is the kind of transformation that is hard to model mathematically or with computer simulations. It usually entails discovering pregnant possibilities in materiality, which are hard to predict because they escape the "space of functions" used to describe the agent before the transformation takes place. The few models that exist of this process of progressive "sensitization" of organized matter in a system are, unsurprisingly, material models. One of them is the famous demonstration by Gordon Pask of the growth of a sound-sensitive network structure (an ear) within an initially undifferentiated ferrous solution in a container with submerged electrodes (Pask 1959; see, also, Bird and Di Paolo 2008; Cariani 1993). A "training" regime is used involving the production of a loud sound and the administration of direct electrical current to some of the electrodes. As electrons flow through the ferrous solution, they leave metallic threads

---

7 Some models however consider a related question: what kind of sensorimotor dynamics can lead to new of sensitivities. For instance, Sato, Iizuka, and Ikegami (2013) have looked into how the body interacts with tools and objects and develops sensitivities to the properties of an object whether they are sensed directly or via a tool. In a simulation model (later tested with real adult participants), they studied a system able to determine the number of vanes on a windmill by touching them with a simple simulated arm. The system achieved this discrimination despite being designed without any a priori distinction between sensory and motor processes (a single interface variable acts as both sensor and motor of the simulated arm's movements). It relied instead on an implicit distinction (at the level of internal dynamics) between self-generated and externally (windmill) induced movements. The authors then placed a second windmill beside the first so they moved as gears. They successfully trained the system to recognize the number of vanes in the second windmill using the first one as a tool to move it. In this way, the model shows the emergence of new mediated sensitivities to distal objects through active sensing.

between the electrodes, forming a network, which is precarious because these threads dissolve unless "reused." Pask's device grew networked threads close to the vibrating walls of the container and could respond to vibrations in the 50Hz frequency by closing a circuit and turning on a light. Pask was also able to make the system discriminate between *stimuli* of 50Hz and 100Hz, and, apparently, other physical couplings such as magnetic fields.

In the late 1990s, a team of researchers led by Adrian Thompson at the University of Sussex, pioneered work on evolvable hardware. They often found themselves outwitted by the relatively simple systems they tried to evolve. Using artificial evolution on reconfigurable chips or motherboards to perform simple functions often resulted in systems that could perform the function that was required of them, but in ways no one could have foreseen. For instance, in one of the first examples of this work, Thompson evolved the connections between many gates in a reconfigurable chip (a Field-Programmable Gate Array or FPGA) so that the chip as a whole could discriminate between two inputs of different frequencies (Thompson 1997). On examining the successfully evolved circuits, Thompson noticed that some of the gates were disconnected from the circuit that linked the input and the output of the chip. So he clamped them, assuming they played no function and re-tested the chip. The chip stopped working. Thompson realized that even though there was no electrical current that could connect to these gates (the only way the gates in the chip were *meant* to link with each other), artificial evolution had made use of electromagnetic field couplings within the material chip, so that seemingly disconnected gates were playing a functional role after all. The chip worked by developing an unexpected sensitivity. A similar example by the same team proved equally surprising. Paul Layzell, a student of Thompson's, attempted to evolve oscillations in an evolvable motherboard using the same methods. Some of the circuits achieved high fitness, but on examination, the apparently robust oscillations became unstable. Moreover, the circuit seemed to behave erratically depending, for instance, on whether there were people working at the computer stations nearby. Upon examination, the researchers realized that the evolved circuits were amplifying radio signals generated by neighboring computers to generate the required oscillations, even using part of the printed circuit board as an aerial. Unwittingly, Layzell had evolved a radio (Bird and Layzell 2002).

These examples vividly demonstrate what we mean when we say that the intercourse between agent and environment is more fundamental and infinitely richer than an informational exchange. It only make sense to analyze sensory and effector processes in informational terms *after* the necessary transformations have settled into a quasi-stationary mode—after the motherboard has turned into a radio—but not *during* the process; not while the space of functions is undergoing transformations, as it regularly does during the kind sensorimotor equilibration discussed in Chapter 4.

It is through processes of this kind that differentiated sensor and effector functions emerge during development and evolution and become progressively attached to specific (neuro)anatomical structures. However, this differentiation does not overwrite the co-constitutive relation between sensory and effector moments in an adaptive response. The entanglement of sensory and motor processes is at the root of the very notion of adaptive regulation, and so at the root of agency, whether we are speaking of motile flagella used to detect an obstacle or the arm we extend ahead when walking along a dark corridor or the contacts at the tip of a blind woman's cane.

It should be clear by now that when we speak of sensorimotor integration, or mastering the laws of SMCs, we are not starting from separate processes external to each other—the sensory and the effector processes—which we then bring together into an explanation of action and perception. We start from already co-defined pairs (i.e., from moments of a same adaptive engagement between agent and environment). Distinguishable anatomical structures may become specialized during development and evolution, and so become reused and recombined into various different adaptive engagements (this permits the dynamical descriptions we have used in Chapters 3 and 4.) We call these structures sensors and effectors, but this is a reification that hides the fact that what makes them so is the co-constitutive role they play in the adaptive enactments of an agent.

5.7 **A remainder of indeterminacy**

In earlier chapters we have made use of notions of agency and normativity without fully grounding them operationally. This could be seen as a potential risk for our task of elaborating a non-representational theory of action and perception because these are precisely the kind of concepts that typically beckon some variant of representational discourse. As we have shown in this chapter, it is possible to propose an alternative approach, one based on the examination of the dynamic organization of complex processes of life and mind, and arrive at naturalized, non-representational concepts of agency and normativity. Their formulation may still see further elaboration—we will see how the picture gets more complex in the following chapters.

The three requirements we have identified—self-individuation, interactional asymmetry, and normativity—are individually necessary and jointly sufficient to determine whether a system is or is not an agent. Our definition of agency not only meets these requirements, but it does so in a way that integrates them into mutually entailed relations. With this concept of agency, we can consider candidate systems and decide on their status as agents, based on what we observe about their coupling with the environment and what we know about how such systems are constituted. We can ascertain that the small rocks falling down the slippery mountain slope in our opening example are not agents because their interactions with the environment are neither asymmetrical nor normative, and, in addition, they are not constituted as rocks by an ongoing process of self-individuation. The lamb that follows the falling sheep is an agent and is successfully acting as one in our example. Apart from being self-individuated, the lamb's coupling with the environment is regulated by breaking the fall and orienting his trajectory toward safer patches of grass. The sheep that fell down the slope was unable to regulate her coupling with the environment successfully. But this does not make her less of an agent than the lamb because in addition to self-individuation, we can observe that the movements of the sheep's body are normatively oriented toward achieving the asymmetric regulation that would tend to bring her out of her predicament, although she does not succeed in this attempt.

We have discussed some minimal examples of agency, for instance, the case of chemotactic bacteria. The idea has been to spend some time in simpler biological manifestations of agency to ground our understanding of this concept. We saw that the ongoing self-individuation of unicellular organisms provides the normative framework for their asymmetrical regulation of the environmental coupling (in the form of active regulation of exchanges through the membrane, motility following gradients, and behaviors that are more sophisticated).

But we have also noticed that the mechanisms through which the regulation of the coupling with the environment is achieved may function with some independence of the processes of metabolism. Such is the case of more traditional research of chemotaxis in *E. coli*, where the signaling pathways linking the sensing of chemical attractants and the likelihood of running and tumbling can be conceived as relatively independent of metabolism. Such pathways are honed by evolution, and they take opportunistic advantage of species-typical environmental relations. But we observe that it is not quite the case that adaptive regulatory processes are entirely decoupled from self-individuation when we consider the metabolic-dependent aspects of chemotaxis discussed earlier.

A question nevertheless emerges at this point: What is the relation between behavior and its underlying mechanisms, on the one hand, and the intrinsic norms of self-individuation, on the other? If the regulatory mechanism may become at least partially decoupled, underdetermined by the metabolism, are the actions induced by such mechanisms normative only with respect to biological self-maintenance? Do they cease to be actions at all? What about that remainder of indeterminacy?

We know empirically that at least in sufficiently complex organisms not every behavior relates directly to organic viability. Some regulations make more sense when we adopt an evolutionary perspective, for instance, any behavior related to sexual reproduction or parental care. But even if we keep our focus on the individual and not its evolutionary history, the agent as it is at this moment in its organization and its immediate coupling with the environment, it is clear that the potential for indeterminacy between the logic of metabolism and the logic of behavior may open the possibility for other forms of agency. We explore this possibility in Chapter 6.

Chapter 6

Sensorimotor agency

To trace the development of mind from the earliest forms of life that we can determine, through primitive acts which may have vague psychical moments, to more certain mental acts and finally the human level of "mind," requires a more fertile concept than "individual", "self" or even "organism"; not a categorial concept, but a functional one, whereby entities of various categories may be defined and related. The most promising operational principle for this purpose is the principle of individuation. It is exemplified everywhere in animate nature, in processes that eventuate in the existence of self-identical organisms; it may work in different directions, and to different degrees; that is, an organism, proto-organism, or pseudo-organism may be individuated to a low or high degree, in some respects but not in others, and anomalies of individuality—double-headed monsters, parabiotic twins, as well as properly semi-individual plants and animals—may arise by imperfect or by normally only partial individuation. Under widely various conditions, this ubiquitous process may give rise to equally various kinds of individuality, from the physical self-identity of a metabolizing cell to the intangible but impressive individuality of an exceptional human being, a Beethoven or a Churchill, who consequently seems "more of an individual" than the common run of mankind.

Individuation is a process consisting of acts; every act is motivated by a vital situation, a moment in the frontal advance of antecedent acts composed of more and more closely linked elements, ultimately a texture of activities. The situation, uniquely given for each act (and therefore not amenable to specific description), is a phase of the total life, the matrix from which motivation constantly arises.

—*Susanne K. Langer (1967, pp. 310–11)*

6.1 The plot thickens

Bandit cannot metabolize rubber. Yet he spends a good part of the day chasing and chewing rubber balls. He swallows small fragments of hard rubber when he succeeds in splitting them apart at a rate of about one new ball every couple of days.

Bella has no interest in seafood and she hates it when her owner wants to give her a bath. Yet twice a day she runs after the receding waves on the beach and stays there, paws sinking in the wet sand, expecting the moment she will get soaked in the cold salty water.

Why would these dogs be committed to activities with no apparent consequence for their biological survival? And committed they are. Try to take away Bandit's ball and you will get a growl. Call Bella to come out of the water and she will not even look in your direction. Why do they behave like this? "She's always been that way," Bella's owner might say, "loves getting wet and rolling over in the sand. No matter how much I try to clean her afterwards I always end up walking home with a schnitzel on a leash."

In referring to who they are, we often describe the way an animal or a person typically behaves. And in trying to explain what they ordinarily do, we often find no better reason than to say "that's just who they are." Perhaps this is what Susanne Langer intends when she says that individuation consists of acts?

In Chapter 5, we built a concept of agency upon the bedrock of the concept of biochemical self-individuation in living systems. In a way, we have proceeded as expected from an ontology that demands that entities must exist first and only after can they relate to other entities. This ontology started to break down a bit when we saw that behavior (what agents do) and self-individuation (what they are) can intermingle in intricate manners, along a historical dimension of co-definition.

But even these bold forays have so far kept us relatively safe within known territory. A bit of nonlinearity here, a bit of self-organization there, nothing that will shake any ontological edifices. In this chapter we are going to suggest that self-individuation, and even agency, can occur at the sensorimotor level. The processes that individuate a sensorimotor agent are *acts themselves*. It is acts—the acts of an agent—that constitute and reassert a new kind of agency, one that is enabled and constrained, but ultimately underdetermined, by biology. It is literally a case of explaining who you are by referring to what you do, and explaining what you do by referring to who you are.

If there is a sensorimotor level of agency different from basic biological agency, it will have to fit our definition of the term. We will already find important clues in how certain sensorimotor schemes become self-sustaining, as in the case of habits. But we will also need to look at the structure of everyday action. In everyday activities, sensorimotor schemes relate to each other in complex, adaptive ways, forming webs of mutually supporting relations. These networks of acts develop and change as processes of differentiation and integration between schemes (including the disappearance of schemes or the emergence of novel ones). Ultimately, these sensorimotor networks can achieve operational closure, in other words, their own individuation. The norms that guide action at this level are not only those given by biological self-individuation; some of them are also sensorimotor norms.

The processes that make up a sensorimotor life are the very acts that it performs (as well as the embodied and environmental structures that enable and constitute these acts). This may cause a few cases of ontological dizziness, but the idea is to follow closely the operational method of the enactive approach, while relating our findings to other theories and supporting evidence as well as phenomenological insights, to make certain at each step that the path we are laying down is firm.

6.2 **Habits**

In Chapter 5 we used the living cell as a concrete example of the emergence of individuation and normativity out of physicochemical processes. The three conditions for agency are clearly manifested in the cell's material embodiment and behavior. But we can envision the emergence of autonomous agency also at levels far removed from the specific processes of the biological substrate. Much of the behavior we observe in animals and persons, while remaining biologically viable, is clearly underdetermined by the conditions for biological autonomy. Many actions acquire value "on top of" their organic functionality, sometimes even in tension with it. Movements can be dexterous; postures awkward; a gait elegant; and so on. Some actions are as effective as others are in terms of their biological purpose, but they are preferred because they are habitual and comfortable. Also, complex agents may sometimes pursue goals that have the tendency to undermine long-term biological viability, such as the rewards gained from risky situations (extreme sports), or as in the case of substance abuse. Bandit's taste for chewing rubber cannot be said to contribute to the maintenance of his metabolism and eventually may even put him at some risk of poisoning. But we still intuit in his activity the hallmarks of an agent in action: a motivation, a commitment, and compensatory acts when something deviates him from his goal. Similarly, a person suffering from Parkinson's disease might not be concerned mainly with physiological consequences if she fails to grasp the glass of water. The very same frustration would arise if the glass were full of wine. In a sense that is yet to be properly spelled out, what is challenged in the patient is a sensorimotor identity, the capacity to assert herself as an agent through everyday safe and effective action.

Any theory of agency beyond the biological level has to provide an account of how autonomy and adaptivity arise in the domain of interest. For the case of sensorimotor agency, this is the behavioral domain.[1] But before considering the question of whether we can conceive of an autonomous organization at the behavioral level, it is helpful to introduce the concept of "habit" as a kind of minimal self-maintaining sensorimotor entity.

The received view about habits is an impoverished version of what the idea once represented (Barandiaran and Di Paolo 2014; Carlisle 2014). Reduced to reinforcement-modulated stimulus-response pairings by behaviorism, the concept all but disappeared from scientific discourse with the advent of cognitivism. Alternative, more organic

---

1 The definition of agency may be applied to other domains too. In the current context, a form of agency that is relevant to mention is social agency. We will only discuss some elements of this in Chapter 8 but not in sufficient detail. The reason is that it would make the exposition of the core ideas of this book too difficult if we kept another kind of agency in mind while simultaneously attempting to establish whether sensorimotor agency is even a coherent idea. In reality, we can expect social factors of different kinds to play important roles in the development of sensorimotor agency. Processes of interpersonal equilibration can scaffold sensorimotor capabilities in the developing infant and new kind of norms emerges during social interaction processes (Di Paolo 2016a; De Jaegher and Di Paolo 2007). Without ever becoming disembodied, forms of social agency take more complex forms beyond the regulation of face-to-face interaction, regulating also other kinds of social encounters, events, and activities, and eventually institutionalizing community-wide norms that enable new kinds of embodied capabilities and sensitivities, what Cuffari, Di Paolo, and De Jaegher (2015) call *linguistic bodies*. These skills are social, some of them are linguistic, but they always also involve sensorimotor elements and biological ones as well. The picture of human agency, in particular, will not be complete until the interaction between all these kinds of agency, corresponding to the three dimensions of embodiment, is properly understood, which is beyond the scope of this book.

conceptions of habit can be traced back to Aristotle, German idealism, pragmatism, and phenomenology. For these traditions, habits are complex and plastic patterns that integrate body and environment. According to a recent proposal, habits constitute self-sustaining behavioral life forms (Egbert and Barandiaran 2014). Following this idea, it will be useful to think of habits as *self-sustaining precarious sensorimotor schemes.*

To be more precise, we propose that a sensorimotor scheme as we have defined it in Chapter 4 is self-sustaining, or habitual, when the elements that support it (muscular dispositions, neural connectivity patterns, spatial arrangement of objects and tools, etc.) depend for their structural stability on the exercise of the scheme (Barandiaran 2008). Habits, thus understood, extend the concept of sensorimotor schemes by adding the notion of precarious self-maintenance: if the habitual scheme is not enacted with sufficient frequency, the structures supporting it start to lose the properties that enable it. Eventually, the capability to enact the scheme degrades and disappears.

Notice that with this definition the concept of habit is different from what we could more generally call "drill." A repeated enactment of a given behavior can transform the agent's support structures so as to predispose future repetitions of the same behavior. But if the support structures remain unchanged after having been established, then the predisposition toward the scheme also remains unchanged, even if repetitions cease. We demand a stricter condition to call a scheme a habit, which is that the support becomes structurally unstable in the absence of frequent enough exercise of the scheme. This is what we mean when we say that a habit is precarious.

This approach captures the idea that repeated enactments of a given scheme can induce plastic changes in the processes that structurally support it—both in the agent and in the environment—and keep these processes from degrading. This is analogous to the situation when people repeatedly walk across a lawn in a park, which leads, after some time, to the formation of paths where the grass is prevented from growing. This in turn encourages further walking along the paths, which continues to "sediment" the path network. Similarly, a habit "calls" for its exercise and its exercise in turn reinforces its durability. In Piagetian terminology the sensorimotor scheme $A \times A' \rightarrow B \times B' \rightarrow \ldots$ is a habit when the stability of the coordination support structures and processes A, B, ... A', B', ... depends on sufficiently frequent enactment of the whole scheme $A \times A' \rightarrow B \times B' \rightarrow \ldots$ (Figure 6.1). The mutual stabilization of a scheme and its support structures may involve neural mechanisms such as Hebbian-type synaptic strengthening during task performance. But our definition also allows for plasticity to occur in the rest of the body (e.g., through adjustment of muscle tone or posture) and the environment (e.g., ordering of habitual workspaces).

The individual habit already provides a first approximation of a sensorimotor conception of identity and normativity (Barandiaran 2008; Di Paolo 2005; Egbert and Barandiaran 2014). A habit can take on a "life of its own": it is both condition and consequence of its own enactment. This form of recursion (a kind of closure) individuates the habit. The dependence of habitual behavior on the precarious brain-body-environment structures supporting it defines a set of viability conditions, and with it, certain normative

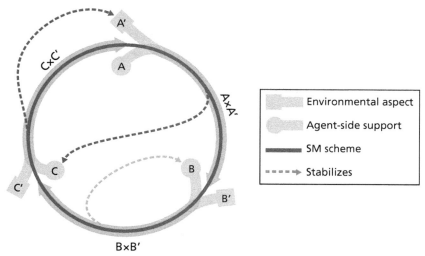

Figure 6.1 Habitual sensorimotor scheme. The circular organization (compare with Figure 4.1) of the three sensorimotor coordination events A × A′, B × B′, and C × C′ (light gray) forms a single sensorimotor scheme (dark gray). The supporting structures A, B, C and A′, B′, C′ in each coordination can be stabilized by the enactment of the scheme as a whole or a particular sensorimotor coordination. The process of stabilization is symbolized by dashed arrows. Only some possibilities are drawn here: for example, a sensorimotor coordination may stabilize its own support structures as illustrated by the light arrow leading from B × B′ back to B. Also, the enactment of the whole scheme may stabilize a support structure in the agent, as indicated by the dark arrow leading to C, or in the environment (arrow leading to A′).

constraints: the environmental conditions within which it can be enacted, the required frequency of repetition at which the habit self-reinforces without vanishing, and so on. One can picture those viability boundaries for specific cases, for instance, the period required between cups of coffee to keep a coffee-drinking habit alive. A coffee every morning may be more than sufficient, but an abstinence of a few months may extinguish it (see Box 6.1 for an example of how habit formation can be modeled).

Various conditions should be in place to sustain a habit. But these conditions may change over time, particularly as a consequence of a history of enactments. It is conceivable (indeed, it happens often) that habits become "hardened," that is, they lose their precariousness (or the timescales of required re-enactments grow too long). We remain predisposed toward such "habits" even if we stop enacting them. This need not be the case with all habits; it is a possibility, not a certainty that habits become strict automatisms. The loss of precariousness (and metastability) in such cases is problematic for our proposal because there is no intrinsic need for the self-sustaining pattern to remain dynamic and adaptive. This relates to the so-called double law of habit (Ravaisson 1838/2008), which links increased ease of performance with progressive loss of sensitivity. Habits that move in this direction involve less effort, but also a receding awareness (loss of precariousness

Box 6.1 Modeling habits

In a model relevant to some ideas in this chapter, Egbert and Barandiaran (2014) study the formation of habits using a "deforming sensorimotor medium." In a simulated robot, sensorimotor trajectories induce plastic changes such that similar trajectories become more probable. The idea relates to the paths formed on a lawn by people's repeatedly walking on them. The process can be visualized as a vector field in sensorimotor space where vectors are rearranged and reoriented in the direction of state trajectories passing by. As the robot is coupled with an environment, self-maintaining metastable patterns ("habits") appear. Habits depend on the agent's body and environment and seem to vary in stability, depending on how the structures that appear in the deformable medium "resonate" with the behavior in the environment. For instance, patterns seem to be most stable if they form a regular interaction with environmental features. Owing to its meta-stability, the system can also exhibit switching between habits, and between periods of exploration (of new sensorimotor patterns) and exploitation (performance of existing patterns). Higher level organizations also seem to occur at times, e.g. the "super-habit" of switching between two "subhabits" at more or less regular intervals.

The model is congruent with our theory of sensorimotor agency, since it treats closed-loop sensorimotor coordination patterns as the fundamental units of development. An interesting aspect of the model is that the existence of stable sensorimotor patterns is precarious. The deformable medium is constructed so as to gradually relax to its default state, such that existing patterns intrinsically fade away if the states constituting it are not revisited frequently enough. Because of this (simulated) precariousness, only through regular enaction can a habit maintain itself.

implies that adaptive regulation, and, as a consequence, sense-making, are no longer necessary). When this occurs, the habit becomes an automatism and stops being a habit in the technical sense we give to the word, although they are still called habits in other contexts.

6.3 From habits to activities

Habits provide us with some powerful elements to address the issue of agency (e.g., precarious self-individuation; a notion of norms at the level of the habit itself). But we do not find ourselves all the time enacting a single habit or another. Normally, we are involved in more complex activities that necessitate sophisticated and fluid relations between several sensorimotor schemes, for instance, the activity of cooking a meal or the activity of shopping for groceries. Moreover, the normativity associated with individual schemes often originates not in their own precariousness but in the way they relate to each other. The efficacy of chopping skills is "evaluated" in the context of related activities, for instance, in terms of avoiding uneven cooking times. We therefore need to examine the origins of the relations between schemes that make up such activities and what distinguishes one activity from another.

In analogy with autocatalytic networks (see Section 5.2.1), we can move beyond a single self-reinforcing habit (comparable to a single autocatalytic reaction) to identify closure and adaptivity at the level of many schemes in mutually self-sustaining interactions (*reflexively* autocatalytic in Stuart Kauffman's terms). This was already suggested by William James (1890), who conceived of animals as "bundles" or "ecologies" of habits. Habits do not stand in isolation as egotistically self-sustaining behavioral patterns. On the contrary, habits (or schemes) are nested in hierarchical, sequential, and ultimately networked relations in a kind of ecosystem, whereby a given scheme calls for, reinforces, inhibits, or subsumes others. Despite the possible loss of precariousness when a habit gets hardened, interdependence makes schemes more metastable (richer in potentialities) and adaptive than the traditional picture that associates habits with automatisms. John Dewey recognized this point:

> Strict repetition and recurrence decrease relatively to the novel. Apart from communication, habit forming wears grooves; behavior is confined to channels established by previous behavior. In so far the tendency is toward monotonous regularity. The very operation of learning sets a limit to itself, and makes subsequent learning more difficult. But this holds only of a habit, a habit in isolation, a non-communicating habit. Communication not only increases the number and variety of habits, but tends to link them subtly together, and eventually to subject habit-forming in a particular case to the habit of recognizing that new modes of association will exact a new use of it. Thus habit is formed in view of possible future changes and does not harden so readily" (1929, pp. 280–1)

This "communication" between habitual schemes, as Dewey calls it, is easy to see if we focus on *activities* and *behavioral genres* (cooking, eating, cleaning, riding a bicycle, commuting in the city, working at the construction site, playing sports, etc.). In these settings or *microworlds* (Varela 1992), the linkage between schemes is inherently meaningful. This is because schemes relate among themselves (e.g., as a sequence of preparatory and central actions) or because adaptive changes in one scheme are necessary to effectively perform another. In his study on dexterity and development written in the 1940s, Nikolai Bernstein (1996) provides several examples of such relations. One is the case of a smoker lighting up a cigarette:

> A smoker takes a cigarette pack out of his pocket, opens it, selects a cigarette, kneads it, and puts it between the lips; then he opens a matchbox; takes out a match; glances at it to check if its head is intact; turns the matchbox; strikes the match once or several times, as necessary, until it ignites; turns it so that the flame flares up; if necessary, protects it from the wind; moves it closer to the cigarette; sucks the match's flame into the cigarette; extinguishes the match; throws it away; and eventually puts all the things back where they belong.
>
> (Bernstein 1996, p. 147)

This series of twenty or so schemes forms a micro-network. Some schemes must be enacted before others are even possible. Others (e.g., taking a pause from walking in order to light up the cigarette) may facilitate other schemes that are otherwise possible but hard to perform. But the activity is not a strict rule-based sequence. If we observe this process many times, we will notice that the whole activity shows adaptive variability, as Bernstein remarks. Protecting the flame from the wind may or may not be required, if a match breaks while being struck, the process must go back a few steps, although if it breaks having also been ignited maybe it can still be used, but if it falls to the ground it must be put out, and so on. Someone could use a lighter instead of a match, and the activity would change a bit, but there is a sense in which similar functional relations obtain and both cases would be related as instances of a same genre. What determines these relations between schemes is a normativity proper to the activity itself (lighting up a cigarette cannot be said to be done as a way of improving organic viability).

In this example, the structural relations among the component schemes are very explicit and easy to see. But this is not always the case. Many complex activities are composed of webs of sensorimotor schemes whose deep structure does not reveal itself until disrupted. Such is the case, for example, in Kohler's (1964) experiments on adaptation to visual inversion goggles discussed in Chapter 4. Recall that recovery of normal vision in this case proceeds in a fragmented fashion. Some "perceptual habits," such as the ability to avoid obstacles, are restored before others (the recognition of numbers on license plates). Only when these individual skills are brought back into alignment is vision as a whole restored. For Kohler, through rehabituation experiments like these one can "penetrate more deeply into the structure and mutual interweaving of habits than would be possible in any direct way" (1964, p. 139). In view of this, we should aim at a concept of sensorimotor agency not so much based on an aggregation of individually self-sustaining habits, but on a network of "mutually interweaving" schemes.

In agreement with this idea, psychologists and biologists interested in behavioral development have long realized that behaviors not only become increasingly differentiated during ontogeny, but that they also integrate into more complex coordinated sequences or, more generally, into hierarchical organizations. John Fentress stresses that the neuroethologist seeks to understand behavior by analyzing how "each act fits into the context of others," and that she must "both find ways to separate behavior into its basic parts and define the rules by which these parts are joined together" (1983, p. 939). Fentress sees behaviors as organized in nested webs of rich interconnections at several levels such that "tugs on any one strand will have ramifications elsewhere, while preserving the web's overall structure" (1983, p. 941). Equally, Kurt Fischer refers to the "complex interconnections among skill components and domains" in a networked "developmental web" (Fischer, Yan, and Stewart 2003); Kenneth Kaye (1979) describes the hierarchical embedding of Piagetian schemes; and Michael Arbib and colleagues speak of complex networks of interdependent schemes where "each finds meaning only in relation to others" (1998, p. 44).

Like Bernstein and Kohler, these authors highlight that actions are linked into functionally coherent ensembles. Because of these links, some actions increase or decrease together in frequency or intensity; some actions compensate for the unwanted effects of others; other actions compete with each other as behavioral options; and so on.

As an example, consider the grooming behavior of rodents, which Fentress analyzes in great detail. The structure of grooming is hierarchical. At the top level, it consists of a cluster or set of related activities that can be clearly distinguished from other clusters, such as locomotion or feeding. At the next level, it can be divided into regions of bodily contact, e.g. the grooming of face, belly, and back, usually in that sequence. Facial grooming itself can be further divided into different stroke types, which themselves exhibit different kinds of groupings and sequential structures. This organization of behavior into hierarchical clusters of schemes makes it possible also to describe how the transitions between activity clusters develop. For example, the switching between activities is more regular than the switching between individual acts within a given activity; and one can describe in some detail the changes in the sensitivity of the animal to external stimuli when immersed in different activities (Fentress 1983, p. 948; also Fentress and Gadbois 2001). Similarly detailed analyses of the networked organization of behavior have been conducted, for example, for food caching and ritualized fighting in canids (Phillips et al. 1990; Moran et al. 1981), the compositional complexity of transition relationships between phrases in bird song (Marler 1981; Berwick et al. 2011; Sasahara et al. 2012; Weiss et al. 2014), and the development of infant locomotion (Muchisky et al. 1996).

Related to this networked view of activity clusters, Piaget's theory of equilibration, as we saw in Chapter 4, suggests that accommodation proceeds not only by creating new schemes when the current repertoire cannot assimilate a new situation, but also by *integrating* these schemes into the already existing network, through mutual equilibration between schemes. Depending on history and context, the changes undergone by different sensorimotor schemes and coordination patterns can give rise to a novel behavior. For example, in a longitudinal study of the development of reaching in infants Thelen, Corbetta, and Spencer (1996) demonstrate that reaching emerges from interactions and modifications of other non-reaching patterns such as bringing the hand to the face or moving arms rhythmically. Depending on the infant's preferred movement speed (which in turn depends on individual factors such as body size), successful reaching may require learning to control fast movements to improve accuracy or learning to expand and accelerate short, slow movements so that the hand can reach the goal. The developmental integration of previous schemes and coordination patterns depends strongly on individual history and preferences, so certain schemes may emerge from others with considerable developmental variability.

These examples show that sensorimotor schemes do not exist or develop in isolation, but always in mutual co-dependence on other schemes (not unlike new species evolving as part of an ecological community). While the concept of a habit indicates the existence of precarious self-sustaining organizations at the behavioral level, an account of sensorimotor agency needs also to consider the networked organization of sensorimotor schemes, their grouping into activity clusters, and their co-dependent development.

6.4 **Sensorimotor networks**

Our main proposal in this chapter is that the interconnectivity between schemes we have just described may serve as a substrate in which a new form of operational closure can develop. We now examine whether a network of interrelated sensorimotor schemes can satisfy the conditions of individuation, asymmetry, and normativity required for agency.

6.4.1 **Sensorimotor individuation**

Let us first consider the question of whether a network of sensorimotor schemes can individuate itself. In the case of the living cell, two elements contribute to active self-individuation: the operational closure of molecular transformations and the construction of a semipermeable membrane. We suggest that a sensorimotor network of mutually enabling schemes may also exhibit closure. In Figure 6.2 we illustrate the kind of relations that may hold between sensorimotor schemes, as well as between schemes and their support structures in the agent and the environment. Comparison with autocatalytic sets (see Figure 5.1) suggests a similarity in these relations. Remember that closure in the latter case involves mutual interdependence between two levels: reactions and molecules. In the autocatalytic network, each molecule is the product of a reaction, and each reaction is

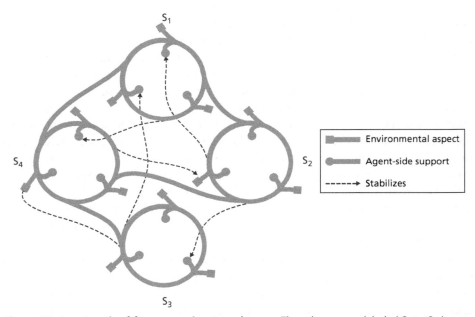

Figure 6.2 A network of four sensorimotor schemes. The schemes are labeled S_1 to S_4 (see Figure 6.1 for details regarding individual schemes). Sensorimotor schemes are interconnected in two ways. On the one hand, predispositions exist that link schemes "horizontally," such that they usually are, or can be, performed in a particular sequential, hierarchical or parallel fashion. This is indicated by the thick gray lines connecting the circular schemes. On the other hand, the support structures in one scheme may depend for their structural stability on the frequent enough exercise of other schemes, indicated by thin, dashed arrows. In this way, the sensorimotor schemes are interconnected also via structural ("vertical") dependencies. Note that we have drawn the support structures of the agent and the environmental aspects as each contributing to a single sensorimotor scheme, but in reality support structures in the agent and the environment may contribute to several different schemes.

catalyzed by molecules produced by other reactions in the network. Analogously, in the kind of sensorimotor network we propose, each scheme depends on at least one support structure in the network, and each support structure, in turn, is stabilized by at least one sensorimotor scheme.

These dependencies may happen as reinforcement of support structures at the level of body and brain (muscle strength, increased cohesion in a neuronal cell assembly, dendritic growth, etc.) or at the level of the environment (the progressive shaping of a shoe, the organization of folders in a computer, or the arrangement of utensils in a kitchen). In the example of the schemes involved in opening a door, the neuromuscular synergies used for reaching for door-handles act as agent-side support structures. We can imagine that the continued stability of these synergies may depend on the repeated exercise of opening doors and/or other activities that involve reaching (say, reordering books in bookshelves). In this sense, precarious supporting synergies can be stabilized by several sensorimotor schemes in the network. At the same time, of course, the schemes for door-opening, book-holding, etc., depend on the presence of appropriate neuromuscular synergies. In this way, schemes and support structures can become mutually dependent on each other at the network level (and not just the level of individual schemes and their *own* support structures, as in the case of habits). This is what we mean by closure at the sensorimotor level.

Figure 6.2 also suggests that closure at the sensorimotor level may go further than the mutual, networked dependence between support structures and schemes, since sensorimotor schemes also relate to each other in various ways. For example, some schemes act as preconditions for the exercise of others, and some situations require the joint, parallel exercise of several schemes for the desired outcome. In other words, complex behaviors often depend on a particular linkage of several sensorimotor schemes. Moreover, this dependence between schemes may be malleable: the usual enactment of scheme S_1 followed by scheme S_2, in a Hebbian-like manner, can make future exercise of the same sequence more likely. Experience may thus shape predispositions that orient some schemes toward the "anticipation" of others. We therefore propose that closure at the sensorimotor level involves a network of mutually enabling conditions both "horizontally" among sensorimotor schemes, as well as "vertically" between schemes and their support structures.[2]

---

[2] The network view of interrelated sensorimotor schemes may still sound abstract, and one may wonder if and how one could identify such a network in real systems. In this regard modeling studies may be one way forward. For example, Randall Beer and colleagues have extracted "functional networks" of stable sensorimotor states, or more specifically, state machines mapping the transitions between different stable states, from the dynamics of a simulated learning agent (Phattanasri et al. 2007; Izquierdo et al. 2008). From an empirical perspective, Paul Cisek and John Kalaska review neurophysiological evidence for a view of the brain as having evolved primarily to support interactive sensorimotor behaviours. In this view, closed sensorimotor loops relate to each other by competing in parallel for their enactment (Cisek and Kalaska 2010).

We may ask where exactly this network of schemes resides. Can we identify a physical boundary that confines it, like the membrane in the case of the cell? Is it located in the brain or perhaps in the biological body? The short answer must be negative. While we may be inclined to point to an organism's body as the locus of sensorimotor agency, it is important to stress that sensorimotor schemes, and networks of these, constitutively involve both the organic body and its environment. Of course, a lot of the support structures, potential for plasticity, and a capacity for selective enactment and equilibration rest within the organism (and in particular within its nervous system). But a network of schemes equally depends on its environmental support. Scheme selection may in some cases rely on resonance of the network with specific aspects of the environmental situation, and equilibration may involve plastic changes in structures of the environment (as we mentioned in Chapter 4, see Agre 1997; Kirsh 1995, 1996). Sensorimotor schemes are by definition modes in which structures in the agent and structures in the environment meet and mutually stabilize. As such, it makes no sense to try to identify their physical boundaries. We may inspect the anatomical and physiological properties of a human body and at best, we will be able to risk a very general guess as to what kind of sensorimotor agency it contributes to instantiate.

It should therefore be clear in particular that we do not propose individuation of a sensorimotor organization as equivalent to closure at the level of neuronal activity, as Maturana and Varela have proposed in their early work (Maturana and Varela 1980; Varela 1979; see Barandiaran 2016 for a critical discussion). Rather, the kind of closure we envision involves both the biological organism, in the form of enabling neurodynamic patterns and bodily configurations, as well as the environment, through its contribution to the closing of each sensorimotor scheme's loop.[3]

But if we are going to propose that the sensorimotor agent is constituted at the relational level of sensorimotor schemes, which in turn involves the environment, how can it be truly individuated (i.e., distinct from the environment)? While this may seem logically impossible, it is only because we have not been very precise in the use of the word "environment." It is important to realize that the environment for the biological organism is different from the environment for the sensorimotor agent. Strictly speaking, in dynamical systems terms an environment is always defined relative to a system. Just like

---

[3] The agent-side support structures for sensorimotor schemes are the most familiar to contemporary researchers in behavioral and cognitive sciences. Our emphasis on the relational nature of schemes does not imply that we should not study their neuronal dimension. In fact, it is reasonable to assume that some of the complexity of the network activity relates to the brain. Functional and structural dynamics in the brain enable sensorimotor schemes and are in turn shaped by their enactment. We should highlight, however, that sensorimotor schemes and the networks they form will rarely depend on specific brain regions, but most probably involve forms of large-scale brain coordination, and that these patterns depend on sensorimotor events, that is, they could not appear in the absence of the right agent-environment coupling (Aguilera et al. 2013; Barandiaran 2016). Some relevant properties of sensorimotor networks might well be pictured as transition networks between large-scale brain patterns (see Aguilera [2015] for an exploration of this idea with robotic models).

biological agents, a self-sustaining network of interdependent schemes brings forth its own domain of interactions. For example, we could argue that a virus is part of the environment of the biological organism, in so far as it triggers in the organism immune reactions and other defense processes. But the same virus, as such, cannot automatically be said to be a part of the sensorimotor agent's environment; we do not perceive the virus floating in the air, we cannot smell it or touch it, nor does it typically affect the relations between sensorimotor schemes. In fact, if the immune response is successful we do not even notice any effects related to the presence of the virus. It is only when the immune response is not quickly successful, as it often happens in the case of the common cold, that the support structures of several schemes are affected and the effects are noticeable at the sensorimotor level too (symptoms such as dizziness, fatigue, etc.). But the same effects could be provoked by other causes (e.g., overwork). However caused, the changes in the support structures themselves (bodily or external) constitute the immediate environment of the sensorimotor network.

The sensorimotor agent's environment, in other words, is constituted by all those aspects that can influence the web of sensorimotor schemes. Conversely, any external factors that in principle cannot influence the dynamics of the sensorimotor network are not part of its environment. The closed network of sensorimotor schemes thus specifies a new domain of what is or is not relevant, and it is in this sense that it distinguishes itself from its own environment.

We can relate the issue of individuation further to the creation of interactional "boundaries." In the first instance, as just described, the particular constitution of a sensorimotor network specifies what are the possible, relevant interactions for the agent. But as the cell produces a membrane for increased adaptive control over its material and energetic exchanges, so a sensorimotor agent may assemble functional boundaries to protect its particular way of life. Now, it might be tempting again to identify such boundaries in physical space. For example, we could point to the notion of territory and note that many animals exhibit a tendency to protect their immediate habitat. This indeed constitutes an easily identifiable boundary. But even a territory is not just a map. It is not simply an objective spatial enclosure, but involves a network of actions such as marking, sheltering, nesting, giving alarm calls, and so on. In the same way, boundaries of individuation are often actively and selectively created through specific actions or inhibition of actions (e.g., not calling a person that systematically challenges my suggestions, arranging my surroundings to be able to exercise a particular activity without disturbance, preparatory actions done for safety reasons, and so on). These are functional boundaries, manifested at the level of relations between sensorimotor schemes.

Last but not least, we want to emphasize the active nature of sensorimotor individuation. A sensorimotor agent—when it emerges as such, not all agents reach this kind of closure, as we discuss later—is not only self-distinguish*ed* but also self-distinguish*ing*. The closure of the sensorimotor network and its boundaries are not just established once and for all, but need to be actively sustained. It is, ultimately, through regulation of the network's coherence[4] (i.e., the degree to which schemes are mutually equilibrated as we saw in Chapter 4) that the sensorimotor agent becomes distinct. The particular way in which a sensorimotor network realizes itself, that is, its own way of life as a sensorimotor agent, is specific to the system's developmental history. Sensorimotor self-individuation is hence a recursive process in which an organization of schemes reinforces and shapes itself through the actions it generates. The system's doing and being are intimately intertwined.

6.4.2 Sensorimotor normativity and interactional asymmetry

The self-individuation of a network of sensorimotor schemes is related with the network of biochemical processes that constitutes the organism, but it is in fact a different kind of system. We already mentioned that these two operationally closed entities have different environments, as well as different ways of self-producing and self-distinguishing. It should then come as no surprise that the norms that emerge in each case will be also related, but different. In general, as we have already hinted at, we can postulate a relation of dependence between the closed sensorimotor network and the organismic body. Processes in the organism (metabolic, physiological, neuromuscular, etc.) enable and constrain all of the sensorimotor schemes in a network individually and in terms of how they relate to each other. However, there remains certain indeterminacy in this enabling relation, which is to be expected if the sensorimotor network can truly achieve its own autonomy.

We have already seen a hint of this indeterminacy at the end of Chapter 5 when we mentioned that the sensorimotor mechanisms in simple organisms such as *E. coli* can become "partially decoupled" from the logic of ongoing metabolism. This decoupling is never complete, as the sensorimotor processes still need to be sustained metabolically. Indeed, in many cases it is shown that taxis is performed contextually on metabolic states in *E. coli* and other bacteria (see Alexandre [2010] for a review on so-called "energy taxis"). Still, the functioning of the sensorimotor pathways that regulate chemotaxis is not entirely determined by current metabolic needs. This metabolism-independence assumption has dominated the studies of bacterial behavior for decades, since it was shown that in *E. coli* the capability to metabolize a reactant is neither necessary nor sufficient for taxis (Adler 1966). From our perspective, independence is an incorrect way to describe this relation. It is more accurate to speak of under-determination.

Despite the possibility of "partial decoupling," many sensorimotor schemes (most if we consider non-human animals) are nevertheless linked to the satisfaction of some vital norm (see, e.g., Barandiaran and Moreno 2008). This is in general explainable in

---

4 By *coherence* we here refer to the use of the term in dynamical systems theory, especially as applied in theories of development (Smith and Thelen 2003; Lewis 2000). It signifies the emergence of stable (coherent) forms, or patterns of coordination, in complex systems of many interacting processes. In particular, a network of sensorimotor schemes is coherent to the extent that its schemes are stable patterns given the system's current environmental context. The network is incoherent to the extent that any of its schemes, or the organization as a whole, becomes unstable; that is, when in Piagetian terms the system is facing disequilibrium and requires accommodation to a new situation.

evolutionary terms. Even if the function of sensorimotor processes is not entirely regulated by metabolism here and now, evolutionary processes still play a role in fine-tuning this wider space of sensorimotor functions so as to satisfy both the norms of self-individuation and the norms of effective reproduction. For the purpose of the current discussion, we may refer to these two sources of norms (organic and evolutionary) using the single label *biological normativity* (despite the well-known fact that many biological conflicts originate precisely in the clash between individual survival and the success of a lineage).

In the human case, at least, but also possibly in other species, not every detail of a successfully enacted scheme matters from this wide biological perspective. I need to hydrate in order to survive, but whether I drink water or ice tea, whether I use my favorite cup, or whether I enjoy doing it with my friends is underspecified by biological norms. These options are regulated by other sources of normativity; we are not indifferent to them or we would not notice them as options. Some of these norms are social, but we will not focus on them at the moment. Apart from these, we propose that specific norms emerge at the sensorimotor level and that they are related to what is needed to sustain a viable organization of behaviors, dispositions, attitudes, etc.

If we keep in mind the linkages between schemes in Figure 6.2, we notice that a particular sensorimotor engagement brings forth a normative dimension over and above biological normativity because of how the implications of its enactment can ripple through the network. We know from Chapter 4 that circular schemes in the Piagetian framework are assumed normative in themselves in that the closure of the cycle of sensorimotor coordination is the measure of successful enactment and the basis on which equilibration works. When schemes are linked via enabling relations, as in Figure 6.2, normative implications spread through these links. Not only can a scheme be a precondition for a subsequent scheme—a "good" or successful reach for the door handle is, among other things, one that allows me to subsequently turn the handle and open the door—but its successful or unsuccessful enactment also bears vertical consequences for its own support structures and those of other schemes. Every action we perform sends waves through a network of structural and functional relations. When these relations become operationally closed, every enacted scheme in the network can have positive or negative consequences for the viability of the whole. The set of structural and functional dependencies between schemes defines the viability conditions for the ongoing maintenance of the sensorimotor network, very much like molecular self-individuation defines how food concentration, pressure, or temperature affect the viability of cellular life.

We should note that it now is possible to relax the condition of circularity of sensorimotor schemes that we introduced in Chapter 4 following Piaget and return to our broader definition in Chapter 3 of a scheme as an organization of sensorimotor coordination subject to normative constraints. Schemes may involve "open," non-circular chains of coordination patterns, which are not required to lead back to the initial coordination. What is required of a scheme within a closed sensorimotor network is that it leads to an appropriate transition to another sensorimotor scheme according to the conditions of viability of the network. What identifies schemes as behavioral units is often resolved at the level of activities, in which sets of sensorimotor coordination patterns form larger units that may be adaptively exchanged or recombined.

Normative relations between schemes become apparent in the cases of activity clusters, as in Fentress' studies of grooming in rodents or Bernstein's vignette of lighting up a cigarette. The relations between schemes within a cluster are adaptive. This is seen in another example provided by Bernstein (1996, p.111). A boy is running and while running he jumps and picks up an apple from a tree. Picking the apple demands various complex coordination patterns (eye, arm, and hand movements; a firm enough grip; a sharp pulling movement to yank the apple from the branch; etc.). But if the branch is too high, the success of the activity depends on the running and jumping schemes, themselves not normally part of the picking-an-apple scheme. In general, the primary "goal" of an activity demands the enactment of some particular schemes but it also necessitates the availability of background corrections to other supporting schemes in "cooperative and harmonious interaction" (Bernstein 1996, p.111). The idea of cooperation or harmony is precisely the kind of normative notion that emerges at the sensorimotor level. Other sensorimotor normative dimensions include efficiency, robustness, adequacy, dexterity, elegance, and coherence.

The felt enjoyment of movement and the flow of action are also indicative of a normative dimension in sensorimotor engagements beyond their biological functionality. The kinesthetic experiences that accompany all movement are clearest in physical play, dance, and exercise. What is felt in these cases is not the potential benefit of these activities, but the embodied dynamics of the activities themselves. "Pleasure or fun in running, chasing, laughing, jumping, beating, and so on, is quite literally pleasure or fun in the flesh. It is not an accessory to a main event, but the main event itself" (Sheets-Johnstone 2003, p. 415). The affective aspects of these "forms of vitality" (Stern 2010) sustain the senses of flow and immersion (Csikszentmihályi 1990), which, in our terminology, could be explained in terms of coherent, long-range relations between integrated sensorimotor schemes.

Conversely, sensorimotor normativity is also clearly manifested in cases of breakdowns. For instance, if we injure our dominant hand so that it cannot be used at all, most of us will still be capable of many everyday activities, such as brushing our teeth, getting dressed, preparing breakfast, or taking money out of the wallet. But we will be painfully aware of an overall clumsiness, lack of familiarity, and frustration that accompanies these activities simply because of the fact we must perform them in a different, less comfortable (or less enjoyable) and less efficient way. The same can happen if we are suddenly placed in an unfamiliar environment or we suffer from some illness affecting sensorimotor schemes.

Let us now consider the requirement of interactional asymmetry and whether an operationally closed sensorimotor network can meet it. How should we understand a network of sensorimotor schemes asymmetrically and adaptively regulating the coupling with its environment? This environment, as we have said, consists of all those aspects outside the network of schemes that may influence the enabling relations between sensorimotor schemes. In particular, this environment includes events or situations that may destabilize or perturb the coherence of the sensorimotor network. Adaptive regulation, therefore, is the network's ability to be sensitive to and to counteract such perturbations and seek opportunities that help reassert the agent's sensorimotor individuation.

We have seen in previous examples that during specific activities, schemes relate to each other adaptively with respect to the norms proper to the activity itself and the network as a whole. Thus, there are "correct" and "incorrect" orderings of some schemes in time or transitions between schemes that occur too fast or too slow with respect to the activity's goals. There is, for instance, a "correct" period in which to move the ignited match toward the cigarette in order to light it up. These acts relate in an adaptive manner in the sense that they depend on factors that may disrupt how schemes link to each other. They are regulated according to these factors (e.g., the presence of wind or an injured dominant hand affect how a person lights up a cigarette). A sensorimotor agent may also regulate her coupling with the environment by adaptively shaping the environmental factors involved in the whole activity. Adaptive regulations of this sort include actions taken to rearrange one's environment in order to facilitate certain engagements. Consider, for example, expert workers such as carpenters or kitchen chefs. In order to ensure successful and efficient execution of their work, they become accustomed to a specific arrangement of tools and utensils, arrangements that they themselves have developed over time. They often prepare themselves for the task at hand, by wearing the right shoes, goggles, aprons, etc., not just for safety reasons but also to work better. Ensuring that everything is in its proper place and other acts *aimed at further acts*, are asymmetric regulations of the primary activity that is precision woodwork or preparation of a sophisticated dish.

Asymmetric regulations of the conditions affecting the network of schemes may also involve the kind of internal adaptations we have mentioned in Chapter 4 as examples of equilibration. For example, a child whose experience with animals is limited to interactions with dogs may have developed certain schemes and skills to engage in with four-legged, furry animals. When this child first encounters cats, which fit the same description but do not participate in the fetching game, assimilation by some schemes fails. This may lead to a process of accommodation in which new schemes for cats are developed (possibly also affecting the child's behavior in the presence of dogs). The child has modified the way she interacts with four-legged furry animals by reorganizing her sensorimotor schemes. This change in the organization of the sensorimotor network induces asymmetrical changes in the agent-environment relation, in this case, by splitting an assimilated sensorimotor category (furry four-legged animals) into subsets of interlinked schemes associated with two new categories (dogs and cats).

We then conclude that it is possible for operational closure to obtain at the level of networked relations between precarious sensorimotor schemes. The self-individuation of this network entails its own normative space: the conditions that affect its viability and could potentially destroy its closure. It is conceivable that this sensorimotor network is able to change adaptively with sensitivity to how it currently stands with respect to its viability and that regulations are enacted according to these conditions—in other words, the network may also exhibit interactional asymmetry. Since the three requirements for agency in principle can be met by a network of interdependent precarious sensorimotor schemes, we conclude that sensorimotor agency is a viable concept. Several examples support this idea, which would help us explain a variety of phenomena that would otherwise remain obscure if we stuck to a single form of agency. We now turn to exploring this concept of sensorimotor agency by looking at some developmental considerations and further evidence.

6.5 Becoming a sensorimotor agent: networked micro-identities

In arriving at the idea of sensorimotor agency as a precarious self-sustaining network of schemes, we have followed the implications of the theory of agency, allowed them to roll, and found that they point to a possible solution to a conundrum posed by the theory itself: is sensorimotor life (and with it action, emotion, and perception) reducible to the operational logic of the self-individuating organism? In other words, is psychology reducible to biology? Since evidence seems to point to a negative response, the theory would be in trouble if we were compelled to answer in the affirmative. But we are not.

The proposal is out in the open and we need to see whether it can fly. This is going to be the concern of this and the next sections, and largely also of the next chapters. Our first stop is to look at the idea of a closed network of sensorimotor schemes in more detail and over longer timescales—developmental ones.

6.5.1 Weaving a developmental web

In Chapters 3 through 5, we have heavily relied on a dynamical systems perspective, in which phenomena are understood as emerging from interacting processes involving brain, body, and environment. It seems natural to employ a similar approach to examine the developmental processes that underlie the formation and maturation of a network of sensorimotor schemes. Luckily, we do not have to reinvent the wheel. A long tradition of research, drawing on ideas from nonlinear dynamical systems theory (Waddington 1977; Kugler, Kelso, and Turvey 1982; Thom 1983; van der Maas and Molenaar 1992), motor skills and synergies (Bernstein 1967; 1996; Kelso 2009, Latash 2008; Sporns and Edelman 1993; Turvey and Carello 1996), and dynamical interpretations of the work of Piaget and others (e.g., Fischer 1980; van Geert 1998) has converged on a view of behavioral ontogeny not as the deterministic unfolding of a prescriptive developmental plan, but as the emergence of stable patterns in a complex system of multiple interacting levels, from the genetic to the cultural (Kuo 1967; 1970; Gottlieb 1992; Oyama 1985; Fischer and Bidell 2006; Thelen and Smith 1994). We cannot here review this field as a whole, but we shall make use of some its key ideas to sketch how a sensorimotor network may arise from a history of agent-environment co-determinations.

A recurring metaphor for developmental change is Waddington's epigenetic landscape (Waddington 1977). The underlying assumption of this metaphor is that processes at the level of genes, cells, and the organism as a whole, including its behavior, interact in complex ways such as to produce more or less coherent and coordinated patterns that change in stability over time. Development, in this view, consists in a "series of evolving and dissolving patterns of varying dynamic stability, rather than an inevitable march towards maturity" (Smith and Thelen 2003, p. 344). In Figure 6.3 we visualize a particular inter-pretation of Waddington's metaphor: an ontogenetic landscape of changing behavioral patterns and stability, as proposed by Michael Muchisky, Esther Thelen and colleagues. Time, here, proceeds from the top to the bottom of the figure, and each horizontal curved

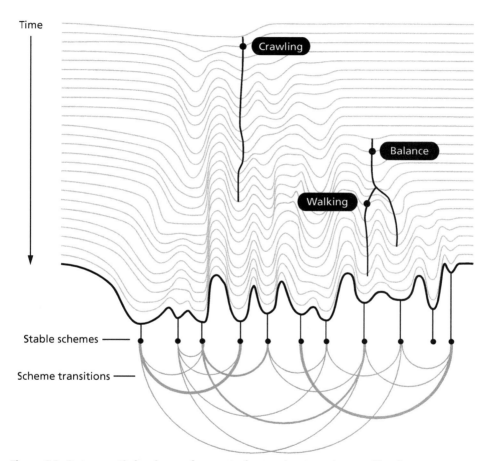

Figure 6.3 Ontogenetic landscape for a set of sensorimotor schemes. The diagram shows the development of sensorimotor schemes as a process of differentiating patterns in a (hypothetical) ontogenetic landscape. Developmental time flows downward. Each curved horizontal line corresponds to a snapshot of the stable schemes at a particular point in time. Valleys correspond to stable sensorimotor schemes. From an initially almost uniform state (top), the landscape becomes increasingly differentiated. As an example, crawling may be the only stable locomotion pattern in the early stages of infant development. When the infant learns to balance in an upright posture, walking emerges as another. At the bottom an arc diagram indicates possible relations between schemes (behavioral transitions).

line represents a one-dimensional snapshot of stable patterns at a particular time. These patterns could correspond to behavioral capacities, traits, or, more specifically, sensorimotor schemes. The depth of a valley in the landscape indicates the stability of the corresponding pattern. Accordingly, development starts from an initially undifferentiated, highly synergistic attractor and proceeds by creating and dissolving stable states, with the changes in the shape of the landscape (over time, i.e., down the figure) indicating rates of developmental change. In dynamical systems terms, development is the succession of bifurcations that create and destroy attractors with varying degrees of stability. This is a useful way of describing processes such as the one depicted in Chapter 4 (Figure 4.2), in which equilibration results in one scheme differentiating into two new ones.

Consider the development of infant locomotion (Thelen and Smith 1994; also, see Muchisky et al. [1996], for a detailed ontogenetic landscape in this case). At a certain stage, when infants have developed sufficient strength to support themselves in a quadruped posture (which itself depends on the prior exercise of stepping and kicking behaviors that are already present), crawling is a coherent and stable pattern of locomotion (see top of Figure 6.3). In other words, the infants' specific muscle strength, skeletal configuration, and neural circuits, at this stage and in the context of an appropriate supporting surface, all contribute to a particular dynamic movement pattern, which remains stable for several months. Once infants develop the balance and strength required for an upright posture, however, they will learn to walk. In addition to crawling, patterns for bipedal standing and walking now become stabilized. In Figure 6.3 this corresponds to the emergence of a stable attractor for upright balance and its subsequent differentiation into walking.

The ontogenetic landscape metaphor captures some important aspects of development quite intuitively (e.g., the emergence, dissolution, and stability of coherent patterns), but it fails to capture others (Newell, Liu, and Mayer-Kress 2003). For our current purpose its main limitation is that snapshots at given points in time only show which behavioral patterns are potentially present. But it does not show how the patterns relate to each other nor to how their stability changes at behavioral timescales so that the system may move from enacting one scheme to enacting another.[5]

At the bottom of Figure 6.3, we have included hypothetically existing behavioral transitions between the different stable patterns. For these to occur, if we wished to keep the landscape metaphor, certain currently stable patterns would have to disappear temporarily, along with the boundaries separating them, such that the system may transition horizontally toward a different stable state. We would thus have to imagine the landscape

---

[5] In part the limitations of the ontogenetic landscape metaphor are a result of a reinterpretation of Waddington's original landscape. The original idea was meant to capture the developmental path of, for example, different tissue types in the embryo. The development of each type of tissue corresponds to taking a single path through the epigenetic landscape (illustrated by a metaphorical ball rolling down through it). The result corresponds to the selection of one of the various final stable states. In the reinterpreted form of the landscape, in contrast, the stable states are meant to exist all at the same time as potentially available behaviors. Clearly, the one-dimensional layout cannot do justice to how the stable schemes are functionally related to each other.

changing also at faster timescales, such that patterns may become transiently unstable, disappear, and reappear. In other words, we have to distinguish between the schemes that are *structurally* available in general, and the schemes and transitions between them that are *functionally* accessible right now.

We suggest that a better model for these ontogenetic aspects is to view sensorimotor development as the growth of a network of stable patterns and the relations between them, very much like John Fentress's "behavioural network" (1983) or Kurt Fischer's "developmental web" (Fischer and Bidell 2006) previously mentioned. Figure 6.4 is an attempt to visualize this idea. Each network should be interpreted here as corresponding to a particular point in developmental time of the ontogenetic landscape in Figure 6.3, but we now expand the arc diagram at the bottom of this figure into two dimensions to make its network structure explicit. Note that the dynamic nature of the picture does not change. Each node in the network here corresponds to a stable sensorimotor scheme, which itself is the result of various support processes in the organism and its environment interacting and mutually stabilizing each other. In other words, each node corresponds to schemes such as those shown in Figure 6.2, but we have now "zoomed out" further and in the process abstracted some details regarding the schemes' shared support structures. The connections drawn in the network figure represent both the "vertical" links (stability dependencies between schemes and their support structures), as well as the "horizontal" connections among schemes themselves (transitions, preconditions, etc.).

The network view makes explicit that behavioral ontogeny involves processes of differentiation and integration. Differentiation is the creation of new stable sensorimotor schemes and sometimes their dissolution. The formation of new attractors in the ontogenetic landscape metaphor corresponds here to the appearance of new nodes in the network and the loss of an attractor corresponds to the removal of a node. Integration, in turn, is the connection of new schemes to the existing repertoire (the establishment of enabling relations, disposition, transitions, and so on). In the network this corresponds to the changing pattern of connectivity between nodes. Such connections may usually be created between related schemes, for instance, between those that usually follow one another (in a Hebbian-like manner) or between those that play similar roles in a given context (i.e., generally between nodes of the same color in the figure). Links may also be established between remote schemes (nodes of different color). Note that transitions between schemes, as represented by the links, may sometimes occur spontaneously as in the case when the termination of one activity leads naturally to the initiation of the next, and in other cases, it may involve environmental triggers, such as the ringing of the phone while one is immersed in preparing a meal. Our pictorial representation of the network does not distinguish these differences.

Over developmental time the processes of differentiation and integration (the latter driven by preferential connectivity between schemes involved in similar situations) lead to the formation and change of highly integrated "behavioral clusters." These clusters (nodes of the same color) should be interpreted as regional activities, for example, those related with locomotion, with feeding, with interacting with others, and so on. Initially,

Figure 6.4 Networked sensorimotor schemes. A (hypothetical) sensorimotor
network is depicted at three different stages of development. Initially (top), the network is
relatively small, but different sensorimotor schemes already appear connected, forming clusters
that correspond to types of activities. Individual clusters may themselves be
organized differently, e.g., mostly hierarchically (pink) or sequentially (turquoise). Some pairs
of clusters are more strongly connected than others. Over time (middle and bottom) clusters
get progressively interconnected and differentiated as the network develops. A wider range of
transitions (links) is available, activity bundles are split into smaller subunits, and so on.
(See color plate.)

the connectivity between and within clusters may not be well differentiated. But over time, the functional grouping of actions becomes more precise. Fentress (1983) has observed, for example, that young mice often switch between unrelated activities when the component acts are similar. If during locomotion a paw passes along the face, for example, the mouse may suddenly switch to a short grooming sequence. In the network model, this may be represented as a case of strong connectivity between schemes in different clusters that share similar conditions for their enactment. During development, as activity clusters become more clearly separated, such "motor traps" tend to disappear.

As the figure illustrates, some nodes have connections to more than one cluster. These nodes can be interpreted as schemes that support more than one activity. For example, the scheme for grasping cups may be involved both in having breakfast and in organizing the kitchen's cupboard. Also within each activity, some schemes may be more central in the sense that they are involved more often in that context. In the network figure, this corresponds to the size of nodes, which more specifically indicates the amount of connectivity between a scheme and others.

The network perspective could shed new light on certain phenomena in (human) ontogeny. For example, it is known that both micro-level and macro-level processes contribute to developmental change, the former consisting in the construction of new skills to be exercised in specific situations, and the latter in their consolidation, generalization, and integration (Fischer and Bidell 2006). In the network view, this distinction may correspond to small and local changes in the structural connectivity on the one hand, and large macroscopic changes on the other, such as the emergence of clusters and connections between large components. The same observation may provide a means for better understanding the presence of developmental "levels," in other words, the synchronized co-occurrence of many discontinuities in developmental change, or "spurts" of growth (see e.g., Fischer 1980; van der Maas and Molenaar 1992; van Geert 1998). "The clustering of discontinuities in macrodevelopment arises not from a mysterious underlying stage structure but from the dynamics by which people build skills through the integration of earlier components in a gradual process with constraints" (Fischer and Bidell 2006, p. 364). An intuitive explanation for how such discontinuities arise from otherwise small and gradual changes may lie in the dynamics of network growth and the resulting transitions in network connectivity patterns. Phenomena such as the consolidation of previously weakly connected schemes into a strongly connected cluster, the establishment of long-range connections between remote clusters given a novel context, and, generally, phase transitions that result in the sudden emergence of new network properties (small-worldness, hierarchies, etc.; Dorogovtsev, Goltsev, and Mendes 2008) may underlie the occurrence of sudden developmental spurts.[6]

---

[6] Neuroscientists will wonder at this point how the network structures and phenomena that occur at the sensorimotor level relate to similar structures observed in the brain. The analysis of structural, functional, and effective networks in the brain is a growing field (see, e.g., Bullmore and Sporns 2009), and it would be a natural next step to investigate the links between these two levels. It would also

Compared with the landscape model, not only does the network description (see Figure 6.4) better capture the development of the complex structure linking individual acts, it also allows us to describe how the interconnections between schemes may become operationally closed, and therefore autonomous. In this view, development starts with the early emergence of simple sensorimotor support structures that enable the organism to engage in basic sensorimotor couplings with the environment. The formation of these structures may originate, for example, in anatomical and functional constraints (e.g., retinotopic wiring; Kirkby et al. 2013), the fixation of self-organized patterns (e.g., spinal circuits driven by spontaneous firing; Marques et al. 2013) or the chaotic exploration of bodily constraints and their sensorimotor regularities (Kuniyoshi and Sangawa 2006). For example, basic kicking, stepping, or arm flailing patterns are present at birth, and constitute part of an initial repertoire that allows the infant to start exploring more complex sensorimotor capabilities and regularities (Thelen and Fisher 1983). During the early stages of development, the differentiation and organization of sensorimotor schemes may be directed primarily by biologically adaptive signals and anatomical constraints together with social and environmental scaffolding. For example, the development of balance and an upright walking pattern depends on the infant having sufficient muscle strength, motivation (e.g., the desire to reach a toy on the table or to follow her mother), and parental support.

Over time, in some species and given typical developmental conditions, a nested web of sensorimotor schemes appears that becomes progressively more independent from biologically adaptive signals and more dependent on the coherence and stability of the organization of sensorimotor schemes themselves. This is inevitable when under internal tensions individual sensorimotor schemes can no longer be accommodated in isolation. For instance, basic patterns of locomotion become important parts of many other activities, including games, dancing, etc., and their relevance in these contexts is no longer determined by the kind of original needs that motivated their creation. The network as a whole becomes increasingly reliant on higher order stability dependencies between individual schemes, and the interactions that they sustain with the environment. The stability and coherence of the organization as a whole depends on the kind of higher level forms of equilibration mentioned in Chapter 4. Eventually, the adaptive regulation of behavior to preserve the consistency and stability of the network of schemes becomes a central organizational principle. It is at this point, we propose, when a sensorimotor network

---

not be a great conceptual leap to identify neuronal cell assemblies (see e.g., Buzsáki 2010; Huyck and Passmore 2013) and processes of rhythmic coordination and synchronization (Varela et al. 2001; Engel, Fries, and Singer 2001) as supporting parts of sensorimotor schemes and their relations. One may also attempt to find correlated phenomena in behavioral and neural development (Fischer and van Geert 2014). But, at this stage, we refrain from hypothesizing about specific mechanisms at the neural level and how they contribute to the sensorimotor network (although they are no doubt integral to its functioning), and limit ourselves instead to a description of the cognitive organization necessary for sensorimotor agency at a mesoscopic level.

achieves autonomy from its underlying organismic self-maintenance, that one may speak of *sensorimotor life.*[7]

6.5.2 **Sensorimotor networks in action**

When seen over developmental timescales, we tend to emphasize changes in the sensori-motor network depicting structural relations (e.g., emergence or disappearance of nodes and links). But we have already suggested with some of the above examples that the net-work metaphor is also useful to understand how behavior is enacted in concrete situa-tions and how schemes relate functionally, as well as structurally, to each other, as we have seen in Bernstein's example of the activity of lighting a cigarette.

The functional relations between schemes involve the context-dependent ways in which the structural enabling relations, so to speak, are "navigated" at the timescales of behav-ior.[8] Thus as one or more schemes become active, linkages between schemes may become potentiated or inhibited. This effect may be path-dependent, that is, regulated according to the history of how the agent engages in a particular activity. One could imagine a sin-gle node in a cluster lighting up as symbolizing a current enactment of the correspond-ing scheme. This is illustrated in Figure 6.5, the red node indicating the currently active scheme. As a consequence, several of the links in the cluster and in the network in general get momentarily strengthened while others are momentarily weakened (depicted in the figure as a change in the thickness of links).[9] In the example of lighting up a cigarette, we can imagine all the actions depicted by Bernstein as belonging together in a functional cluster. However, the transitions between schemes in this cluster are not arbitrary; they follow a set of constraints. Some transitions, such as striking the match before taking it out of the matchbox, are impossible. Other transitions are possible but "incorrect,"

---

[7] Some of us have previously used this very characterization for the more general or abstract notions of *cognitive autonomy or mental life* (Barandiaran and Moreno 2006; Barandiaran 2007, 2008, Barandiaran and Di Paolo 2008).

[8] We use the terms *functional* and *structural* in the neuroscientific sense in analogy with the way in which brain connectivity constraints but not fully determines (functional) interactions at short time scales, with the latter in turn modulating structural connectivity over longer timescales (see e.g. Sporns 2013; Byrge, Sporns, and Smith 2014). This usage should not be misunderstood in terms of schemes necessarily fulfilling a functional role in a given architecture. In specific conditions, it may be possible to interpret the enactment of some schemes in this way, of course (scheme A achieves certain results that facilitate scheme B). But, more generally, such interaction links between schemes are not neces-sarily defined in terms of functions as such, but in terms of dispositions, inhibitions, pre-activations, etc., some of which may not play any functional role (e.g., certain quirks and mannerisms induced by recently enacted schemes).

[9] Fentress (1984) already emphasized that at "each moment the animal or person performs within the context of actions that *have* occurred, *will* occur, and *might* occur" (p. 120, original emphasis), and compares the dynamic balance of different actions using a center-surround model to account for their relations. Similarly, Cisek (2007) argues that the brain structures involved in several currently available actions can be activated in parallel while competing for enactment by mutually inhibiting each other.

Structural Network   Functional Network

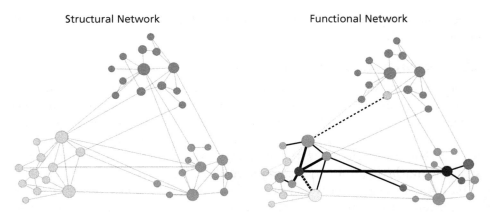

Figure 6.5 Structural and functional sensorimotor networks. Depicted is a small part of a network of sensorimotor schemes, consisting of three clusters of activities. On the left, the structural connectivity is shown. It consists of all vertical and horizontal links between the included schemes. On the right, the functional connectivity at a particular instance in time is shown. The currently active scheme is highlighted in red. Its activation makes transitions to some other schemes more likely (thicker solid lines) or less likely (thicker dashed lines). Through the transitional links, some neighboring schemes may already become pre-activated ("primed") or inhibited (shown here as a change in the intensity of the node's color), which in turn may lead to modulation of their own transitional links. Note that this is a simplified picture. In reality there can be more than one hotspot of concurrently active schemes at the same time, and their activation not only involves the influence of neighbors (which in addition may be path dependent), but also the world. (See color plate.)

for instance, lighting the match and blowing it out before lighting the cigarette. This defeats the purpose of the activity. Other transitions have equivalence (e.g., blowing out the match or putting it out by shaking it). Others have some flexibility (e.g., using the same ignited match to light up more than one cigarette). And others simply co-occur or are facilitated without any particular function or meaning for the activity itself, although they may fit the overall style (e.g., adopting a Humphrey Bogart or Lauren Bacall attitude). In addition, some schemes in a cluster tend to act as entry and exit points, due to their functional connections to schemes outside the cluster (e.g., putting away the matchbox to resume walking). Thus one does not enter the activity of lighting up a cigarette, so to speak in the middle of it, nor is the activity typically abandoned halfway.

That these functional relations follow certain normativity suggests that activities are microworlds, to use the apt term proposed by Varela (1992). They are frames of significance inhabited by a micro-identity formed by tightly connected structural and functional relations between schemes. Phenomenologically, once an agent inhabits a microworld, there is a certain readiness-to-act and a certain set of sensitivities implied by the possibilities and demands inherent in this activity. In other words, activities are to a greater or lesser extent *absorbing*. This does not mean that the agent is insensitive to possibilities and demands outside the current activity. Hearing someone urgently calling for help can freeze a person between picking a match from the matchbox and striking it. But even these possibilities show some structure. It may be easier to break the flow of an activity between some schemes but not between others, it may even be harder to break the flow within schemes. Even if I am expecting an important telephone call while I am also repairing a chair, I am not likely to freeze my activity if the phone rings, say, the instant I have begun a strike of the hammer. In other words, there is a degree of commitment that is manifested differentially as the agent is drawn into the different stages of enacting a microworld.

Microworlds are realms of significance, where whatever happens is either readily interpreted within the frame they provide or else it induces a kind of breakdown that can bring the agent outside the frame. If events allow an interpretation within a microworld, they are assimilated into it. Thus, if I am repairing a chair, small breakdowns in my activity are dealt with by the adaptive variability already in place within the cluster of schemes. I may pick up a nail but realize when I start hammering that it is too thin or too short, and so I must stop hammering, remove it if it is partly driven into the wood, go back to the box of nails, and select a different one. There is a concrete focus of my sense-making, and even a sort of style in my actions, which is given by the kind of relations that have developed historically between schemes clustered together. In the absence of major breakdowns, the activity will flow naturally toward its "exit nodes."

In contrast, if an event leads to a major breakdown, the activity is interrupted. This could be an irrecoverable breakdown within the activity (I hurt my thumb while hammering) or it may be something external to the activity: an alarm bell or the sudden recollection I had promised to be somewhere else at this time. These induced breaks in the activity demonstrate that absorption is not absolute and that sensitivity and readiness of a larger scope remain present even while I am quite committed to actions in a microworld. Again the network metaphor is useful here. We can see the analogy between activities and densely connected clusters of schemes, which nevertheless may still connect to schemes in other clusters. There may even be some loose sense of hierarchies (clusters of clusters) or *genres* ("a texture of activities" in Susanne Langer's happy phrasing in the opening quote of this chapter). Take the various activities involved in making repairs around the house; they can themselves form a wider microworld with its own realm of significance, sensitivities, and readiness. So for instance, realizing I should repair the electrical light installation before it gets dark, can make me abandon temporarily my actions on the chair, while I remain involved within a home improvement genre.

Still, what happens if a breakdown is so severe that the agent is not, so to speak, "caught" within any particular activity or genre? There is likely at this stage a hiatus of deep disorientation, of simultaneous partial abandonment and retention of the old frame of significance. We may find ourselves still involved in some of the previous schemes, only that they do not seem to make much sense now. In fact, until the situation is resolved and a new microworld emerges, we are *world-less*. This is a time between moments where *enaction happens*: "at the hinge between one behavioral moment and the next" (Varela 1992, p. 443). At these junctures, the frame of sense-making is reconstituted, brought forth, possibly as a transition to some different existing frame or even as the start of a totally novel one. To begin to enact a microworld is to bring ourselves into the adequate genre and the adequate activity such that we can re-establish the coordinates of significance that were lost during the breakdown.

Microworlds—dynamic clusters in the sensorimotor network—resonate with theories in psychology that characterize regions of interconnected behaviors according to environmental or social settings. One important affinity is with activity theory, the Soviet school of psychology built on the work of Lev S. Vygotsky, Alexei Leont'ev (1978), and others. Activity is the category used to describe a set of interrelated actions that confront a subject with a situated, culturally mediated, often social series of objects, tools, and goals. The daily running of a bakery, the writing of a joint report, and the care of a sick person are examples of activities. The concept is introduced in an attempt to move from an exclusive focus on the "inner" categories of psychology (motives, beliefs, etc.) and explain them as embedded in networks of actions that do not assume a hard distinction between psychological and socio-technical processes. The actions that make up an activity are not unitary in the sense that each of them must satisfy a specific goal to fulfill a specific need. Their conditions of satisfaction are oriented toward the goals of the whole activity. Activities thus involve patterns of coordination between actions, normative orientation toward the environment from which tools and objects are selected and used, and coordination with other actors. Activities are highly dynamic and context-dependent; new motives and goals can emerge in the course of an activity, which is then reoriented accordingly. These properties fit well with our clustered sensorimotor network model; with the difference that activity theory is also oriented to describing the sociocultural and technical organizations. Our network, in contrast, can be interpreted as a projection of these larger macrostructures onto the sensorimotor organization of a single agent. From our perspective, we follow the agent over time, who may transit between activities.

Another useful projection of the concept of activity, this time not onto the organization of the agent but onto the structure of the environment is the notion of behavior settings. The idea, developed by Roger Barker in the 1940s is deeply influenced by Gestalt psychologists like Kurt Lewin and Fritz Heider, with whom Barker worked closely. Behavior setting theory (Barker 1968; Schoggen 1989) associates recurrent behavioral patterns with structures in the environment that constrain and guide these behaviors. An example of a behavior setting is a waiting room. The arrangement of sitting places, the general entrance and the entrance to some inner space, the temporality of events such as the time to admittance, and so on, all are environmental factors that massively influence the flow of activities, including the enactment of sensorimotor schemes that accord to the setting. In this way, behavior setting theory studies activities in terms of the objective environmental processes that enable behavior and how these processes are sustained, partly by the very enactments they enable.

The idea of sensorimotor agency as a dynamic self-sustaining network of schemes is in general compatible with activity theory and behavior setting theory (about the potential links between enactivism and the latter, see McGann [2014]; see also the discussion of affordances as relative to the situated normativity of a form of life in Rietveld and Kiverstein [2014]), although as we note, their emphases are complementary. Activity theory's emphasis is on the sociocultural embedding of activity and its influence on psychological processes. Behavior setting theory highlights environmental structures and spatiotemporal relations in the milieu as strong determinants of behavior.

We notice that since we are dealing with a network of sensorimotor *schemes*, nothing of what goes on is this network is entirely determined either by the support structures in the agent or in the environment on their own. Environmental settings can funnel activities into a few constrained behavioral options, but they do not wholly remove the possibility of an agent breaking away from those activities. In fact, it may be necessary to introduce a further distinction between the general environment of a sensorimotor agent, and the more specific environment as part of a microworld (activity theory and behavior setting theory could provide useful tools for developing this distinction). The same general environment, a street, becomes radically different in its significance depending on the microworld we are currently enacting (e.g., whether I am strolling aimlessly or returning home from shopping with heavy bags). Thus, we provide only partial information when we describe what the general environment affords an agent if at the same time we do not also describe the activity that frames her current microworld within that environment.

6.6 **Kinds of agency, sensorimotor bodies**

The definition of agency is sufficiently explicit to be operational. We have demonstrated this by using examples of simple life forms, showing that even single cells exhibit agency in the minimal sense. We should equally be able to determine which sort of system exhibits *sensorimotor* agency. For example, can we say whether non-human mammals, like cats or dolphins, are sensorimotor agents? What about "simpler" animals? Do fish have the required sensorimotor organization? Do ants or flies? How about a bee colony or a tree?

To answer these questions, we must test whether the three requirements are satisfied in each case. In other words, does the system in question form a network of sensorimotor schemes with stability dependencies among schemes as well as between schemes and their substrates? Is the system organized such that its activity regulates the coupling with its environment so as to normatively preserve its particular sensorimotor organization? In what sense does the sensorimotor network establish a functional "boundary" that separates and protects its mode of being?

These questions are tricky because agents in the minimal sense tend to engage in overt behaviors, often displaying the frequent use of lawful sensorimotor regularities (coordination patterns, schemes, and even series of related schemes). In short, we may find creatures able to enact sensorimotor schemes of various degrees of complexity, but in itself, this is not yet a sign that all of the requirements for *sensorimotor* agency are fulfilled in this kind of agent.

We can propose a rough progression in complexity from minimal to open sensorimotor agency.

1. *Minimal agency.* The fulfillment of the three requirements in the minimal sense, without specifying any restrictions on the sensorimotor domain. All living organisms are minimal agents.

2. *Minimal sensorimotor engagement.* The minimal agency of many organisms involves what we describe as sensorimotor dynamics with specialization into sensory and effector processes. Many of these organisms can enact a variety of sensorimotor schemes. These schemes are coupled to metabolism, grounding their normativity in the organism's survival. Yet the schemes themselves are more or less fixed, showing only limited scope for adaptive change in their organization.

3. *Minimal sensorimotor learning.* Of the creatures capable of minimal sensorimotor engagements, some can exhibit learning and flexibility in individual sensorimotor schemes. Each scheme can equilibrate, but not necessarily in relation to others. Schemes relate to each other via organismic agency (i.e., coupled through metabolism).

4. *Minimal sensorimotor agency.* In some organisms, precarious sensorimotor schemes are organized into a network of enabling relations. At this stage, the sensorimotor network becomes operationally closed in itself and not only through the organismic closure; it becomes individuated in the sensorimotor domain.

5. *Open sensorimotor agency.* Some sensorimotor agents have the adaptive capacity to learn new sensorimotor schemes in an open manner and integrate them in the overall network.

According to our definition, only creatures at or above level 4 of this progression are proper sensorimotor agents. As we move up through the lower categories, we find examples that are not yet sensorimotor agents, but which gradually move toward increasingly complex forms of biological agency. For example, even the bacterium *E. coli* exhibits sensory-guided behaviors and is able to adaptively switch between different behaviors in response to metabolic demands. But for exactly that reason, the operation of the bacterium remains firmly rooted in its organismic agency, and it is not able to flexibly adjust its schemes according to norms at the sensorimotor level. We would thus place *E. coli* in category 2 (i.e., as an agent capable of minimal sensorimotor engagements).

Next, consider the nematode worm *C. elegans*. It has a nervous system, exhibits a whole range of different behaviors, such as chemotaxis, thermotaxis, and avoidance of noxious stimuli, and is able to adapt behaviors to new circumstances. For example, it will learn to associate the presence of food with a certain environmental temperature if the food is reliably found at this temperature along a thermal gradient. And it will unlearn this association when starving and food is no longer found at the learned temperature. *C. elegans* could be said to have a set of sensorimotor schemes and the ability for some of these schemes to adapt. But while there are some linkages between the different schemes (as demonstrated by their associative flexibility), these relations do not yet necessarily form an operationally closed network.[10] Moreover, the driving force behind the adaptations within or between individual schemes remains the metabolic requirements. We suggest that *C. elegans* is an example of category 3, an agent capable of minimal sensorimotor learning.

Sensorimotor agents in the full sense (category 4) also have a flexible organization of sensorimotor schemes, but in addition, this organization forms a closed self-sustaining network of relations between precarious schemes. This means that we would expect to find in sensorimotor agents that some aspects of their behavior make sense according to sensorimotor norms apart from strictly metabolic needs. Many animals fall in this category in more or less obvious ways. For instance, all species capable of exhibiting play behavior are clear cases of sensorimotor agents because they can show behavior that is not directly linked to biological norms and yet are capable of normatively regulating their schemes within the play-activity frame. The same is true for species susceptible of showing maladaptive and compulsive behaviors in the absence of a sufficiently engaging physical and social environment (e.g., rats; Alexander, Coambs, and Hadaway 1978).

Since we considered the question of open-endedness in Chapter 4, we may also distinguish basic forms of sensorimotor agency from open ones (category 5). The behavioral repertoire of open sensorimotor agents is able to change in unpredictable, historically and culturally influenced manners. This is not only the case with humans and primates, or songbirds with very flexible song repertoires, but also with what we could very broadly describe as "trainable" animals (e.g., domestic pets, crows solving problems using tools).

The five categories above are merely indicative. It is likely that they share many gray areas and that suggesting that an organism is an agent of one kind and not of another will be highly dependent on current knowledge about it. The categories are also not perfect in that they move simultaneously along different dimensions: flexibility, plasticity, organization of a sensorimotor repertoire, and whether norms are strictly biological or both biological and sensorimotor. We could attempt to separate these dimensions or even to add other ones such as, for example, the degree to which an agent engages in active niche construction, or the forms of spatiality implied in the agent's movements (animals move differently from plants), bodily reflexivity, or the degree of sociality involved in enabling a sensorimotor repertoire. We also have not said much about the

10 The relative independence of sensorimotor schemes in *C. elegans* is revealed by ablation studies (Rose and Rankin 2001) and specific mutations (Zhang, Lu, and Bargmann 2005) that induce selective inhibition of behaviors while maintaining the rest of the behavioral repertoire comparatively unmodified. Also high degrees of modularity and decomposability in *C. elegans* (Reigl, Alon, and Chklovskii 2004) may considerably reduce the role of self-organization at a large scale and thus the precarious interdependence between scheme support structures. Nevertheless, we are still far from understanding the full complexity of *C. elegans'* sensorimotor dynamics (Izquierdo and Beer 2013), and the kind of agency it embodies is something that deserves a further empirical study. The virtue of our approach to sensorimotor agency is precisely to make possible the generation of hypothesis to be tested regarding levels and kinds of agency.

developmental aspects that could inform these distinctions either: a particular organism may not fall within one of these categories at a given developmental stage but do so at another. And to make matters more complicated, there is no reason why we should not also look at some collectives, such as ant colonies, as potential sensorimotor agents, if we could suitably define what we mean by sensorimotor closure, asymmetry, and normativity in such cases.

These problems could be addressed and we could come up with better ways of categorizing kinds of agency. Our point is simply to indicate some broad differences along some dimensions that tend to be of general interest.

Related to these issues, the question as to whether a robot can become an agent finds a straightforward answer within our proposal: insofar as the robot is capable of supporting the emergence, maintenance, and adaptive regulation of a network of *precarious* sensorimotor schemes it is a sensorimotor agent. Note, however, that current examples from autonomous robotics are not agents of any kind in this classification since they do not self-individuate. Still they can "behave" in functionally more or less complex ways. These behaviors, however, do not quite go beyond being analogous to level 2 of agency. Most often, behavioral layers are independently defined and later connected. Few (if any) of the mechanisms involved can be said to be precarious. The challenge for robotics is to create agency directly at the sensorimotor level. It is an open issue whether this is possible or not. It would require the design of "architectures" that can instantiate a plastic web of precarious "sensorimotor" processes, with the additional capacity to organize themselves into finely tuned hierarchical and horizontal dependencies to achieve global coherence. If such an operationally closed dynamic architecture could be instantiated, it would probably suffice to bootstrap the normative requirements to anchor the activity of sense-making (processes at that point would become properly sensorimotor, without scare quotes).

In order to achieve artificial sensorimotor agency in real environments self-organized processes of accommodation and assimilation within and between schemes need to be in place, and we are yet to find a general-purpose approach capable of producing this (although see discussion on open-ended equilibration in Section 4.6). A promising avenue for engineering artificial sensorimotor life could be to explore the evolution of developmental processes whose deployment in a specific bodily and environmental setup leads to the emergence of a sensorimotor scheme network with regulatory capacities. Some connected ideas have been modeled (Di Paolo 2000b; Egbert and Barandiaran 2014; Iizuka and Di Paolo 2007), but the principles for open-ended sensorimotor agency have yet to be implemented artificially. While strict material self-production may possibly be sidestepped, open-ended attunement will nevertheless demand that the body also be subject to material evolution and self-organization, otherwise it cannot be open-ended. The environment will need to be sufficiently complex too, and open enough to afford selective and self-organized support of the sensorimotor scheme network. An example of this kind of material and environmental openness are the cases of evolvable hardware and growth of artificial sensors mentioned in Chapter 5.

The flipside of the question of artificial sensorimotor life without material self-production is the question of how biological and sensorimotor agency relate in a same organism. We may sketch some of the expected relations between the two forms of agency, especially since in many cases we see an overlap of self-asserting organizations in a given organism, which could potentially aspire to agency status. This is a complex area that deserves more elaboration so we cannot go further than some very broad considerations.

The first thing to note is that there are not simply just two forms of self-individuation: organismic and sensorimotor. What we have broadly described as organismic or minimal agency, based mainly on the adaptive, self-sustaining identity of metabolic processes, is in itself in many cases constituted by more than one level of autonomy. This has long been recognized by proponents of the organizational approach to the study of living processes. Varela (1979) proposes apart from autopoietic closure, the autonomy of the immune system, as well as other self-sustaining loops, such as the relation between cellular activity and the extracellular matrix (Varela and Frenk 1987), as genuinely autonomous systems. These many levels of autonomy do not immediately translate into forms of agency, of course. In some cases, only one of the three requirements would seem to be warranted, that of individuation. But other cases could potentially be shown to fulfill the other requirements too (we could speculate for instance that self-affirming immune networks may be a possible candidate; see e.g., Varela et al. 1988; Stewart and Varela 1991).

Similarly, as we have seen, sensorimotor agency involves self-sustaining sensorimotor schemes that are themselves equilibrated in higher order organizations. Even within an operationally closed sensorimotor repertoire, we can find other kinds of closure. Micro-identities may emerge as subnetworks of tightly bound schemes within the broad sensorimotor repertoire, like the clusters in our network description. It may well be the case that we enact different forms of sensorimotor agency, for instance, in different social situations, and that at the psychological level, these are seen as different, sometimes conflicting aspects of a personality.

Given these complications what can we say about the relation between biological and sensorimotor agency? The general answer is that there is no general answer, and that the relation between these types of agency depends on the specific relations between the processes involved (and sometimes on the particular history and circumstances of the organism in question).

We can make a few observations, though. It would seem that in some sense, metabolic/organismic identity is more fundamental that sensorimotor identity, at least as presented to us empirically by every example of agency known to us. Thus, organic norms play the role of constraints on sensorimotor norms because organic self-individuation reliably enables the existence of the sensorimotor agent. But enabling is not the same as fully determining. There are many ways in which organic constraints can be met, but only some of them seem to be preferred, a fact already noted by neurophysiologist Kurt Goldstein (1934/1995). There are therefore some degrees of freedom within which sensorimotor agency establishes its own normativity. Sensorimotor norms can even be in tension with organic norms as we have noted. This tension may be sustained in time. What cannot happen is that sensorimotor agency in living systems sustains itself beyond the ultimate barrier of organism's viability. At that point both forms of agency cease to exist.

The above describes one "arc" of the relation between biological and sensorimotor agency. But the description would not be complete if we left it at that. In many ways, once sensorimotor agency is at work, it can offer the organic body many possibilities to sustain its viability that may not have been available to it before. Sensorimotor agents explore their environments and modify them, changing in so doing the conditions that affect their organic selves. And in turn this can lead to the organic identity also changing in accordance to new possibilities afforded by sensorimotor agency. At some point in this history of mutual changes, however, the organic level may not only be "helped" by the sensorimotor level, but actually become *organizationally* dependent on it. Such is the case with practically all animal life. The energy budgeting of animals (different from that of plant life) allows the development of nervous systems that control rapid movement and permit mobile forms of agency where the whole body is able to uproot itself in controlled manners. This in turn affords a variety of possible sources of nutrition. But this "gain" in opportunity is at the "cost" of organismic dependence on sensorimotor success because the energy levels required to sustain mobile agency cannot be achieved in any other way (see Moreno and Lasa 2003; Barandiaran 2008). We see then that while it is clear that organismic agency *reliably* enables sensorimotor agency, the historical/evolutionary development between the two forms can *sometimes* reach situations where sensorimotor agency in turn also enables specific ways in which biological agency is organized. These ways may eventually be such that it could become impossible for the organism to constitute itself materially, to produce its body, other than as a sensorimotor agent.

In this sense sensorimotor life can be reabsorbed into the material constitution of the organism. Sensorimotor agency becomes anchored in the body. It is no longer a case of a purely organic body, which later, in addition, enters into the relations from which sensorimotor agency emerges. It becomes a *constitutively sensorimotor body*, a second nature.

As we have said, the schemes forming a closed and precarious network are relational in nature, and therefore cannot be conceptually confined within any particular physical boundaries. Yet in everyday life, it is of course the most natural thing to identify the "sensorimotor ecologies" surrounding organisms with their physical bodies. How do we justify this association, if schemes constitutively involve both agent and environment structures? One answer may lie in the way that agent and environment support structures relate among themselves. For while environmental aspects involved in different behaviors do not generally need to be connected in any specific way, the agent-side support structures depend for their existence not only on the exercise of certain sensorimotor schemes (as per our definition of closure at this level), but of course also on a common physiological basis or substrate that is itself maintained at the level of organismic autonomy. It is through this kind of scaffolding of the different levels of autonomy that each higher level can become anchored in the levels below it. And it is this anchoring, in turn, which justifies our intuitive association of the sensorimotor agent with its living body.

This anchoring in the body is the dialectical resolution of the tension between the two kinds of agency. The double normativity of sensorimotor engagements (they are the regulatory engagements of a biological agent, but they are also the way in which a sensorimotor agent reasserts its identity) results in tensions that can only be overcome actively and sometimes partially by the ongoing (re)integration between the two kinds of agency and their historical co-dependencies. Thus, we are justified in associating sensorimotor agency with a body, only this body is not just the organic body it is anchored to. There is also a sensorimotor living and lived body. This body allows for non-organic elements (prosthesis, glasses, hearing aids, and tools) to become part of it, to be *incorporated* into it whenever they participate in sustaining the closure of the sensorimotor network. The sensorimotor body can also be extended beyond direct contact with the organic body, while remaining functionally anchored in it. It extends into the nest, the territory, the vehicle of the expert driver, etc. This occurs, again, whenever the (structural or functional) stability of extra-organic elements also faces precarious circumstances, and by becoming involved in the closed network of precarious sensorimotor processes these external elements become stabilized.[11]

These briefly sketched mutual constraining and enabling relations do not exhaust the interdependence between organic and sensorimotor agency. In their call for a radically embodied approach to the neuroscience of consciousness, Thompson and Varela (2001) highlight the fact that the brain is not only the mediator of an agent's sensorimotor embodiment, but also the point where behavior is organically regulated. The organic body and the sensorimotor body in fact engage in direct and mutual interactions at various levels, mediated by links between the autonomic nervous system and the limbic system via the hypothalamus, as well as through contributions of the endocrine and immune system. Emotional states and homeodynamic metabolic processes thus regulate not only specific functions such as sleep, wakefulness, or arousal. As Panksepp (1998) notes, the interaction between the neurodynamics of basic emotional circuits and neural schemas related to bodily action plans might also be the origin of affective feelings of pleasure, anger, desire, etc. Disruptions to sensorimotor agency can be emotionally distressing (see Box 6.2). Adaptive regulation through interaction of the organismic and sensorimotor levels

---

11 Some cases of active niche construction fall within the category of extended sensorimotor bodies, but not all. Many of them have the added complication of involving collective and intergenerational effects. Examples of more or less obvious sensorimotor extensions or incorporations include spider webs, the building of amplifying singing burrows by mole crickets, and the trapping of air bubbles for underwater respiration in insects (see Turner 2000). A nice example is that of the Weddell seal in Antarctica. She maintains a breathing hole in the ice so she can access the water below where there is food and protection from blizzards. The breathing hole is constantly shrinking as ice builds up; hence, it can be said to be a precarious structure. The seal invests a lot of energy and time trimming the ice around the hole with her teeth. This prevents the hole from freezing over (Castellini, Davis, and Kooyman 1992). There is, in this case, a relation of co-dependence between biological individuation and sensorimotor agency, and between sensorimotor agency and the structuring of the environment, all of them precarious networks of processes that stabilize each other.

Box 6.2 Sensory deprivation and sensation seeking

Sensorimotor agency is precarious; a systematic prevention from sustaining the networked relations between schemes and support structures can, over time, result in minor, traumatic, or even fatal damage to sensorimotor autonomy.

Such is the explicitly stated purpose of techniques described by the 1983 *Human Resources Exploitation Manual* used by the CIA to train (para)military personnel in Latin America. These techniques include manipulation of the environment to disrupt familiar sensorimotor patterns and induce disorientation, physical weakness, and dread. Ultimately, the goal is to provoke a "psychological regression in the subject by bringing a superior outside force to bear on his will to resist. Regression is basically a *loss of autonomy*, a reversion to an earlier behavioral level" (Schutz 2007, p. 155, our emphasis).

Studies in sensory deprivation in the 1950s—covertly funded by the CIA (McCoy 2006)—revealed that apart from the need for "the sensory stimulation of the normal complex environment" during growth, "the *integrity* of the mind at maturity continues to depend on that stimulation" (Hebb 1980, p. 96, our emphasis). Sustained isolation from patterned stimulation (often inseparable from limitations to action possibilities) leads to visual disturbances, anxiety, mood swings, apathy, and cognitive impairments.

Attempts to explain these effects on autonomy and integrity often involve the notion of stimulus hunger: a need for an optimal level of sensory stimulation (e.g., Hebb 1955), although this does not clarify why deprivation leads to some effects in particular and not others, nor does it make explicit what need is served by satisfying this "hunger." Given the chance to voluntarily trigger a stimulus even if it is meaningless (e.g., pushing a button to show some colored stripes), sensory-deprived subjects will seek these sensory sources (Zuckerman and Haber 1965). This sensation-seeking behavior is explained as originating in an inherent personality *trait*, which is determined by the neurochemistry of reward systems in the brain and modulated by social and occupational factors (Zuckerman 1979).

We can formulate an explanation that moves beyond considering stimulus seeking as a trait by exploring its underlying logic as a disruption to sensorimotor subjectivity, i.e., as threats to the integrity and autonomy of sensorimotor networks. The need for stimulation is literally a *hunger*; not so much a trait that just happens to be modulated by neurochemical and social factors, but a well-defined, contextual sensorimotor demand that fits the logic of self-maintenance of sensorimotor networks (as hunger for food fits the logic of metabolic self-individuation). The clue is that, if sought after, stimulations are not meaningless. On the contrary, they afford (or fail to afford) the enactment of certain sensorimotor coordination patterns and even whole schemes (e.g., following the direction of the colored stripes with the eyes, or crossing them over repeatedly; even patterned color flashes can pre-activate schemes), so they are meaningful according to sensorimotor norms. "Sensation" seeking is a longing for the action/perception *enactments* needed to stabilize precarious relations between schemes and their support structures.

of agency may even underlie a "basic self-affection or a core consciousness of one's bodily self-hood. Thus, processes of life and processes of mind are inseparably linked. Every conscious state is rooted in the homeodynamic regulation between brain and body, and, in a sense, integrates the present state of the organism as a whole" (Fuchs 2009, p. 5). Similar points are elaborated by Giovanna Colombetti in her book *The Feeling Body*, where she proposes an enactive, dynamical theory of affectivity and emotions. What she calls primordial affectivity is a property of the whole organism as a sense-maker. This affectivity belongs to living organisms, even those that are not themselves sensorimotor agents. For the latter, primordial affectivity takes the shape of moods and emotional episodes, which she describes as "self-organizing configurations of the organism" (Colombetti 2014, p. 82). Emotions act as emergent high order constraints to the dynamics of muscular, neural, and autonomic processes that integrate the affective experience of the whole body engaged with the world.[12]

6.7 **Sensorimotor subjectivity and mastery, again**

If we take the definition of agency in Chapter 5 seriously, then evidence of actions that follow "non-biological" norms also counts as evidence of some other kind of self-individuation happening at a level that corresponds to these norms. But exactly *what* is being self-individuated? To answer this question, we have explored whether individuation and agency at the sensorimotor level are coherent ideas.

Self-sustaining precarious patterns of sensorimotor activity do occur; we call them habits. They are metastable relations between organic and environmental processes poised between blind automatism and unpredictable spontaneity. Habits confirm that the indeterminacy of sensorimotor processes with respect to metabolism allow the possibility of self-individuation in a different space from that of biochemical processes. The circularity between the enactment of a habit and the stabilization of its support structures is extended as an oblique relation between several schemes and their support structures. In this case we are dealing with a self-individuating network of sensorimotor schemes.

Sensorimotor networks of this kind can meet the requirements of self-individuation, normativity, and interactional asymmetry. As we explore this idea in detail, we profit from the network metaphor and find that through regional developmental differentiation and integration of schemes, integrated subnetworks can be formed that correspond well to notions such as activities, microworlds, and sensorimotor genres. The structure of the

---

[12] Here we should remind ourselves of Hans Jonas's reflections on the passage from basic forms of affectivity in metabolizing life to the co-emergence of action, perception, and emotion with the mobility typical of animal life (1966). For Jonas, not only are action and perception possibilities afforded by movement that arise as developments of basic forms of sense making (in our terms), they are also inextricably linked with a complexification of primordial affectivity as novel emotional dimensions are afforded by animal mobility. Thus, fear is an emotion that integrates temporal and spatial elements (e.g., sensing now the approaching predator in contrast to sensing now the pain of an injury).

sensorimotor network reflects the history of the agent. In humans the possibilities are open-ended and path-dependent, as we would expect, leading to a way of characterizing otherwise vague concepts such as sensorimotor styles (an idea that merits further exploration).

Sensorimotor agency, a non-homogeneous network of interrelated micro-identities, fends off the frequent criticism of enactive cognitive science reducing mind to biology. This is a clear example of the emergence of autonomous domains, in this case a sensorimotor domain, which does not contradict life–mind continuity, but on the contrary is only understandable through it, as a dialectical development of the relations between organism and environment. In this case, in particular, it is thanks to the operational concept of agency that developed at one level that we may find a different kind of agency at another. The distinction between biological and sensorimotor levels corresponds to the ontological leap afforded by distinguishing between manifestations of autonomous entities at different domains. There is a sensorimotor body that extends beyond the organic body, but at the same time, it is anchored in it. The organic body in turn becomes dependent on the sensorimotor body, not only in the direct sense that behavior is required for survival in sensorimotor creatures, but also in the sense that organic self-individuation is organized in ways that are enabled by the sensorimotor body.

We will continue to explore some of the implications of sensorimotor agency in Chapters 7 and 8. Before we do, we briefly return to Thompson's (2005) critique of the sensorimotor approach to perceptual experience. According to him, sensorimotor contingencies theory does not provide a sufficient account of perceptual experience as it obviates reference to the subjective aspects of this experience. Could the concept of sensorimotor agency help in addressing these concerns? We think so.

Myin and O'Regan (2002) argue that the sensorimotor approach can account for fundamental properties of perceptual experience, viz., forcible presence (sensory experience imposes itself on us), "ongoingness" (this experience happens in the here and now), ineffability (experience is always richer than what may be described in words), subjectivity (sensory experience is an experience for me, the subject). The latter property would obtain simply by the fact that there is a subject whose resources are put into constituting her conscious experience. According to Thompson, to repeat what we said in Chapter 2, the sensorimotor account of these properties is enriching but incomplete, as it requires a more compelling account of selfhood and agency as well as an account of pre-reflective bodily self-consciousness (Thompson 2005, p. 417).

Thompson reminds us of the difference between simple agents like bacteria that meet our definition of agency and systems that do not, like O'Regan and Noë's example of the missile guidance system (2001, p. 943), which they claim has some kind of mastery of sensorimotor regularities that allows it to operate adequately. Perhaps it can be argued that this system is able to alter its coupling with its environment in an asymmetric manner (for instance, by activating new sources of information contingently on its current state). But the system is individuated only by convenience of design and use, not as a consequence of its own activity. Moreover, while its operation is subject to norms, these norms are strictly external to the individuation of the system—see a similar argument by Jonas (1966) about goal-seeking control in the guidance system of a torpedo. This is not an agent. Guiding a missile to the right target or missing it with the concomitant tragedy of so-called "collateral damage" is not something the guidance system cares about. There is no risk and no opportunity, in fact, nothing to "perceive" by this system that is of any interest to the system itself. Its "individuation" does not put the system into that kind of Jonasian relation of needful freedom with its surroundings.

In a sense, a fundamental property of perceptual experience that has so far been missed by the sensorimotor approach is that perception is inherently meaningful for an agent—even when we perceive something nonsensical, this character is precisely a locus along dimensions of meaningfulness (see e.g., Froese and Cappuccio 2014). There must be an agent that does the perceiving, and there must be a meaningful relation between agent and world in order to speak of action or perception at all. What sensorimotor agency adds to this is that this meaningful relation is not only established between the bio-logical body and its environment, but involves also the new forms of self-individuation that obtain in the sensorimotor spaces opened by basic kinds of agency; it involves the sensorimotor body.

We should keep in mind that for the enactive approach to have a meaningful relation to the world is to enter into the activity of sense-making. What is meaningful is that which the agent is sensitive to and adaptively capable of regulating. In the case of the sensorimo-tor agent, the world consists of those factors that the agent is able to identify as relevant to the ongoing individuation of its closed sensorimotor network of schemes, or more concretely, as relevant to its effective action in the world. These include factors relevant for the normal enactment of schemes, which are needed to reaffirm the agent's identity and factors perturbing this enactment, potentially impeding it.

Some confusion may arise from the fact that by definition a sensorimotor scheme is constituted by the interaction of agent-side and external support structures (the A, B, C, ... and A', B', C', ...). If we think of the schemes involved in picking an apple from a branch and taking a bite of it, it is obviously the case that the apple in this situation is meaningful. It is meaningful for the biological agent but it is also meaningful for the sensorimotor agent. Although related, these meanings are not identical. What matters from the sensorimotor perspective is not the nutritional value of the apple, per se, but the effectiveness at performing a series of actions, the relations between the schemes involved (stretching the arm and body long enough to reach the apple, gripping it firmly enough so it does not slip, pulling it with sufficient force, etc.) and also the relation to other schemes. The apple provides several properties that enable these schemes, e.g., its solidity and its graspable size. These are the A', B', C' ... elements that co-constitute the different acts, while the neuromuscular activity, the body posture, the pressure put on the grip, etc. are the A, B, C, ... agent-side support structures. What is meaningful to the sensorimotor agent are not these elements in themselves but the apple's potential for providing these sensorimotor supports, of affording certain schemes, as well as the body's own capacities and skills involved in performing them. The sensorimotor agent will be sensitive to the apple's affordances and to the bodily possibilities and this includes factors that disrupt or alter the successful enactment of the scheme and therefore impede its role in asserting the sensorimotor network that makes up the agent. Thus, it will be meaningful for the sensorimotor agent if the apple is too high and out of reach. If the agent is a small child, the apple may be too big to grasp with one hand or too firmly attached to pull off the branch. But these sensorimotor breakdowns (obstacles and lacunae) may be adaptively equilibrated; after a few failed attempts, the child may jump, grab the apple with both hands, and pull it with her whole body. Equilibration in this case shows the presence of sensorimotor norms at play, that is, the regulation of relations between schemes so as to overcome sensorimotor barriers and restore the agent's effectiveness in the world.

Thinking of a sensorimotor network as closely connected schemes appropriate to certain activities and behavior settings—clustered into what Varela refers to as micro-identities—brings a new angle to the notion of mastery. In Chapter 4 we have linked mastery to the equilibration of sensorimotor schemes. The way that sensorimotor agency is organized—as a non-homogeneous network of enabling and functional relations between schemes—allows us to see mastery more concretely. Not only do we master a given sensorimotor scheme as we undergo progressive equilibration such that it better fits a given situation, we also master the regional relations between schemes and their adaptive variability in the concrete context of an activity (the effects shown in Figure 6.5). Our picture of development shows that new clusters may emerge through processes of differentiation and integration between schemes. Achieving cluster-level mastery is equivalent to saying that the elements of the cluster become progressively integrated and adapted to each other. Enacting a single scheme in a cluster involves the influence of several other non-enacted schemes (cf. the quotation by Mead on this point in Chapter 2).

This is why some of the examples of perceptual know-how given in the sensorimotor literature make intuitive sense. An object like a tomato is seen as voluminous because I may change my angle of vision or grasp it or turn it, and these schemes—which are not necessarily enacted—inform my visual perception. But why do they? And why these schemes and not others? My visual perception of the voluminous tomato does not appear to be that much informed by many other things I could potentially do, such as smashing it with a hammer, touching it with my forehead, or covering it with a newspaper. This is because what we master is strongly colored by the clustered schemes that correspond to concrete activities, in other words, the subnetworks of schemes in Figure 6.4. Thus seeing a tomato, grasping it, and moving it so that I now see it from a different angle are all schemes that belong together as part of activities such as cooking or shopping for groceries. And this is the reason why these closely networked schemes affect the enactment of a single one of them, such as looking at the tomato. The structural and functional links that connect schemes in a cluster mean that I rarely just look at a tomato in the abstract (an issue we will come back to in Chapter 8), but instead I look at it in order to start preparing the salad, for example. If the looking scheme belongs to clustered activities, looking is always already *looking-in-order-to* as well as *looking-because-of*. In other words, mastery of sensorimotor contingencies is *regional*, something that is only implicit in the sensori-motor approach to perception.

There is operational closure at the whole network level as well. Thus, we speak of a single sensorimotor level. However, the high integration of regional, closed or quasi-closed sensorimotor clusters is phenomenological relevant in questioning the extent to which there is a unified self at this sensorimotor level. As Varela (1992) has noted, we transit through microworlds of significance (walking aimlessly in the park or hurriedly to catch the train) in which we literally act as micro-identities. Enaction is the transit between microworlds, the time between moments.

If this is so, there could be, apparently, not a single *sensorimotor* subject, but many, which are nevertheless interlinked, and also integrated by the same sensorimotor body and same sociohistorical-narrative embeddedness. Varela (1991) has spoken of the various forms of self-individuation that make up an organism as "a meshwork of selfless selves." By this he means to picture an organism somehow as the ongoing integration of many kinds of autonomy (metabolic, immune, neural, sensorimotor, cognitive, social, personal, etc.) with no single substantial locus where one could find a unifying "self." The subnetworks of schemes that make up a sensorimotor agent add to this picture, as they make even one level of selfhood (sensorimotor) into a potential multiplicity of micro-identities.

It would be tempting—but incorrect—to conclude from this insubstantiality and multi-plicity that the notion of selfhood should be discarded, that such a thing does not exist. We have done nothing so far but show that concepts of this kind can be perfectly grounded in natural science. A self that like many other concepts (energy, force, dynamics, etc.) is not an intrinsic property of matter but a relational one, does not for this reason have any less reality or efficacy. The enactive approach is not skeptic about the self, in any of its manifestations; it has precisely reinvigorated the idea of selfhood while at the same time demonstrating that it is a multiple, dynamical, relational, historical, and situated concept.

So, to come back to the sensorimotor approach, sensorimotor agency, as we have pro-posed it, not only places the agent at the center of the engagements that constitute a per-ceptual act, it also provides us with a deeper understanding of context-dependent mastery and activity-dependent selfhood. This helps us address only the first of Thompson's worries.

> Adding an enactive account of selfhood to the dynamic sensorimotor approach goes only part way toward addressing the body-body problem [the problem of the relation between the subjectively lived body and the living body of the organism]. In addition we need to include subjectivity in the sense of a phenomenal feeling of bodily selfhood linked to a correlative feeling of otherness.
>
> (Thompson 2005, p. 419)

In Chapter 7 we intend to move into this direction by exploring whether our account of sensorimotor agency can help us elucidate at least one aspect of the phenomenal feeling of bodily selfhood: the sense that we are the agents of our actions.

Chapter 7

The sense of agency

As I navigate my way along the path up the hill, my mind totally absorbed anticipating the difficult conversation I'm going to have at my destination, I treat the different features of the terrain as obstacles, supports, openings, invitations to tread more warily, or run freely, etc. Even when I'm not thinking of them these things have those relevances for me; I know my way about among them.

—*Charles Taylor (2002, p. 111)*

7.1 A sense of a meaningful relation with the world

In the epigraph to this chapter, Charles Taylor invites us to consider that in everyday situations we often experience the world around us without consciously reflecting on it. The objects of my experience, in this case, such as the rocks, streams, or bridges along my path, already figure for me in the kind of actions they afford or solicit, and their relevance for the satisfaction of my desires and meeting of my goals, even when I am consciously thinking about something unrelated. It is this fundamental kind of sense-making, of already meaningful engagement with the world, that qualifies us (and other animals) as agents.

How I experience *myself* in situations such as the one described by Taylor is a controversial topic. A lively debate surrounds the question of whether and how I experience myself as being the author of my actions during absorbed bodily coping. Hubert Dreyfus (2005, 2007a, 2007b), for example, initiating the now famous debate between himself and John McDowell, has claimed that in certain cases of both expert performances as well as everyday behavior, I literally lose myself in the flow of my activity. According to this view, during skillful, absorbed coping, my activity becomes automatic and reactive, bodily responses being drawn out of me by environmental solicitations. McDowell (2007a, 2007b), in contrast, rejects that agency and subjectivity should be lost in these situations. He argues that when we react to environmental affordances—and we do so normally in a context-sensitive manner (i.e., taking into account our own current state of affairs)—this *always* involves conceptual capacities and is therefore permeated with rationality. According to McDowell, this rational orientation toward the affordances involved in embodied coping recovers the subject in the act, and with it, the possibility of a minimal form of mindfulness.

We are not the first in proposing that there may be an overcoming of the opposition between these views, a description of skillful intentional action that is characterized neither by the total absence of mindedness suggested by Dreyfus, nor by the always-rational mindfulness proposed by McDowell. The alternative is to consider the possibility of other ways in which a subject can be involved in the process of acting on environmental solicitations. As we will see, the concept of sensorimotor agency developed in Chapter 6 offers exactly this. On this account, we can act intentionally, that is, in ways that can succeed or fail, not only as a result of prior, conceptually manifested reasons, but because we are non-conceptually involved in the process of shaping the dynamics that lead to the engagement and control of particular sensorimotor schemes. We argue that this kind of motor intentionality, as Merleau-Ponty (1945/2012) would describe it, opens up the possibility for minimal, non-conceptual forms of subjectivity and self-awareness in action.

Consider as another motivating example the case of trail running:

> I'm trail running in the pouring rain in midwinter. The trails have become a network of creeks. The water flows commandingly down the trail, yearning through exposed roots, striving to reach the lowest point. As I find my rhythm on the trail, I become aware of the fluency of my movements. Compensating with my right foot on a root. Balancing with my left arm to avoid tree branches. The more insistently the water flows, the more the creek appears to move with determined force. On some occasions, my own actions seem to join the stream. I don't reflect on the difference between moss and ice under my feet. Miles of the trail disappear from memory and I seem to have vanished. At one stream crossing, the water becomes too wide to cross with a single leap. I step back and detach from what I'm doing. I decide to rest and suddenly my concerns overwhelm me. Maybe I should be grading papers or reading books with my son. I look for a way to cross the creek. With one foot on a fallen tree, the other on a jutting rock, I hop to the other bank. I am back in the flow of running.

> (Dow 2015, p. 2)

Not only, in this case, do I perceive the affordance of a fallen tree or a rock as allowing me to cross the creek, as Taylor already suggests in his example. I also act on these affordances and purposefully choose and carry out carefully targeted movements with my body, allowing me to traverse complicated terrain, all the while, I am completely absorbed in the flow of my running. Immersion in my activity does not mean that I lose control over my actions, or that my actions are not directed at my immediate goals. Nor do I sense that the trail is the source for my actions instead of myself, as Dow points out (2015, p. 2). *I* choose whether to tackle a boulder by stepping on and over it, or by running around it. *I* adapt my running style to the condition of the terrain, taking into account whether it is wet or dry even when I am not consciously reflecting on the boulder or the terrain. All of these are intentional actions, and my non-conceptual involvement in their selection and control, we propose, allow me to experience myself as the author of my running activity, or put another way, to realize myself as a runner.

We will show in this chapter that this idea not only applies to cases of skillful and absorbed performances. Rather, we believe that the phenomenology associated with an agent's intentional sensorimotor engagement contributes to the pre-reflective sense of agency in general.

In our daily lives, when we engage in activities that do not necessarily involve particular skill or immersion, we do not always reflect about our own actions consciously either. We carry out many activities habitually (e.g., operating a door handle when leaving the room), or more generally without our conscious involvement or decision to do so (catching an object unexpectedly having been thrown at us). Yet, arguably, when we engage in this manner with our environment, our experience is permeated by the diffuse sense that it is we who are initiating and controlling our actions. In other words, these actions are accompanied by a phenomenological experience—sometimes thin, at other times richer—of our own agency. Where does this phenomenology originate? Can we account for its different qualities and dimensions in the same way that the sensorimotor approach aims to explain differences in *perceptual* experience? In this chapter we aim to show that the sensorimotor approach, when extended with an enactive understanding of subjectivity arising in embodied agents, may point the way toward a new understanding of the sense of agency.

Our proposal is markedly different from existing theories aiming to explain the experience of agency, the majority of which can be described as "in-the-head" approaches. These traditional explanations emphasize computational processes occurring in the agent's brain that instantiate internal models and which are used to draw comparisons between intended and actual states of the body and world. In this view, experiencing oneself to be the agent of an action is first and foremost a question of verifying whether these states match or not. These theories are subpersonal and internalist. We propose an enactive account of agency experience that, in contrast, is world-involving. It aims to link the personal-level relation between agent and environment to the dynamical coupling of subpersonal processes in the two systems. It does not see the sense of agency as merely an epistemological problem (i.e., as the question of how my cognitive system computes that my sensor and motor activities correspond to an action I am performing), but rather assumes that it is an intrinsic aspect of how sensorimotor schemes are organized and enacted in the world. A sense of the bodily self as an agent, in this alternative view, corresponds to what we experience during the ongoing adventure of establishing, losing, and re-establishing meaningful relations between the world and ourselves. The enactive account of the sense of agency endeavors to further unpack this view by articulating in operational terms what these meaningful relations consists of, as well as what it means to establish them or to lose them.

Building on our extension of the basic concept of agency into the sensorimotor realm, described in Chapter 6, we propose that the various phenomenological aspects of the sense of agency relate to both the intrinsic as well as the relational (meta)stability of the action/perception schemes that together constitute the sensorimotor level of agency. These intrinsic and relational aspects always involve the world in some nontrivial sense and do not require internal comparison between neural signals as the epistemic signature of a controlled act. Instead, the enacted schemes "belong" to the agent to the extent that they assert her agency in the first place. This is manifested in different forms as feelings of action initiation, of action control, of effort and control exertion by the various ways and degrees in which an enacted scheme is met with, and surpasses (or not), obstacles and resistance both internally within a given act and relationally between acts.

Our account of the sense of agency is situated at the meso-level between neurodynamics and personal experience. In developing it, we follow the general research direction of neurophenomenology (e.g., Varela 1996, 1999), according to which phenomenological descriptions of lived experience and naturalistic explanations are contrasted in order to uncover their mutual constraints and compatibilities. As a matter of convenience, we will often talk about "explanations" or "accounts" of the sense of agency. The use of this terminology should not be understood as an attempt to crudely reduce experiential aspects to naturalistic processes, but as an effort to provide a neurophenomenologically coherent story. Nor, despite the label, will such explanations be restricted to neural processes only, but will generally involve complex linkages between brain, body and world. In short, they will be world-involving. We discuss toward the end of the chapter, and again more generally in Chapter 8, how our proposal can incorporate the fact that human agents are also (perhaps primarily) involved in a social world, in which many agents interact.

7.2 **The phenomenology of agency**

As we hope to have illustrated, when I engage in intentional actions, these are usually accompanied by the experience that I am their author or initiator, in other words, by an awareness that the actions are mine, and that I have caused them. Upon further reflection, and based on empirical data, some of which we summarize below, this sense of agency is not, however, a unique and unified sense. Rather, one can distinguish different levels of action awareness, and various aspects of one's agency at which this awareness can be directed.

7.2.1 **Pre-reflective and reflective sense of agency**

To start with, at a very general level, we can distinguish between a *pre-reflective* and a *reflective* self-awareness in action (Gallagher 2007, 2012). The latter, also referred to as the *judgement* of agency (Synofzik et al. 2008a, 2008b), is the experience of ourselves as agents when taking an introspective stance that is detached from our ongoing activity. For example, when deliberating and planning actions we are about to take, or when explicitly monitoring the success of our actions, we may judge ourselves to be responsible when the actions are consistent with our personal beliefs, or when the task results in the achievement of a goal I have set myself. This reflective sense of agency is usually conceived as a retrospective and conceptual attribution. It also takes a transitive form. In other words, a form with subject-object structure, insofar as it is an experience of which I am both the subject as well as object: I experience myself as he who is acting.

The judgement of agency is best illustrated by cases in which normal agency is manipulated in some way. For example, when subjects receive distorted or delayed visual feedback about their own movements, they are less likely to conclude, when asked retrospectively, that they have caused the observed effect (Tsakiris and Haggard 2005). At least in such cases of distorted or disrupted agency, then, people may attempt to infer agency from observation of movements and their effects. This highlights another important feature of the interpretative sense of agency, namely the fact that it can be non-veridical. I may both judge that I am not the author of a perceived action when in fact I am (as in the case of distorted feedback), as well as judge to be the agent of a movement when I am not. The latter may occur, for example, when a group of people jointly aims to bring about the same effect, such as the movement of an object on a Ouija board, or in the case of table turning (Wegner 2003). Another example is an experiment carried out by Wegner and Wheatley (1999), in which two participants (one in fact a confederate) jointly control the movement of a mouse-like control device, which in turn moves a cursor on a screen depicting a series of random objects. The subjects are asked to move the cursor in slow sweeping circles, to stop more or less midway through a 10 second music clip played every 30 seconds through headphones, and to then judge whether they had personally intended to stop (because the stop could be initiated by either subject independently at random points during the clip). In some trials, the real participant hears a word that corresponds to an object on the screen, while the confederate is instructed to move toward the same object and to stop at it during the music clip (unbeknownst to the subject). Interestingly, the subject is more likely to report to have intended to stop in these cases the less time has passed since the presentation of the word, even when it is most likely he or she has not been responsible for the stop at all. Wegner and Wheatley argue that this sense of agency relies on the perceived contingency between mental states and observed action effects. As long as the action outcome is congruent with one's intention (consistency), follows the intention within a certain time window (priority), and there is no other conspicuous cause (exclusivity), then the action can be experienced as intended and effectuated by oneself. This is in line with Graham and Stephen's account of the reflective sense of agency, which they describe as the construction of "self-referential narratives," or as the result of our proclivity to explain retroactively our actions to ourselves in terms of our beliefs and desires (Graham and Stephens 1993).

The pre-reflective sense or *feeling* of agency (Synofzik et al. 2008a, 2008b), in contrast, is the experience of agency that accompanies my actions when I am immersed in my activity, without paying particular attention to or consciously reflecting on the details of what I am doing at the moment or why (such as the experience of myself when trail running while thinking about the day ahead). At this level, my agency is not given to me explicitly as an object of my experience; in other words, unlike the reflective sense of agency this is not a transitive experience of I, as he who is acting. Rather, agency here is implicit in the unperturbed flow of my action and the egocentric perspective underlying it. It is the basic, diffuse feeling that I am carrying out an activity, an "I do", and it is a kind of feeling that is intrinsic to what we do when we act in the world. This form of agency experience is also phenomenologically recessive, in the sense that in normal circumstances I am primarily aware of *what* I'm doing, rather than the fact that it is I *who* is doing it. Often we become consciously aware only of the absence of the feeling of agency when being interrupted while immersed in a task, or when unexpectedly failing in some way.

For example, as I sit at the desk writing the text before you, my awareness is primarily focused on the content of my writing. As a relatively skilled touch typist, I do not have to pay attention to the fact that I am typing on a keyboard to express my thoughts on the screen. However, while I may not be consciously aware of my typing itself, for example of the precise finger movements involved, I nevertheless experience myself as being engaged in a certain activity—which happens to involve typing on an unconscious level, so to speak. Now, there is nothing esoteric about this. The fact that I am in "typing-mode," without paying attention to this part of my activity itself, may, for example, have the very real consequence of inhibiting me from engaging in other activities with my hands, such as doodling. And on a certain level I am diffusely aware of the different constraints and opportunities, whether bodily or cognitive, that are the result of being engaged in the typing activity. Also, while the feeling of being immersed in this kind of skilled performance is diffuse, and difficult to describe precisely, it is nevertheless different from the feeling we have when moving unintentionally. No matter how absorbed I am in the flow of my performance, it never feels the way it does, say, when my limbs are moved passively by a physiotherapist without my intention or intervention. When I act purposefully and skillfully, I am never in doubt as to who is the source of my activity, without having to reflect on the question of authorship explicitly. But this feeling of agency is not merely characterized negatively by the absence of a feeling of "non-mineness" (i.e., as in contrast to passive movements). As we unpack in the following sections, a positive and distinctive phenomenology may derive from the fact that in realizing my (not necessarily conscious) intention to perform skill X, I affirm myself as an agent able of "X*ing*." In other words, the feeling of agency may primarily result from processes involved in forming, selecting, and realizing meaningful sensorimotor schemes, through which I re-assert myself as a skillful agent. In the successful realization and harmonious flow of my typing, I thus assert myself as a skilled typist, and implicitly experience myself as such.

At first it would seem that understanding the pre-reflective sense of agency in this way implies that, unlike the judgement of agency, it cannot be non-veridical. In other words, it would appear that I am able to experience a feeling of agency if and only if I am successfully deploying the appropriate sensorimotor skills. However, as we describe in the next section, multiple aspects contribute to the feeling of agency, many of which are neither sufficient nor necessary on their own for producing it. Some of these can be susceptible to manipulation or disruption, and therefore may allow for non-veridical aspects in the feeling of agency (e.g., a lacking or disturbed feeling of agency even though I am skillfully performing a particular action). We return to this issue later in the chapter.

Even though we do not usually experience it as such, the pre-reflective and reflective forms of the sense of agency are separable, in the sense that one can enjoy one without the other. This is demonstrated, for example, in anarchic hand syndrome (Marcel 2003; Marchetti and Della Sala 1998; Pacherie 2007a), one of several symptoms involving a disturbed sense of agency. Anarchic hand refers to patients with certain types of neurological damage, who with their contralesional hand, perform complex, goal-oriented, and often environmentally cued actions, such as operating a doorknob or scribbling with a pencil (Frith 2005), yet report that these actions are independent of their will. These patients seem to have no awareness of their own intentions, but feel as if the anarchic action occurred involuntarily. Of particular interest is the case of a man who disowned the anarchic actions his hand was performing, reporting that he was not doing them, yet maintained, "of course I *know* that I am doing it. It just doesn't *feel* like me" (Marcel 2003, p. 79, emphasis ours). Based on such observations, Anthony Marcel argues that the person with anarchic hand "is often clear that their experience of the action as disowned is a 'seeming'" (Marcel 2003, p. 79), which indicates that the first-order feeling of agency and the second-order judgment of agency are indeed separate processes, and that they may come apart, at least in pathological cases.

7.2.2 Aspects of pre-reflective agency experience

Before elaborating on the different aspects of actions that one can be aware of, and which may contribute to the overall feeling of agency, we should separate the sense of agency as described above from the sense of ownership (Gallagher 2000; Synofzik et al. 2008b). The latter is the pre-reflective experience that it is me, i.e. my body, that is moving, or more generally, that a given body part belongs to me. In everyday voluntary activity, these two aspects contribute to a unified, minimal self-awareness for action. That they are nevertheless distinct phenomenological aspects is revealed, for example, in involuntary movements. In such cases, for example when being pushed from behind, I lack a sense of having initiated the movement, or of controlling it, yet I still have the sense that it is me who is moving.

Now, under normal conditions, and in the case of voluntary movements, our own agency tends to present itself as a unified experience. I may or may not feel as if I am the author of my actions, but usually this feeling does not present itself in varying shades or qualities, in the same manner, say, that the visual appearance of a surface may differ in terms of perceived light intensity, color, texture, and so on. However, experimental manipulation and phenomenological analysis reveal that there are multiple different aspects that can contribute to the overall feeling of agency (Gallagher 2012). More specifically we will differentiate here between phenomenological aspects along two qualitative dimensions. First, we can distinguish between intentional aspects involved in the sense of agency and those related to bodily movement and control. Second, we can identify contributions to the sense of agency from processes occurring before overt action (prospective) and those ongoing throughout the action (concurrent).

7.2.2.1 Prospective aspects

Sometimes, events or processes occurring before an action is carried out are already sufficient to elicit an experience of agency. A striking example is the case of patients with anosognosia for plegia, that is, patients with an unawareness or denial of their own paralysis. A subset of such patients sometimes believe to have raised their hand even though they suffer from a condition preventing them from voluntarily initiating any movement with the affected limb. The attempt, or intention, to initiate a movement alone seems sufficient in this case to create a sense of agency, even when that movement is never carried out (and the experience, therefore, non-veridical, de Vignemont and Fourneret 2004). Also, deafferented patients, who do not receive any tactile or proprioceptive feedback about, and therefore have no sense of, their own ongoing movement or body position, know perfectly well whether or not they are moving, and are aware also of the physical effort their movements require (Lafargue et al. 2003). That central processes before overt action may contribute to the sense of agency also provides an intuitive explanation for the experiential difference between voluntary and involuntary movements, as such processes would be operative only in the former but not the latter case.

Awareness of movement initiation. Both intentional as well as movement-related aspects may contribute prospectively to the feeling of agency, and the examples just mentioned cannot discriminate between the two. By movement-related we here refer to processes involved in the initiation of action, but which are neutral with respect to that which is to be achieved by the action, and which occur after the intention to act has been formed. For example, the idea that the presence of motor commands may be one contributor to the sense of agency during movement initiation is supported by a series of experiments reported by Tsakiris and Haggard (2005). Using a paradigm involving the independent manipulation of efferent and afferent signals (e.g., the same arm movements executed either voluntarily or involuntarily), the authors show that the presence of efferent motor signals is important, for example, to attenuate the sensory consequence of self-generated movements (even when the consequences cannot in principle be predicted); to account for the observation that during voluntary movements the awareness of actions and their effects show an attraction in time toward each other ("intentional" binding); or to distinguish between one's own hand passively moving, and that of an experimenter performing the same movement. In all cases, the authors emphasize the importance of efferent signals in particular (motor commands), as different from prior intentions, a view supported by experimental evidence for the involvement of the motor cortex, for example, rather than prefrontal areas normally associated with higher level intentional processes; or by the observation that interrupted intentions, which do not lead to voluntary movement, also do not result in intentional binding (in contrast to what would be expected under Wegner's hypothesis of a reconstructive sense of agency described above). Haggard (2005) associates the presence of efferent signals with the sense of "urge" preceding the execution of an action or the feeling of "being about to do something."

Awareness of prior intention. The above does not mean that prior intentions cannot or do not enter into the phenomenology of agency awareness. In fact, one aspect contributing to the sense of agency is the feeling that not only have I initiated a certain action, but that my initiating it is in accordance with my intention to do so. Patients with schizophrenic delusions of control, for example, seem to have a perturbed sense of the latter kind. The symptoms in this condition are in some respects similar to anarchic hand, in that patients will report not to be the agent of movements they are carrying out. But the two conditions are different in crucial ways. In anarchic hand, patients often carry out actions that are against their conscious will. They will recognize this and try to stop them. Patients with delusions of control, in contrast, often continue to make the movements they intended, such as when following the instructions of an experimenter. They moreover seem to have awareness of the intentions underlying their movements, since they do not try to stop or correct them. Yet, they insist that their movements are being controlled by external forces, reporting to believe, for example, that somebody has implanted a computer in their brain that is making them move, or feeling as if being steered around (Frith et al. 2000b). These patients are aware of their intentions, and of the movement as it occurs (as well as having a sense of ownership), but not of having initiated it. Patients seem to experience a feeling of somebody anticipating their actions (reading their intentions), and initiating them without their intervention. Delusions of control thus illustrate how the willing of an action and its initiation are separable aspects that can be experienced independently.

7.2.2.2 Concurrent aspects

Awareness of being the cause and initiator of action, however, does not exhaust all there is to the phenomenology of agency. Another element that enters into it is the extent to which I am controlling my ongoing movements and their consequences in the external world. Even when I know or feel to have initiated an action, if that action happens to go awry, my sense of agency may diminish or disappear. Again, we can distinguish between intention-related aspects and those deriving from motor processes.

Sense of control. When my movements unfold smoothly and in the absence of environmental perturbations that need to be corrected for, I may experience this as the feeling of being in control of my movements or as a certain fluency in my action. A person walking on ice for the first time, for example, will not experience the same stable and harmonious flow of bodily activity she normally associates with walking. Even though she is aware of her intention to walk, and knows how to do so in different circumstances, she will experience her failure to control her movements as a lack of agency for her walking. Or consider the case of riding a bicycle without freewheel mechanism for the first time, in which case it may literally feel as if the bike is in control of my actions, rather than myself. In short, before yet considering whether my action is efficacious in bringing about the desired result, my movements can be clumsy, awkward, or inefficient, and the extent to which they are this may impact on my sense of being the author of these actions. To clarify, what we refer to as the pre-reflective sense of control relates primarily to unconscious processes of regulation and the exercise of fine-grained adjustments and corrections, to micro-choices about alternative ways to perform a particular action, and so on. Even the most settled skills are not strict automatisms, like reflexes. Rather, skilled performance is characterized by an openness to contextual differences and specific requirements of the here and now, as well as by our non-conceptual, continual involvement in guiding our actions (Fridland 2014; Dow 2015). According to Harry Frankfurt, the "sense of our own agency when we act is nothing more than the way it feels to us when we are somehow in touch with the operation of mechanisms of this kind, by which our movements are guided and their course guaranteed" (1978, p. 160).

Sense of achievement. Intentional aspects enter in agency awareness during ongoing action through my concern for the desired outcome of an action, which results in another form of feeling in control (or not), which we here refer to as the sense of achievement. An inherent property of intentional actions is that they are directed at achieving a meaningful effect, and the extent to which I am successful in achieving this effect may influence the level to which I feel to be the agent of the action. For example, if I play tennis with a racquet different from the one I commonly use, I will still be able to perform and control my swing as usual. But owing to a difference in the string tension, the ball may not react the way it normally does and adversely affect the quality of my play. Without necessarily knowing this to be the case, and while still enjoying a sense of control over my movements, I may nevertheless experience a diminished sense of agency over my play.

Two points should be highlighted here. First, for the experience of the intentional aspect of ongoing action it is not required that I am consciously monitoring the success or failure of my action. As Gallagher emphasizes, when "I reach to grasp a cup in order to take a drink, there is no conscious judgment required—that is, I don't have to constantly verify that my reach is going in the right direction, or that my hand is shaping the proper grasp for the cup" (2012, p. 12). But I may nevertheless have a vague feeling that my action progresses as it should. Second, it is not clear to what extent the sense of accomplishing one's goals contributes to the feeling of actions as being mine and initiated by me (i.e., to the stricter concept of agency). Even when my actions fail, I am not normally in doubt as to whether it was me who was responsible for the incorrect movement. Nor do I feel that the environment made me do it or that it was in fact a passive movement. In such cases, I am in doubt only about whether I am in control of external events that I perceive indirectly, for example, of the tennis ball hitting the desired spot. In other words, I may doubt whether I am in control of the situation as a whole, but not about the nature of my own movements.

Interestingly, many experimental studies fail to discriminate these senses of agency. In some experimental setups, for example, subjects are asked to control events on a screen, while the visual representation of their movements is distorted to varying degrees (e.g., Farrer et al. 2003). In other variations, the observed movements, in some instances, may be controlled by another subject or by the experimenter (e.g., Farrer and Frith 2002). Arguably, if and when subjects experience a diminished feeling of control in these experiments, what they would refer to is a lack of control over what happens on the screen, not a doubt about whether the actual movements they produced were under their control. In other words, I may recognize movements observed on a screen as actions (i.e., as goal driven), and I may doubt whether the observed action corresponds to the action I am performing myself, or whether I control the events on the screen. But it is unlikely that I would feel as if not in control of my own actual movements.

What is the relationship between the different aspects of pre-reflective action awareness? It is clear that, although we can draw a distinction between them, under normal circumstances we do not experience these aspects in isolation. They rather seem to enter into a unified, qualitative and diffuse sense that I am the author of my actions and that I am in control. Moreover, as Marcel (2003) argues, few if any of the individual aspects

Table 7.1 Phenomenological aspects in the sense of agency along two qualitative dimensions: initiation versus ongoing action; and intentional versus movement-related aspects

Prospective/Action initiation	
Intentional aspects	Anticipatory awareness of one's intention (before movement)
Movement-related aspects	Sense of being the initiator of one's movements; sense of urge
Concurrent/Ongoing monitoring and control	
Intentional aspects	Sense of achieving one's intended goal, of completing a task
Movement-related aspects	Sense of currently being in control of one's actions

seem to be strictly necessary for the experience of agency. For example, neither a sense of ownership based on proprioceptive awareness (as in deafferented patients), nor conscious awareness of the reasons for our actions (e.g., anarchic hand) seem to be required in order to feel that we are performing intentional actions. At the same time, most of the different aspects may be sufficient on their own to create a sense of agency. For example, merely being aware of one's intention to move (or of one's attempt to initiate a motor response), can be sufficient to feel that one is engaged in action, as is the case in anosognosia for plegia. Like Marcel, Gallagher (2007, 2012), as well as Synofzik et al. (2008a, 2008b), therefore defends a view of the sense of agency as involving a complex integration of afferent, efferent, and intentional forms of feedback.

Without claiming to provide a complete catalog of the phenomenology of action awareness, we summarize in Table 7.1 what we take to be some of its most important phenomenological aspects. We will later map these different aspects to specific microgenetic processes involved in an enactive and sensorimotor-based theory of agency. The experiential distinctions we have introduced here are not meant to be exhaustive. Pacherie (2007b), for example, proposes a more fine-grained hierarchy involving the experience of intentional causation at different timescales. Christensen et al. (2015) work out further phenomenological distinctions at the reflective level for the case of skilled athletic performances such as mountain biking. Hoffding (2014) provides a detailed phenomenological account of joint musical performance. Nevertheless, the categories summarized in Table 7.1 capture the broad structure of the minimal, pre-reflective sense of agency, the origin of which we aim to elucidate in this chapter.

7.3 **The comparator model**

Current theorizing about the sense of agency is dominated by in-the-head approaches that emphasize computational processes in the brain that produce and consume representations, such as facts about my intentions and the actions I perform. A common model is based on the concept of efference copy, introduced in the early 19th century (Steinbuch 1811) in the context of motor control. The concept refers to the idea that appropriately transformed copies of motor commands can be used to modulate afferent signals, in order, for example, to cancel out sensory signals resulting from self-generated movements (von Holst and Mittelstaedt 1950). An updated version of this proposal (Figure 7.1) postulates detailed internal models that predict the proximal and distal sensory consequences of an action, and which can serve in comparisons with either actual (estimated) sensory feedback or the desired state of the agent (for predictive in-the-head models of cognition in general, also see Grush 2004; Knill and Pouget 2004; Seth 2014; Wolpert, Doya, and Kawato 2003).

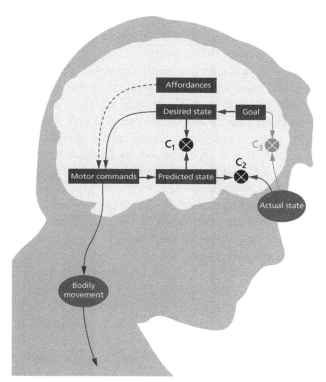

Figure 7.1 The computational motor control hypothesis of action awareness: the comparator model. A person's representations of his intentions or goals are first translated into a desired state (bodily or world). Using an internal inverse model, motor commands are computed that are appropriate for achieving this desired state (motor commands). The resulting bodily movement leads to sensory feedback (proximal and distal), from which the actual state of the body and world are estimated. The result of the comparison between predicted and desired state (C_1) is supposed to underlie the sense of control, while the comparison between actual and predicted state (C_2) allows for self-attribution of sensory events. Awareness of action-effects (the sense of achieving what was intended) involves the comparison between one's goals and the state of the world (C_3). In patients with anarchic hand or utilization behavior, movements are triggered directly by environmental affordances, which override the subject's conscious intentions.

Frith and colleagues (2000a, b; see also Synofzik et al. 2008b), for example, propose that a match between predicted and desired state underlies a sense of being in control (comparator C_1 in Figure 7.1). In anarchic hand, for instance, the patient's movements are triggered directly by environmental affordances, rather than what the patient holds to be her intentions. Predictions made on the basis of involuntarily executed motor commands would thus differ from the desired state as derived from the subject's goals. The resulting discrepancy could hence be experienced by the subject as a lack of control over her own actions. Similarly, in patients with delusions of control, the predicted state is supposed to be unavailable for monitoring. Again, a false discrepancy is thus generated at the comparator, even though awareness of intention and action execution is normal. The absence of an awareness for having initiated actions, so the authors suggest, is tied to an absence of predicted action consequences before movement.

The comparison of predicted and actual state, in contrast, allows for the attribution of sensory events to oneself (comparator C_2 in Figure 7.1). In this scheme, if the actual state matches one's prediction, i.e. if events that have occurred can be "explained" completely by what one has intended, then such events are attributed to one's own agency. The absence of any discrepancies at the comparator manifests itself in the attentive recessiveness of the feeling of agency. By contrast, if actual sensory feedback is incongruent with sensory predictions, external causation of the sensory stimuli can be inferred (ex-afference). According to Frith et al. (2000a), in patients with anosognosia, motor commands normally initiating actions do not lead to any movement (due to paralysis), while predictions are still made based on their motor intention. The disorder, in this case, results not from the corresponding discrepancy between predicted and actual state itself, but from the patient's inability to register this discrepancy.

According to the comparator hypothesis, the pre-reflective feeling of agency is thus a diffuse sense that results from the matching flow of predictions and sensory feedback, which occurs at the subpersonal level and is phenomenologically recessive, in the sense that one normally becomes aware only of disruptions of the process, that is, when one's anticipations are not met (Synofzik et al. 2008a).

In proposing a computational mechanism underlying sensorimotor self-awareness, the comparator model seems to fit naturally into the growing ecology of predictive brain theories (Friston 2010; Clark 2013; Seth 2014).

Despite its appeal, the model is neither sufficient nor necessary to explain the feeling and judgement of agency. Synofzik and colleagues (2008a) present some arguments to this effect. Firstly, subjects can attribute the same comparator mismatch in some cases to themselves and in others to the world. In other words, any potential mismatch has to be registered and appropriately categorized by another process different from the comparator. Also, in order to learn the required internal models, that is, to learn the sensory effects of its own movements, the system already has to somehow know which of its movements are caused by itself and which are not. Based on cases of pathological loss of action awareness, as well as neuroanatomical lesions in areas supposed to be involved in the comparator, the authors reach the conclusion that a much less specific congruence between efferent and afferent signals in general (e.g., between an action intention and a distal sensory effect, see comparator C_3 in Figure 7.1), alone or in combination with certain intermodal congruencies, suffices to explain the feeling of agency. This idea is supported, for example, by patients who feel agency for phantom limbs (Ramachandran and Rogers-Ramachandran 1996), that is, in the absence of proprioceptive feedback as well as efference copy. Similarly, in the "helping hands" pantomime task, passive subjects experience high degrees of agency for movements that are in fact performed by another agent, when only the other agent's hands appears in the place where the subject's hands would normally appear (Wegner and Sparrow 2004), in other words, when there is most plausibly no efference copy and no comparator involved.

Synofzik and colleagues conclude that it is not a specific, unique, and accurate prediction that underlies different forms of action awareness. Rather, all kinds of action-related perceptual and motor information such as efference copies, sensory feedback modalities, and their congruence, are combined in a multifactorial weighting process at different levels of cognitive processing, where the importance of the different authorship cues may vary with task, context, and person.

Apart from the question of whether comparators are necessary or sufficient for different aspects of action awareness, another problem the approach faces concerns the transition from subpersonal computations to an account of personal-level agency *experience*. In other words, how good is the comparator model from a neurophenomenological perspective? Note that it cannot be simply the presence of certain computations and resulting measures of mismatch that constitute the subjective sense of agency. If this were the case, then one would have to grant agency awareness also to thermostats and simple robots. Computations such as those involved in the comparator model may hence play at most an enabling role in the sense of agency. As Synofzik et al. (2008a) have pointed out, whether the results of such computations contribute to the experience of agency or not, must be decided by processes different from the comparator itself. The sense of agency, on this view, is thus not in fact an "*intrinsic* property of the action processing itself" (Synofzik et al. 2008a, p. 222).

The comparator model on its own does not offer an answer as to the experiential nature of agency awareness. To do that, it needs to go a step further, and explain how it is that we have first-person, subjective experiences at all, accompanying (and conditioned by) the proposed subpersonal computations. Although this aspect is less well developed, one proposal is that the result of the comparator enters into the creation of internal representations of "mineness" that become available to other cognitive subsystems (Synofzik et al. 2008b). In the case of the non-reflective feeling of agency, these representations are supposed to be non-conceptual, yet abstracted from ongoing action instances. They classify actions as caused by the self or not, but without explicit attribution. They are not compositional (they are not composed of parts representing the self and the fact of being the cause), and have no object-property structure. The reflective judgement of agency, in turn, is supposed to arise from further processing of these primary representations, now involving conceptual capacities and beliefs, which results in an explicit, propositional, and compositional kind of representation of actions as one's own (Synofzik et al. 2008b, p. 415).

Even if a meaningful interpretation could be given to the idea of non-conceptual representations, it is clear that whether one believes the comparator approach to be fruitful hinges on one's acceptance or not of experiences as fundamentally representational in nature (Metzinger 2000; Dretske 2003). In light of the enactive turn in cognitive science, these doubts and worries warrant the exploration of non-representational alternatives. We do not claim, of course, that all non-representational approaches to action are necessarily compatible with the phenomenology of pre-reflective agency described in Section 7.2. But we believe that the following proposal not only matches this phenomenology well, but may also shed some light on its origin.

7.4 **An enactive approach to the sense of agency**

As we have argued above, cognitivist models aim at a view of action awareness in which the experience of agency arises as an intrinsic aspect of action itself (at least in the case of the pre-reflective feeling of agency). It is arguable whether this attempt could be successful, given that it relies on the explicit construction of internal representations from brain-side computations that may or may not be required for the sense of agency. We propose here an alternative that synthesizes our theoretical developments up to this point, and which views a cognitive agent as essentially an integrated ecology of sensorimotor skills.

7.4.1 **Elements of an enactive proposal**

The premise of computational approaches to agency experience, of which the comparator model is one example, is that this experience is epistemically based on the co-occurrence of actions and their typical sensory consequences (i.e., sensorimotor contingencies). This assumption, combined with a certain sensitivity to non-typical consequences, is the precondition for the sense of agency in these approaches. However, in contrast with the comparator model, the general idea of a dependence of the sense of agency on sensorimotor contingencies (more precisely sensorimotor schemes) can also be developed in a nondualistic and non-representational manner, in which agency experience is truly intrinsic to the performance of action, rather than involving higher level representational mechanisms that consume the results of enacted sensorimotor contingencies.

The idea behind such a development is to apply the lessons learned from the sensorimotor approach regarding *perceptual* experience to the case of *agential* experience. Remember that the sensorimotor approach postulates that both the "content" and form of experience, what one perceives and how, is constituted by the exercise of one's embodied know-how of the relevant SMCs. The experiences associated with the various sensory modalities, or with distinct aspects of the environment (e.g., colors or sounds) differ, because different sensorimotor regularities are involved in, say, seeing and hearing. Since the different qualities of perceptual experiences derive from different modes of interacting with the world, they are *relational* in nature, rather than properties or states of the brain or mind. In particular, the regularities of interaction they derive from need not be available to the agent representationally. The enactment of SMCs is sufficient for an agent to enjoy their corresponding experiential qualities. Pre-reflective agency experience may be of the same nature. We propose that the different aspects of the sense of agency derive from the way that sensorimotor skills are enacted. In this case, however, not from regularities in the stream of sensorimotor signals associated with the performance of a particular sensorimotor scheme, but from the way in which this scheme is selected, initiated, and controlled. Like SMCs in perceptual experience, these processes of sensorimotor equilibration involve the world constitutively, and as such, the sense of agency is relational in nature, rather than something that can be reduced to representations in the head.

But the idea of a world-involving, relational, and sensorimotor-based sense of agency would still not be sufficient to address all of the weaknesses we have identified in the comparator model. This is because the sensorimotor approach explains the origin of differences between sensory modalities or instances of perceptual experiences, but as Thompson (2005) has noted, it does not address the question of why we have experiences at all (i.e., the problem of "creature consciousness" or the problem of what makes an entity capable of being phenomenally conscious). In the same way that a self-guiding missile, despite claims to the contrary (O'Regan and Noë 2001), does not have genuine mastery of airplane-tracking, and hence no genuine perceptual experience of "seeing" airplanes, the sensorimotor account alone is insufficient to describe the kinds of systems that can experience themselves as agents. It is necessary but not enough to claim, as Hutto and Myin do, that "the phenomenological character of experiences must, ultimately, be understood by appealing to interactions between experiencers and aspects of their environment" (2013, pp. 176–7). Defining *agency experience* as a relation between the acting subject and its environment may eliminate the need for an explanation of the otherwise mysterious transition from physical happenings to experiential phenomena. But this only reframes the problem as the question of what an "experiencer" is in the first place, in other words, what is it about such entities that they can have experiences? We therefore need to enrich our sensorimotor-based approach to the sense of agency with the enactive concept of *sensorimotor agency*, introduced in Chapter 6, that is, with a naturalized concept of a sensorimotor-based *subject*, with an intrinsically defined perspective, able to engage in intentional actions subservient to her self-generated desires and norms.

Lastly, we will relate specific phenomenological aspects of the sense of agency, such as the sense of effort or control, to the microgenetic processes involved in the equilibration of an autonomous sensorimotor network, by drawing on our interpretation of Piagetian sensorimotor learning as developed in Chapter 4.

7.4.2 Sense of agency as sensorimotor self-synthesis

We are now in a position to describe the kinds of systems that may experience themselves as agents and to elucidate the origin and phenomenology of the various aspects of the sense of agency. The starting point is the concept of sensorimotor agency. According to the enactive proposal, a sensorimotor agent is constituted by a self-asserting network of

mutually enabling sensorimotor schemes; a network whose stability is constantly challenged by environmental and bodily demands, and which undergoes adaptive processes
of equilibration as a result. Such an organization satisfies the three minimal requirements for agency (self-individuation, interactional asymmetry, and normativity) and
may thus be said to constitute an identity that emerges at the sensorimotor level, bearing its own concerns, acting on its own behalf, and entertaining its own perspective: a
sensorimotor subject. What and how this agent experiences perceptually is determined
by her sensorimotor repertoire. Crucially, we propose that how the agent experiences
her engagements with the world, i.e., whether as self-driven and self-controlled or not,
is determined by different modes of equilibration both within and between sensorimotor schemes, and by how these modes of equilibration re-assert the individuation of the
sensorimotor subject.

For the sake of clarity, and to match the different aspects of the sense of agency mentioned at the beginning of this chapter, we proceed by describing separately aspects associated with the initiation of actions and those associated with their ongoing control.

7.4.2.1 Prospective aspects

As we have already mentioned, aspects of action initiation and control in agency experience are separable, as demonstrated, for example, in pathological cases involving subjects
that report a sense of agency for intended movements that are either not executed at all
(anosognosia for plegia), or initiated without control (deafferented patients). But before
addressing the phenomenology of action initiation, we first need to describe in more
detail the processes we believe to be involved in selecting, engaging, and controlling sensorimotor schemes.

Everyday skillful coping relies on the deployment of networks of interlocking sensorimotor skills as we saw in Chapters 4 and 6. But what do we mean by deployment of a skill?
How exactly does a person engage a particular sensorimotor scheme in his repertoire? For
example, sitting at my desk, what determines whether I reach for the computer mouse
to open my inbox in response to an incoming email, or rather for the glass of water to
quench my thirst? Let us outline a plausible answer. Though constitutively involving the
environment to close the sensorimotor loop, sensorimotor schemes are supported on the
brain side by metastable neurodynamic structures in conjunction with metastable states
in other non-neural systems in the body (see e.g., Tognoli and Kelso 2014; Kelso 2016).
Like dynamic fields (Erlhagen and Schöner 2002; Schöner and Dineva 2007), for example,
such neurodynamic patterns may be characterized by a continuous distribution of activation, with neural assemblies supporting different schemes exhibiting localized peaks in
this distribution (say, one for mouse control and one for grabbing the glass). The formation of neural activation peaks, and more generally the activation of schemes, may involve
various sources, as well as the competition (e.g., through mutual inhibition) between the
different neural assemblies involved. If the activation of any assembly comes to dominate,
the corresponding scheme is engaged, in other words, the corresponding neurodynamics
are coupled with the motor system, and with the world. If the environmental conditions

are supportive, and the motor system is in a compatible state (e.g., not prevented from moving), then the corresponding scheme is enacted.[1]

Now, different sources of activation may prime, potentiate or inhibit individual sensorimotor schemes (or rather their corresponding neural assemblies), for instance, by raising or lowering their excitability. Such sources include internal signals related to volition and readiness (my thirst), as well as environmental affordances (the presence of the glass or computer mouse). In a general sense, the enactment of a particular sensorimotor scheme depends in normal circumstances on a resonance between external and internal conditions related to the agent's desires and needs. Bodily states, history, dispositions, and external factors may resonate with the water-drinking scheme more intensely or sooner than with the email-reading scheme, and it is this resonance between agent and world that breaks a symmetry in the critical state and tips the balance. As the agent is involved in regulating its coupling with the environment, an activity that in turn affects her own states, it is conceivable that she may be able to influence the process of selection of sensorimotor schemes, as a form of higher order regulation. This may take place through environmental mediation (e.g., the arrangement of a work-environment to encourage "good habits"). But it may also involve internal "gestures" that differentially amplify specific neurodynamic patterns. This is clearly the case in agents possessing linguistic abilities and capable of using them for their own self-control (Cuffari, Di Paolo, and De Jaegher 2015). On lower levels this may include the kind of volitional, readiness or appetitive signals already mentioned. In both these cases—via alterations in the world, or via internal gestures—the whole agent is involved in the process by encouraging certain sensorimotor schemes to resonate with the situation while others are avoided or suppressed (see also Fuchs 2011). We propose that aspects of agency experience associated with the initiation of movement, identified in Table 7.1 as the awareness for one's intention, and the awareness for having decided to engage (now) in action, can be related to these dynamics of sensorimotor scheme competition and activation.

Awareness of (prior) intention. For instance, the emergence of one or more preferred schemes (the passage from metastable to more stable neuro- and body dynamics), even if not yet activated at the muscular level, can nevertheless have real covert consequences for the agent. This corresponds to the agent being in an *anticipative state*, in which the activation of a scheme may influence other processes in the agent's cognitive organization. In the network model of interrelated sensorimotor schemes introduced in Chapter 6 this anticipative state corresponds to the activation of one or more nodes, which represent schemes appropriate in the current situation. This is illustrated in Figure 7.2. Because of the networked nature of the schemes, the activation of one of them has the result of

---

[1] The dynamic field analogy is just one possible way of conceptualizing this relation between different metastable neural and bodily states, rich in potentialities, which according to the circumstances can be differentially actualized in the initiation of a particular action. Other models also fit this account in terms of critical states (see e.g. Kostrubiec et al. 2012; Wallot and Van Orden 2012).

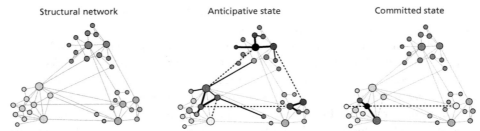

Structural network   Anticipative state   Committed state

Figure 7.2 Stages in the selection and engagement of sensorimotor schemes. Part of a network of sensorimotor schemes (as introduced in Chapter 6) is shown at three points in time. The leftmost network depicts an ideal initial state where the sensorimotor agent is not engaged in any activity at all. It shows all possible (structural) relations between schemes. As a result of internal demands, and/or in response to environmental solicitations, several schemes may become latently activated (intensely colored nodes in middle figure). At this stage the agent has not yet settled on a final response, and various schemes, including those belonging to different activity clusters, compete concurrently for engagement. In this "anticipative state," the agent is sensitive to the various options of response and their consequences. This is shown as the spreading of activity to other (neighboring) schemes in the network. Through mutual excitation and inhibition, a particular scheme eventually comes to dominate (right). Even when not engaged yet, the selected scheme continues to influence the organization as a whole, for example, by priming or inhibiting potential successor schemes. (See color plate.)

inhibiting other competing schemes as well as facilitating or priming others, for example, those that constitute viable alternatives. The activation of a scheme thus sends ripples across the network, changing the balance of which other schemes are likely to be enacted now or in the immediate future. The totality of such changes in an activated scheme's "neighborhood" forms a kind of sensorimotor context through which the agent becomes sensitive to the scheme itself as well as its consequences. In this way the agent is pre-reflectively aware of processes and events about to ensue, and we identify this as coherent with the intentional aspect of action awareness (the awareness of having the intention to move). For example, when a tennis player forms the intention to return his opponent's serve, he is already prepared to move in a certain way (e.g., by setting up required motor primitives and synergies). But at the same he is still sensitive to the different ways the ball may be served, as well as to the different options for his opponent's reaction to his upcoming return. There is a narrowing down of possibilities, which nevertheless remain open, as the selection of the scheme to be enacted is still ongoing. The particular manifold of possibilities and consequences associated with the process of selection of a specific return scheme is experienced by the player as his intention to enact that return.

Awareness of action initiation. The irreversible activation of a scheme, in contrast, implies a commitment to move. Now the agent is less sensitive to the possibilities and tendencies for actions that may or may not occur. Instead, his activity is funneled toward the execution of a single selected sensorimotor scheme, which again may lead to changes

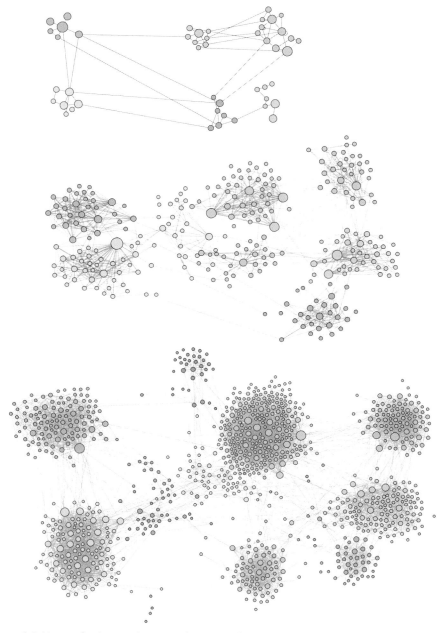

Figure 6.4 Networked sensorimotor schemes. A (hypothetical) sensorimotor network is depicted at three different stages of development. Initially (top), the network is relatively small, but different sensorimotor schemes already appear connected, forming clusters that correspond to types of activities. Individual clusters may themselves be organized differently, e.g., mostly hierarchically (pink) or sequentially (turquoise). Some pairs of clusters are more strongly connected than others. Over time (middle and bottom) clusters get progressively interconnected and differentiated as the network develops. A wider range of transitions (links) is available, activity bundles are split into smaller subunits, and so on.

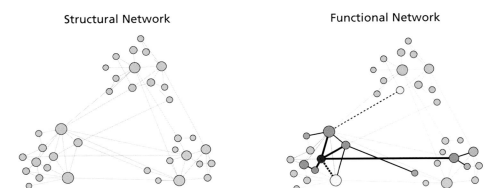

Figure 6.5 Structural and functional sensorimotor networks. Depicted is a small part of a network of sensorimotor schemes, consisting of three clusters of activities. On the left, the structural connectivity is shown. It consists of all vertical and horizontal links between the included schemes. On the right, the functional connectivity at a particular instance in time is shown. The currently active scheme is highlighted in red. Its activation makes transitions to some other schemes more likely (thicker solid lines) or less likely (thicker dashed lines). Through the transitional links, some neighboring schemes may already become pre-activated ("primed") or inhibited (shown here as a change in the intensity of the node's color), which in turn may lead to modulation of their own transitional links. Note that this is a simplified picture. In reality there can be more than one hotspot of concurrently active schemes at the same time, and their activation not only involves the influence of neighbors (which in addition may be path dependent), but also the world.

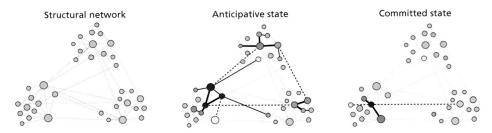

Figure 7.2 Stages in the selection and engagement of sensorimotor schemes. Part of a network of sensorimotor schemes (as introduced in Chapter 6) is shown at three points in time. The leftmost network depicts an ideal initial state where the sensorimotor agent is not engaged in any activity at all. It shows all possible (structural) relations between schemes. As a result of internal demands, and/or in response to environmental solicitations, several schemes may become latently activated (intensely colored nodes in middle figure). At this stage the agent has not yet settled on a final response, and various schemes, including those belonging to different activity clusters, compete concurrently for engagement. In this "anticipative state," the agent is sensitive to the various options of response and their consequences. This is shown as the spreading of activity to other (neighboring) schemes in the network. Through mutual excitation and inhibition, a particular scheme eventually comes to dominate (right). Even when not engaged yet, the selected scheme continues to influence the organization as a whole, for example, by priming or inhibiting potential successor schemes.

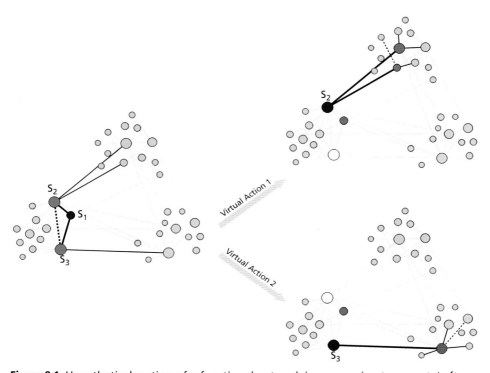

Figure 8.1 Hypothetical section of a functional network in a sensorimotor agent. Left: A single scheme (S_1) is currently enacted, as indicated by its black shading. At the same time, through the influence of S_1 as well as the potential contribution of environmental solicitations, two other schemes are already "primed" (S_2 and S_3, darker shade). This could involve, e.g., the "subthreshold" activation of the support structures in the agent required for these schemes. Though S_2 and S_3 have no overt effect yet, their latent activity may nevertheless already influence yet other schemes in the network (illustrated as thicker connections to nodes in different clusters), which may as a result become more likely to be enacted in the future. These "primed" schemes constitute a strongly anticipative context for the current action, which thereby extends its horizon of influence into the future. The two schemes S_2 and S_3 are also mutually inhibiting and thus they compete with each other for enactment. **Right:** activity shifts to either S_2 or S_3 and one of the two schemes is enacted (top and bottom). Several events in the agent can cause the tipping of the balance between the two competing schemes. This may be the influence of other active nodes in the network on S_2 or S_3 (not shown), an uneven influence of S_1 on the two schemes, or a more holistic act, such as the agent changing the relation to the environment with the result that one affordance becomes more salient than another. All these are examples of virtual actions, and their result is the shifting of activity in relevant parts of the network. Note that the agent-side structures involved in previously enacted schemes may continue to exhibit influence on other parts of the network, even when no longer having any overt effects. Such "lingering" activity extends a currently enacted scheme's horizon of influence into the past, which current activity can resignify.

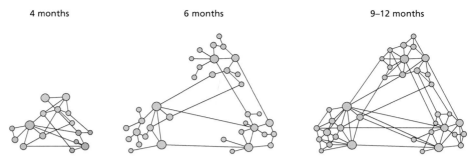

Figure 8.2 Schematic (hypothetical) representation of the scaffolding effects. These correspond to mother-infant interactions on the infant's sensorimotor network during the first year. The first panel shows the highly synergistic schemes that infants of about 4 months of age bring to an object. As the mother solicits bids for attention to herself and other objects, these schemes begin to diversify (middle panel) into new, loosely linked clustered schemes. The mother now encourages integration of these schemes during play by sustaining attention and manipulations on single objects and on relations between objects and persons. This results in the network growing increasingly consolidated as the schemes for attention, object manipulation, shared attention, social acts, etc., get progressively integrated (third panel).

in other parts of the organization, such as the priming of sensorimotor schemes that usually follow the current one (rightmost network in Figure 7.2). The tennis player's engagement of the return action may already have primed his subsequent move toward the net, for instance. In our network model, the selection of a single scheme corresponds to the concentration of activity onto a single node, and to the corresponding collapse of the context that constitutes the agent's sensitivity to the consequences of different action choices. We identify this transition from a narrowing down of dynamical possibilities to the individuation of the selected scheme (the *committed state*[2]), and the corresponding real consequences for the agent as a whole, with the awareness for action initiation. The associated collapsing of possible alternative courses of action is experienced by the player as an urge to perform the return.

We may also picture the temporal dynamics of the selection and activation of a scheme as the emergence and shaping of a particular basin of attraction in the agent's sensorimotor space. The structure of the scheme's basin determines the agent-environment synergies, or constraints on their coupling, that are appropriate in the given task context. This poises the agent in an anticipative state rich in equally valid propensities (paths) to respond, all of which share a common end. This is the intentional aspect of the emerging action. A slight change in circumstances (internal or external) then can be enough to break the system's symmetry and select one among the manifold of possible actions. This is experienced as action initiation. Wallot and Orden (2012) defend an analogous view to explain cases of so-called ultrafast cognition, i.e., cognitive performances with response times as fast as or faster than the estimated physiological limits of information processing. This is evidenced, for instance, in visual or acoustic go/no-go tasks in which participants hold down a key which they release as soon as particular stimulus appears (e.g., a natural scene with an animal, as opposed to one without animals). The minimum response times for reliably correct responses in such studies (some of which include a training phase) can be as low as 200 ms which is estimated to be about the time needed for neural activation to propagate from eye to finger, without any time for information

---

[2] The concept of commitment, seemingly vague, has been measured in simulation models of minimal cognition and indirectly in neuroscientific studies. When a simulated agent must choose between alternative behaviors, one can systematically study the effect of variations in the agent's or environmental states on the choice made by the agent. One can explore this effect over time and find counterintuitively that commitment to a choice is not a monotonic quantity that increases progressively as the time for execution approaches but that instead it can fluctuate in complex ways (Beer 2003). Alternatively, one can "freeze" time just before the simulated agent has made its choice and ask the question: What if things were slightly different?, then quantify the effect of changes to the environment on the decision—the larger the effect, the less the commitment to the action that is nevertheless actualized (Iizuka and Di Paolo 2007; Di Paolo and Iizuka 2008). In the case of humans, the stability of neural cortical patterns may be measured indirectly in single perceptual trials by analyzing the statistical variability corresponding to different cortical regions. Evidence shows that low variability or other measures of coherence between recorded patterns indicating a more stable global state correspond to the emergence of a commitment to a conscious perceptual choice (Schurger et al. 2015; Varela et al. 2001).

processing. Ultrafast compensation is also found in quick adaptive reactions to external perturbations. For example, Kelso et al. (1984) asked participants to say either /baeb/ or /baez/. Without warning, a force is applied to the jaw during the closing gesture of the utterance. This externally provoked displacement of the jaw is compensated within five to ten milliseconds, which is faster than a loop of neurons could have calculated a new sensorimotor configuration. Within this short timeframe, the lower lip moves to a new configuration that preserves the intended utterance.

Such evidence suggests that instead of processing incoming information as a pre-condition for making a decision or articulating an adaptive response, brain and body are already poised in a critical anticipative state appropriate to the current context and whose "symmetrical" stable resolutions are sensorimotor synergies that correspond to the context-relevant behavioral options. The symmetry is broken by the environmental coupling; one of the options is selected; and the corresponding sensorimotor scheme enacted. Cases of ultrafast cognition, thus, count as particular cases of our interpretation of sensorimotor scheme dynamics of selection and initiation. A crucial addition in our account is that the emergence of a sensorimotor scheme's basin of attraction, and the symmetry-breaking, also have covert consequences on other parts of the cognitive organization, which results in the agent's awareness of intention and action initiation respectively.

The general view of a prospective sense of agency arising from processes of scheme selection is also supported empirically by experiments in which agency experience correlates with the dynamics of action selection. For example, Chambon et al. (2014) find behavioral and neural evidence that fluent, uncontested action selection contributes positively to the sense of agency, while conflicts between alternative possible actions (induced in their experiments by priming subjects subliminally with alternative action "instructions") reduces the subjective feeling of control over actions (also, see Wenke, Fleming, and Haggard 2010). In our dynamical view, this corresponds to situations in which two or more equally appropriate schemes compete for activation, and in which this competition is resolved only with great difficulty. One can imagine, for example, two conflicting schemes being equally active until the last moment, when an external trigger breaks the symmetry (the action can no longer be delayed). Such situations can be induced in somewhat artificial situations, such as experimental setups involving little additional context to guide action selection and enforced responses, but may also occur in everyday life. We can speculate that due to the late resolution of the scheme competition their covert consequences for the rest of the network do not play out as they would if the final action had been selected more rapidly or fluently. Because no single scheme emerges early enough, the context of possible consequences that they create through their influence on the remaining network is equally undefined, or unusually broad. The person experiences this ambiguous situation as a lack of awareness for his intention, which contributes to a reduction in the overall sense of agency.

7.4.2.2 Concurrent aspects

Other aspects of pre-reflective action awareness, such as the sense of being in or exerting control or the sense of achieving the desired effect, we can relate to ongoing processes of assimilation and accommodation as the selected sensorimotor scheme unfolds. To motivate this idea, consider the following parallel. The sense of agency is traditionally understood as originating in a kind of self-monitoring, that is, a diffuse awareness of the effect of one's actions on the environment and of one's body. This is analogous to the requirement of adaptivity for minimal agency, namely, an agent's monitoring and adaptive regulation of the consequences of its activities for its own viability. In the context of sensorimotor agency, we have developed Piaget's theory of equilibration as a possible form of adaptivity in Chapter 4. Our suggestion is that some aspects of the sense of agency derive from the modes in which a sensorimotor agent monitors and adaptively regulates its interactions such as to stabilize and deploy a precarious network of sensorimotor schemes. This involves both the metastability of sensorimotor schemes and the metastability of their mutual relations, i.e., aspects intrinsic to individual actions and aspects concerning how schemes relate to each other (e.g., whether they tend to go together, in parallel, or in sequence, or whether they tend to be incompatible, and so on). To be more specific, our proposal is that, rather than originating in particular relations between sensory and motor signals (as in the case in perceptual experience), the aspects contributing to the sense of agency throughout an action cohere with processes of assimilation and accommodation in a sensorimotor agent; in other words, they originate in specific relations within and between sensorimotor schemes.

Ongoing sense of control. For example, we consider the diffuse and attentively recessive feeling of agency to be the experiential consequence of a network of sensorimotor schemes successfully assimilating the current environmental context, i.e., actively engaging with it. In other words, the absence of any perturbations to the dynamic equilibrium of the enacted sensorimotor schemes is the feeling that "everything is going according to plan," (i.e., of being in control). According to the concept of sensorimotor agency, this is not simply a static state of affairs, but an actively sustained relation between sensorimotor schemes that mutually enable each other. The feeling of being in control can be further separated into movement-related and intentional aspects. The movement aspect refers to the "fluency" of my performance, which requires, according to the stability condition of assimilation (see Section 4.5), that the pattern of environmental states and the pattern of the agent's states, when engaged in a sensorimotor coordination, are mutually stabilizing. If this is the case, that is, in the absence of *obstacles* that would lead to perturbation of my interactional stability (and to violation of the stability condition), both agent and environment play their proper parts, and the sensorimotor coordination fulfills its role in the enaction of the scheme as a whole. The absence of obstacles and the corresponding stability of the sensorimotor coordination are experienced as fluency in my interaction, which contributes to the feeling of being in control of my action in this particular moment.

While this sense of agency is phenomenologically recessive, we do become more consciously aware of situations where our interactional stability is challenged and an assimilation fails (or threatens to fail). For example, in the case of obstacles, the environment or my body do not "play along" as they usually do in my enactment of a sensorimotor coordination. When aiming to dribble a ball, for instance, I may try to engage in the usual sensorimotor pattern. But if the ball happens to be somewhat deflated, or my hands sweatier than normal, this might result in an unstable coordination and prevent me from dribbling correctly. In this case I experience a lack of, or a perturbed, sense of control for my action. Failures of assimilation usually rise to the foreground of my attention, which may have the consequence of directing attempts to accommodate the source of the perturbation (e.g., adjustment to a ball with different pressure or wiping one's sweaty hands).

Intentional sense of achievement. Another contribution to the feeling of being in control of one's actions is the sense that I achieve the desired effect, or that my interaction proceeds in the right direction. In the language of assimilation, this means that the *transition condition* is satisfied (see Section 4.5), in other words, that my involvement in stable sensorimotor coordination A leads in time to the production of coordination B, where B is the next stage in the sensorimotor scheme comprising the individual patterns of coordination. In our example of dribbling a basketball, a feeling of achievement or successful procession may result, for instance, when my pushing of the ball toward the ground is followed by the ball returning to my hand in the correct way. This intentional aspect of feeling in control does not require explicit monitoring of goals and comparison with distal feedback about the state of the world. Rather, it results directly and pre-reflectively from the successful closure of the sensorimotor scheme (the feeling of successful dribbling). However, there is here also a potential for failure of assimilation, namely in the case of *lacunae*. For example, I may be successfully engaged in what used to be the correct sensorimotor coordination given my current context (I perform the right arm movements to dribble a basketball). But because of changes in the structure of the environment (say, a loose floorboard), or changes in my own body, my enacting of the coordination does not lead to effects appropriate for achieving the desired outcome; in other words, the transition to the next coordination in the scheme's own organization fails (the ball does not return to my hands, which prevents me from catching it). I will experience this as a lack of agency for the outcomes of my actions, or as a disturbed sense of achievement.

Sense of exerting control. Another aspect of agency awareness can be associated with features of the ongoing process of active equilibration of sensorimotor schemes. Though failure of assimilation may in the first instance raise awareness that I am not in total control of my interactions and their consequences, a different sort of awareness can arise from the *effort* I have to exert in order to re-equilibrate, that is, to counter perturbations and re-establish a "smooth" interaction. In other words, this aspect of the sense of agency corresponds to the amount of self-correcting activity involved in achieving increased equilibration. As this activity diminishes whenever equilibration approaches its maximum condition, the intensity of the effort-related aspect of agency experience also diminishes. If this is the case, a prediction would be that if we were to perform a very repetitive task, such as the folding and stapling of dozens of copies of a paper hand-out for the delegates of a conference, our sense of effort should be greatest at the beginning, since we have to deploy more regulatory resources to establish a stable interaction to begin with. But as we repeat the actions, and as long as we do not suffer from fatigue, these would come to feel more and more automatic. Once the fluent interaction is established, and not challenged by unexpected events, the initially strong sense of effort is thus replaced by a less intense feeling of being in control (and though the actions may feel "automatic," it never feels as if somebody else is in control).

In this respect, Nahmias (2005) and Pacherie (2007b) acknowledge a certain ambiguity affecting the sense of agency in situations of effortless control. Pacherie suggests introducing a distinction between the feeling of being in control and the feeling of exerting control. The latter is what we may call the kind of agency awareness associated with progressive equilibration. As equilibration is achieved, the feeling of exerting control becomes ambiguous. Our prediction of a loss of intensity in this feeling during fluent, repetitive tasks (before we reach a situation of fatigue), we note, is quite different from what would be expected from the comparator model, since in this model we cannot introduce the distinction of being in control and exerting control. The better the match between predicted and sensed signals, the less the error in the comparator, and, therefore, the more reaffirmed or confirmed the sense of agency should become, contrary to what we suggest would happen according to our proposal.

In summary, we can relate the sense of being the initiator of an action to intentional and premotor processes involved in the selection and activation of sensorimotor schemes, and the sense of being in, as well as exerting, control to intentional and movement-related processes of assimilation and accommodation. The feeling of control can be further mapped to the type of assimilation condition being challenged, or the different forms of equilibration. We summarize the relationship between phenomenological aspects and microgenetic processes involved in the sense of agency in Table 7.2.

Table 7.2 Mapping of phenomenological aspects in the sense of agency to microgenetic processes involved in the selection, initiation, control, and equilibration of sensorimotor schemes

	Phenomenology	Processes of equilibration
Action initiation		
Intentional aspects	Anticipatory awareness of one's intention (before movement)	Activation of and selection between metastable SM schemes and the covert effects of the *anticipative state* on other parts of the sensorimotor organization
Movement-related aspects	Sense of being the initiator of one's movements; sense of urge	Engagement of single SM scheme, and covert effect of a *stabilized committed state* on other parts of the sensorimotor organization
Ongoing monitoring and control		
Intentional aspects	Sense of achieving one's intended goal, of completing a task	Satisfaction of the *transition condition*; lacking or perturbed sense of achievement in case of *lacunae*
Movement-related aspects	Sense of currently *being in control* of one's actions; sense of currently *exerting control over* one's actions	Satisfaction of the *stability condition*; lacking or perturbed sense of control in case of *obstacles*; adaptive processes of *equilibration (accommodation)*

Interestingly, the basic feeling of being in control that derives from successful assimilation (both the intentional aspect of achieving the desired effect and the state of fluency), is best described as an *absence* in our account. It is the feeling that my agency is *not* currently challenged by lacunae nor obstacles. This seems to be a good match for what is usually meant by the "attentive recessiveness" of the sense of agency.

Also note that the sensorimotor "congruence" required for this pre-reflective feeling of being in control is not one involving precise matching of predicted feedback, as in the comparator model. Rather, it depends on the satisfaction of the stability and transition conditions for assimilation, in other words, on the fact that agent and environment play the roles required for the enactment of the sensorimotor scheme to be successful and coherent with the rest of the network of schemes. It is thus not the registration of congruence between internal signals in the brain that is crucial, but the appropriateness of the chosen scheme given the current situation in agent and environment. The agent does not need to "find out" whether a sensorimotor scheme equilibrates by comparing signals. It is the enactment of the scheme itself that results in success or failure to various degrees (see discussion on intrinsic normativity in sensorimotor schemes in Chapters 4 and 6). An obstacle or a lacuna is manifested *directly* in the failure to equilibrate within a scheme or between one scheme and another. The manifestation is world involving and personal since it implies the agent as a whole. Also, since sensorimotor schemes are always enacted by the sensorimotor agent owning them, and their enactment in turn re-affirms the organization that constitutes the sensorimotor agent, the "first-person givenness" is already implicit in the processes that lead to the selection and enactment of a scheme. In other words, whether or not I am the agent of my actions, at this level of equilibration, is not a question of reflection and verification. Rather, *it is the enactment of the act that asserts the agency of the agent*.

7.5 Application to pathological cases

We now come back to some of the pathologies we mentioned in the beginning. This will serve to highlight further the differences between our world-involving enactive proposal and the comparator model as the main contender of in-the-head approaches. It also shows how we can derive additional testable predictions from the enactive proposal, even though it is still in need of further work, for example, in order to elucidate the neural mechanisms underlying processes of sensorimotor equilibration.

7.5.1 Passivity and schizophrenic delusions of control

As an example of the applicability of our proposal, consider the experience of passivity in schizophrenic patients with delusions of control. These patients do not disown the intentional aspect of their actions. They successfully follow instructions by an experimenter, and do not try to stop the affected movements. They report, however, not to have been responsible for initiating their actions, i.e., not to have consciously intervened in their execution (passivity). In other words, they have correct awareness of their intentions, but not of action initiation.

According to the comparator hypothesis, this aspect of agency experience results primarily from efferent signals, i.e., from the flow of motor specifications that are responsible for translating intentions into movement and occur before their execution. On one account (Frith et al. 2000a), it is suggested that the passivity experience arises from an impaired predictor of limb position. While the patient is aware of his intentions, and while these are translated into the correct movements and achieve the desired effect, in the absence of predictions at the comparator (both C_1 and C_2 in Figure 7.1), the patient is not aware of the precise specification for movement, and as a result, it is presumed, of his initiating it. The situation at the comparator is as if the movement were passive, and the afferent feedback is thus attributed to external causes rather than to the self.

From our enactive perspective, there is no need for central predictors to account for the awareness of action initiation.[3] As we have argued, awareness of intention and action initiation results from the real but covert effects that the selection and activation of sensorimotor schemes have on the rest of the agent's sensorimotor organization. In particular, the activation of a specific scheme places the agent in a state of that may constrain future actions by differentially modulating other schemes that usually follow (pre-activating some, pre-inhibiting others). Awareness of action initiation may be disrupted, resulting in passivity experience, when a scheme's activation no longer has the right secondary effects on the remaining sensorimotor repertoire, in other words, when the scheme is executed as if decoupled from the greater network. This view is compatible, for example, with speculations by Frith et al. (2000a) regarding the underlying neurological abnormalities in patients with delusions of control. The authors argue that the overactivity in the inferior parietal cortex, as observed in some patients, may be due to abnormal modulation via the anterior cingulate, which in turn is speculated to be involved in attention to future actions and suppression of inappropriate actions.

In short, pathological awareness of action initiation may be due to impaired linkages between different sensorimotor schemes, which in normal circumstances results in anticipative modulation of future actions. Such an impairment would not interfere with the actual execution of a scheme, which is compatible with the fact that schizophrenic patients have no problem with motor control itself. We could predict, however, that rapid execution of (unusual, non-automatic) action sequences might be impaired, since the selection of one scheme does not always correctly prime other relevant ones ahead of time. Though we are not aware of any experiments testing this specifically, there is evidence that schizophrenic patients, though not having problems with anticipative motor control per se, show indeed a deficit in the sequencing of motor actions (Delevoye-Turrell, Giersch, and Danion 2003). The idea that the experience of passivity is related to

---

[3] That is not to say that sensorimotor behaviors or schemes cannot be anticipatory in nature, but rather that each scheme, if required, may be intrinsically anticipatory rather than relying necessarily on detailed internal models (see Stepp and Turvey 2010).

a disassociation between parts of the sensorimotor repertoire, reducing the coordination between an enacted scheme and the network in which it is embedded, is also compatible with empirical findings that links schizophrenia with reduced functional integration between different components of the large-scale brain network (the "disconnection hypothesis"; see, e.g., Friston 1999; Tononi and Edelman 2000; Foucher et al. 2005; Lynall et al. 2010).

7.5.2 **Anarchic hand**

As described earlier, patients with anarchic hand syndrome perform purposeful actions, often triggered by environmental affordances, but they report that these actions are independent of their will. They recognize an action as intentional, and as being performed by themselves, but disown, so to speak, the intention behind the action. Synofzik et al. (2008b), explain the symptoms as a disturbed feeling of agency, accompanied by a false belief regarding the cause, a mistaken attribution of intentionality to the hand itself ("it does what it wants to do").

According to the comparator hypothesis, this pathology is explained by an inappropriate activation of representations in the parietal cortex that refer to objects in a person's environment in terms of the actions they afford. These representations are no longer inhibited by other, intrinsically motivated, actions that the patient holds to be intentional and is aware of (Frith et al. 2000a). Since representations for the predicted state (now derived from the involuntary affordance-triggered action) and the actual state are present (comparator C_2 in Figure 7.1), the patient can attribute the action to himself. But comparison of the intended state with actual state (comparators C_1 and C_3) would lead to discrepancy and the corresponding feeling of his actions not having been intended.

We can provide an alternative account on the basis of our equilibration hypothesis that does not involve explicit predictions. As we have described above, sensorimotor schemes are selected by the agent through their "resonance" with aspects of the current environment, such that those schemes compatible with environmental affordances will become latently activated. Which of these is ultimately performed depends on the agent's internal state, needs, and capabilities. This requires the agent's active intervention in the competition between sensorimotor schemes, which may occur internally and locally via modulation of the neurodynamic structures underlying specific schemes, or more holistically by regulating the overall coupling with the world such that different schemes may or may not resonate in the first place (or more or less so). We argue that it is this aspect of equilibration that may be affected in anarchic hand. The impairment affects the agent's control over part of her sensorimotor repertoire (at least at times) and prevents her active involvement in the selection of which schemes are enacted. As a result, some sensorimotor schemes may rise to prominence and be enacted almost arbitrarily (e.g., because of "subthreshold" fluctuations or environmental affordances; see, also, Schöner and Dineva 2007). The loss of the agent's ability to influence the selection of schemes may involve breakdowns in the communication between them (i.e., in their integration), such that enacted schemes do not influence other processes in the agent's sensorimotor organization as they would "normally" do. As a result the agent is not poised in the anticipative state we associate with the selection of a scheme, and therefore lacks sensitivity to the possibilities and tendencies implied by it. We interpret the lack of this sensitivity as the lack of a subject's awareness of her intention.

Note that the involuntarily engaged scheme(s) may proceed successfully and correctly assimilate the corresponding environmental conditions. The patient's pathology does not involve problems with ongoing motor control as such, it does not seem to affect the lower level of equilibration within a scheme. It seems that it is at a "middle-level" of organization that the pathology lies: in how schemes are initiated as if in isolation, without forming a richer context of secondary consequences through which the subject becomes aware of her intentions. Despite a lacking awareness of intention, and because the lower levels seem to function normally, the subject may nevertheless experience the diffuse feeling resulting from actions unfolding smoothly (though compared to the often frightening experience of "alienness" this aspect may go unnoticed).

A difference between this interpretation and the comparator model, apart from disagreement on the need for explicit predictions and representations, lies in the notion of intention. In the comparator model as it is usually presented, the subject is assumed to have a unique, internally generated goal or intention. But rather than acting on this "true" intention, patients with anarchic hand respond with "unintentional" movements triggered by environmental affordances. It is arguable how well this matches the observed phenomenology. According to the comparator model, patients should have awareness of their "true" intention, and perceive the discrepancies between intended and actualized states as a lack of control. But to our knowledge patients do not usually report to be aware of having intended a particular action for the affected arm, as different from the executed one. They rather seem to lack active guidance of that arm, and are unaware therefore of any particular intention underlying the pathological movement (even if upon reflection or observation of the unwilled movement it may be perceived as contrary to other intentions that the patient is in fact aware of).

In our account, in contrast, there is no such thing as a single true intention. Different sensorimotor schemes, whether engaged voluntarily or not, are all equally (motor-)intentional, in the sense that they are goal-oriented or aimed at satisfying a particular need. But because of a pathological lack of communication between parts of her sensorimotor repertoire, the agent as a whole is not involved in selecting which (intentional) scheme is engaged and lacks sensitivity to the scheme's consequences. This explains the lack of the subject's awareness of intention. This interpretation also allows us to avoid otherwise confusing notions such as movements "controlled by intentions that are not part of the subject's actual intentions" (Synofzik et al. 2008b, p. 422).

In short, while the comparator hypothesis would predict an existing awareness for intention (though of non-actualized intentions) along with a disturbed sense of control because of mismatch between intention and action consequences, our alternative hypothesis predicts an absence of intention awareness, but not necessarily a disturbed feeling of ongoing control. The mistaken attribution of intentionality to the hand itself is the result of the patient's reflective stance toward the absence of the intentional aspects of her agency experience.

7.5.3 **Anosognosia for hemiplegia**

Anosognosic patients are unaware or deny the (usually unilateral) paralysis of their affected limb. While some patients confabulate reasons for the lack of movement (arguing, for example, that they are too tired), more interesting are cases in which patients report to have in fact moved the paralyzed limb (e.g., to have clapped their hands), even though no such movements have taken place (only one hand has moved). In other words, these patients on some level experience agency for movements that have not in fact occurred.

The comparator-based explanation is that in these cases, the representations required for the initiation of movement and its predicted consequences remain intact. Because there is no discrepancy between intended and predicted states, the corresponding comparator (C_1 in Figure 7.1) signals success. Moreover, the impairment in this condition is supposed to prevent the recognition of the discrepancy at the other comparators (i.e., involving the limb's actual position), such that awareness for the movement is restricted to the limb's predicted position only. And on the basis of these predictions the patient must conclude that he has in fact moved. This proposal thus seems to suggest that patients with anosognosia for hemiparesis do enjoy a first-order feeling of agency (which is falsely created by the comparator system), and thus "truthfully" conclude, in their verbal report, to have moved. In other words, given the first-order "evidence" of an awareness of movement (which is non-veridical), their second-order reflective judgement is correct.

Our equilibration-based hypothesis suggests a contrary interpretation. In the absence of actual movement, the sensorimotor scheme, say of clapping one's hands, cannot succeed. Both the stability as well as the transition conditions for assimilation would be violated. We thus have to conclude that these patients would not be able to experience a pre-reflective feeling of *movement-related* aspects of agency. The mechanisms underlying action selection, however, may remain intact in these patients, who consequently experience the *intentional aspects* and a sense of urge. We take their reports of believing to have moved as indication that patients use confabulation to deal with this contradiction between their awareness of movement initiation but lack of movement-related agency aspects, which itself presents the main impairment explaining the patient's denial. The condition would thus have to be described as an incompleteness in the pre-reflective feeling of agency, accompanied by an impaired, non-veridical, reflective judgement of agency. In contrast with the explanation based on the comparator model, our proposal would predict that were reliable physiological markers for the pre-reflective sense of agency to become available, these markers would deviate from the typical case in patients with this pathology. These markers would allow us to test which of the two proposals better captures the pathology.

7.6 **Sensing ourselves, involving the world**

The proposal developed in this chapter fulfills several requirements that we believe should apply to any enactive account of the sense of agency. Let us summarize these points:

1) The experience of agency is *relational* in nature, fundamentally world involving rather than internal to the brain. It is constituted by structures or processes present in our active exploration of the world, by properties or modes of the relation between agent and environment.

2) The pre-reflective sense of agency, on our account, is an *intransitive* experience. It is not an experience of "me" as an object of perception (as defended for example by Bayne 2011) or introspection, but rather the basic feeling of my intentional direct-edness at the world; the feeling of (re-)asserting myself as an agent in meaningful interactions with the world.

3) Evidence provided for the comparator model suggests that some form of *congruence* between different motor and sensory streams is, or can be, involved in the feeling of agency. Our hypothesis accounts for this congruence (satisfaction of assimila-tion conditions), and its breakdown (lacunae and obstacles), in a more general way than matching signals and without requiring precise predictions or internal representations.

4) Our proposal accounts for the fact that the sense of agency presents itself phenom-enologically as a *heterogeneous* collection of different ways or aspects of feeling in con-trol that depends on context, the task, and the person's history and capacities. Since individual sensorimotor schemes are by definition task-specific, and therefore vary, for example, in terms of the sensory modalities involved, the balance of contribu-tions from agent-internal and environmental processes etc., it follows that the sense of agency should vary from situation to situation and, when it breaks down, it does so in accordance with the specific demands and properties of the task and person.

5) Our proposal coheres naturally with the *phenomenological recessiveness* of the feeling of agency as the absence of perturbations or instabilities in an agent's active assimila-tion of its environment. Conversely, the fact that we often become aware only of our loss of agency, when our intentions are thwarted or our sensorimotor engagements unexpectedly disrupted, is explained by the occurrence of violations of the assimila-tion conditions.

6) By satisfying the three requirements for minimal agency, our proposal addresses the problem of who is experiencing by positing that there exists a well-defined sub-ject that is experiencing its own agency, namely a sensorimotor agent constituted by a self-sustaining network of precarious sensorimotor schemes. Such an agent is invested in interactions with its own intrinsic norms, and its very constitution brings forth a domain of interactions that are relevant to this emerging self, and therewith an intrinsic *subjectivity* and perspective on the world (Di Paolo 2005; Jonas 1966;

Thompson 2007). In this sense, our account spans both the subpersonal level of sensorimotor processes, as well as the personal level of the experiencing subject, something the comparator model fails to do satisfactorily.

An additional aspect that distinguishes our proposal from most accounts of the sense of agency (e.g., Marcel 2003; Synofzik et al. 2008b; Pacherie 2007b), is that it conceives of actions and intentions as inseparable qualitative aspects of sense-making (in the sense of Merleau-Ponty's motor intentionality, Merleau-Ponty 1945/2012; also see Gallagher and Miyahara 2012). In our view, all actions are by definition intentional; equally there is no such thing as an abstract intentional "state" as divorced from the action that it requires for its realization, even if for whatever reason such action is not fully actualized. The intentional aspect of an action derives from the dispositions that the agent exhibits when a sensorimotor scheme is selected from the greater repertoire, and from the fact that the selection itself involves the agent's needs and desires. For a variety of reasons, however, an action may not be (fully) realized. In this case the intentional aspect may manifest itself without any overt movement (Di Paolo 2015). This manifestation may nevertheless have real consequences for the agent, and as such may underlie the experience of agency (as well as perceptual experience) in situations that do not seem to involve movement or the world, such as illusions, hallucinations, or in case of paralysis or locked-in syndrome (Beaton 2013; Kyselo and Di Paolo 2015).

In our enactive proposal, we have focused on the concurrent feeling, here and now, of realizing oneself as an agent. We have not said anything about a related aspect of our experience of agency, namely the sense of generally being *capable* of acting in the world. This is the experience of *I-can*, as distinct from the experience of *I-do*, which itself may be a general aspect of experience from which I sense myself as a power for acting in the world, or a particular experience of my capability to perform specific tasks effectively. Marcel (2003) suspects that practical knowledge of the "possibilities and constraints of one's bodily action" contributes to a sense of being reliably effective in one's actions and forms a background for the concurrent sense of agency. Without going into much detail, we can speculate about the connection between these two senses of agency. For example, the feeling of being able to perform a particular task, may arise from the fact that in the corresponding situation a number of appropriate sensorimotor schemes become latently activated, immediately structuring a situation into meaningful possibilities for action. In other words, it results from having a possible set of responses to hand (see discussion on virtual actions in Chapter 8). In contrast, a feeling of being incapable of a task, say, stopping an oncoming vehicle with our bodies, results from the absence of any corresponding scheme that efficaciously fits the meaning of a situation. If, moreover, this awareness of ability is reliably accompanied by the successful realization of one's actions, then over time one may come to sense oneself as being a persistent agent, with trust in the world, and reliably effective at one's will.

Trusting the world and trusting our bodies are two sides of the same coin and the extent of this fundamental form of confidence is modulated by our history of sensorimotor and socio-cultural development as well as our biology. In her famous 1980 essay *Throwing like a girl*, Iris Marion Young suggests that it is precisely a systematic curtailing of bodily intentionality that lies at the root of a general lack of confidence of women in their own physical capabilities (her observations, of course, involve certain generalizations which are valid for a given historical and socio-cultural context). One of the factors examined by Young is a tendency of women to underuse their physical strength and sensorimotor skills: "feminine bodily existence is an *inhibited intentionality*, which simultaneously reaches toward a projected end with an 'I-can' and withholds its full bodily commitment to that end in a self-imposed 'I-cannot'" (2005, p. 36, original emphasis). Up to a certain age, boys and girls show practically no difference in motor skills, movement, or spatial perception. These differences tend to increase with age. However, inhibited intentionality is not always found in adult women. This suggests the presence of systematic cultural factors at play that affect how sensorimotor networks develop according to normative expectations about the behavior of boys and girls in male-dominated societies.

In our proposal, the inhibition of neighboring schemes always affects how a selected scheme is enacted. Thanks to this, the current act is historically situated and aimed toward a contextual, activity-dependent fit. Inhibitory and other interactions between schemes help outline motor intentionality beyond the initial "move" in a series of skillful engagements. Without inhibitory interactions, then, our sense of the I-can would be rather limited to certain basic and widely used motor schemes. However, the inhibited intentionality analyzed by Young is not of the same kind as the *networked* relation between enacted and inhibited schemes. It could correspond, instead, to a systematic partial *self*-inhibiting of each scheme at the stage of its selection or enactment, and/or to ambiguous, unresolved transitions between alternative schemes, factors that would both result in hesitant sensorimotor engagements.

Another aspect of bodily confidence and trust in the world is related to the experience of being capable of sustaining persistent projects. If the sense of capability derives from the latent activation of sensorimotor schemes, its presence would depend on the proper organization of the sensorimotor repertoire as a whole (ensuring that the right schemes emerge in appropriate circumstances). This may lead us to speculate further about its connection to higher levels of equilibration. For instance, equilibration occurring to address incongruence between a sensorimotor scheme and the environment here and now, as we have said, may underlie the instantaneous feeling of agency. In contrast, adaptive processes aiming to resolve mutual conflicts between schemes, or between a scheme and the repertoire as a whole (see Section 4.4), may underlie the feeling of generally possessing a coherent and sufficient set of sensorimotor skills to deal with one's everyday life. In this sense, higher forms of equilibration may be necessary for the experience of oneself as a unified, persisting, and coherent source of intentional activity over time.

In this chapter we have assumed a certain account of the phenomenology of agency experience, in particular the distinction between a pre-reflective feeling and a conceptual judgement of agency. However, this phenomenological account is by no means universally accepted. Indeed, it is not uncommon to question whether there is any distinctive feeling of agency at all, or at least its status as a distinct kind of experience. Without reviewing in detail all existing opinions about the epistemological and ontological status of agency experiences, we can highlight a few alternatives and their relation to our proposal.

Bayne (2011) provides a useful categorization of models of agentive experience that distinguishes between three kinds of approaches: the doxastic model (agentive awareness as beliefs about one's actions); the thetic model (agentive awareness as a form of perception); and telic models (intentional actions intrinsically enjoying experiential character). For proponents of the *doxastic* model the "sense" of agency is not a type of experience at all, but rather a species of belief. On one line of thought, for example, the sense of agency is presumed to derive from retrospective conceptual inference about one's intentions and observed events (Alsmith 2015; Stephens and Graham 2000; Wegner and Wheatley 1999). We agree with Bayne that everyday intuition, the results of experimental manipulation, and observations from pathological cases provide sufficient evidence for the separability between the experience of agency and conscious judgements or cognitive inferences about it. As we have said in the introduction to this chapter, the normal experience of agency is clearly not exhausted by one's conscious beliefs about one's actions.

Another strategy for denying the existence of a distinctive experience of agency is Dreyfus's (2007b) account of skillful absorbed coping, also mentioned in the introduction, in which he claims that in such situations not only is there a loss of conscious awareness of and reflection on what one is doing, but also a complete loss of any non-reflective experience whatsoever: "In general, when one is totally absorbed in one's activity, one ceases to be a subject" (Dreyfus 2007b, p. 373). All there is, on this account, are automatic happenings in response to dispositional sensitivities that have been tuned in the process of mastery of the relevant skill. "In fully absorbed coping, there is no immersed ego, not even an implicit one. The coper [...] only needs to be capable of entering a monitoring stance if the brain [...] sends an alarm signal that something is going wrong" (Dreyfus 2007b, p. 374). We agree with Zahavi (2013) here, in that Dreyfus seems to be going too far in rejecting the role of subjective awareness in absorbed coping by implying that there can only be either conscious monitoring or no subjectivity at all, while not considering the possibility of an implicit first-personal givenness in experience (as well as subpersonal monitoring and regulation, which we believe would still be occurring in the highly skilled, if absorbed, activity he refers to). As Zahavi notes, at least in this point Dreyfus seems to be at odds with the phenomenological tradition with which he associates himself. This first-personal "givenness," in our view, partly manifests itself in the phenomenological recessiveness of the sense of agency during skillful coping. This, as we have said, may best be described as a feeling of absence (of friction or challenges impeding my coping). Importantly, a feeling of absence is not the same as the absence of feeling.

Defenders of the *thetic* model, in contrast, while not denying the experience of agency as such, claim that the sense of agency is but another form of perception, namely a transitive experience of certain things being the case about one's intentional actions; in other words, involving representations of certain facts about my actions that are available to my awareness. Such states are understood to have a mind-to-world direction, in so far as they are verified as true when they fit the world, and are otherwise non-veridical (see Bayne 2011, for a more detailed discussion of this school of thought; as well as de Vignemont 2013; and Bermúdez 2015, for a related discussion in the context of sense of ownership). The comparator model squarely falls into this category, which we reject for reasons already given in Section 7.3.

The view we defend in this chapter is best described as a *telic* model: the sense of agency is understood as an intransitive experience that is implicit in intentional action, i.e., actions themselves are considered to have phenomenal properties that do not need to be inferred (see Searle 1983). Telic states have a world-to-mind direction of fit. Their aim is to bring about certain changes in the world. They are satisfied when an intentional action is successfully realized (assimilation), or otherwise fails and remains unsatisfied or frustrated (requiring accommodation). According to our proposal, the pre-reflective feeling of agency is of this kind. It is tightly linked with the reaffirmation of an agent as an agentive system through her actions.

Since ours is a meso-level proposal, we have not said much about how sensorimotor scheme equilibration and selection is supported in the brain and other body structures, other than that we envision the involvement of metastable neuro(body)dynamic patterns. Sensorimotor schemes recruit structures distributed across many brain and body regions, involving different sensory modalities, motor control areas, etc. Premotor areas and what are usually considered higher level cognitive areas, such as prefrontal cortex, are expected to contribute to the selection and mutual priming of sensorimotor schemes. Thus, the fact that we find evidence for neural processes in support of the comparator model (see David, Newen, and Vogeley [2008], for a review) does not immediately contradict our proposal. The different stages of scheme equilibration, selection, and activation may well involve activity in similar brain areas as those associated with stages of the comparator model. We have presented our proposal in contrast with the comparator model, which we find neither necessary nor sufficient for explaining the sense of agency. But we do not claim that similar neural processes may not support sensorimotor schemes and their equilibration. Chambon, Sidarus, and Haggard (2014), for example, show the involvement of the angular gyrus in processes underlying a prospective sense of agency based on fluent action selection (which is compatible with our proposal, see Section 7.4.2). The same area has previously been associated with mechanisms in the comparator model (Farrer et al. 2008). Equally, it may well be that a scheme's satisfying of the stability condition, for examples implemented in some cases through a comparison of different neural signal streams. But such a comparison would only play an instrumental role in the more fundamental dynamics of scheme equilibration, which could also occur through different processes.

Another important element that we have not addressed concerns the social aspects of the sense of agency as well as the constitution of social forms of agency. While the picture given in this chapter is altered in such cases, it changes following along the same enactive principles, i.e., we do not need to invoke representational explanations or higher forms of inference or simulation (as in, e.g., Apperly 2010; Goldman 2006; Leslie, Friedman, and German 2004).

It has been amply demonstrated that in situations of social interaction people extensively coordinate with each other at various levels, from physiological variables, posture, distance, gestures, speech acts, and affect (see, e.g., Abney et al. 2014; Konvalinka et al. 2011; Streeck, Goodwin, and LeBaron 2011; Tschacher, Rees, and Ramseyer 2014). These coordination patterns are part of how individual sensorimotor schemes are enacted. Social agency is characterized by what McGann and De Jaegher (2009) call "self-other contingencies." Action and perception in the social domain are a matter of coordinating the behaviors, emotions, and intentions of the agents involved, in and through the coordination of movement (including utterances). Self-other contingencies are different from SMCs in a number of ways. Social interactions are interactions between agents, each of whom is maintaining their own autonomy. The condition of asymmetry between agent and environment is more complex (able to change over time along different dimensions), since the regulation of social interactions is not completely down to either individual (De Jaegher and Froese 2009). As a consequence, interactions with other social agents are far less predictable than those with (most) objects. In this way, the equilibration of sensorimotor schemes in interactive situations may not obviously arise from a single agent but could in principle be co-authored, leading to ambiguities, for instance, like controlling one's actions without exerting control (like we described above for the case of repetitive tasks). One example of this could be the attribution of agency (or lack thereof) on the basis of tracking the effects of action in situations involving several participants acting at the same time. As mentioned earlier, the movements on a Ouija board may be judged incorrectly as non-self-produced (Wegner 2003). When we consider the interactive element of this situation (several actors attempting *not* to move a common object but to let themselves be moved by it), it is the summation of slight individual forces that produces movement. This movement is not likely to correspond to any sensorimotor scheme by any of the participants. For this reason, and since each participant's contribution differs from the resultant movement, all of them at some point could simultaneously feel that they are passively following a mysteriously moving object.

Social forms of agency can also be manifested in solitude. In their approach to linguistic sense-making, Cuffari, Di Paolo, and De Jaegher (2015) distinguish increasingly complex forms of social agency, from basic coordination in interaction, to partially complementary social acts (e.g., the act of giving/accepting), to acts of mutual control, mutual and self-interpretation, and social self-control (the application of social regulatory acts to the self, as in behavior directed by inner speech). Some of these socially constituted forms of agency can be manifested in the form of top-down impositions of control and norms. The reflective sense of agency we have mentioned at the beginning of this paper, which involves an introspective stance and the explicit conceptual attribution of agency to myself, as well as narrative skills in some cases, may derive from counterparts of social skills such as self-interpretation and social self-control described by Cuffari and her colleagues. A full account of the sense of agency in human beings will have to distinguish the biological, the sensorimotor, and the social forms of agency but also combine them through a core explanatory continuity.

Our proposal is a world-involving, non-representational, meso-level account based on how actions and dispositions are organized as a network of precarious, mutually stabilizing sensorimotor schemes. A given act contributes to the ongoing regeneration of this organization to different degrees or fails to do so. It is the self-asserting logic of this network that determines whether an act belongs to the agent or not. Conversely, the ways in which an agent acts in the world individuate her as the agent she is constantly becoming.

This *meaningful* relation between agent and world that can be established, lost, and regained, is what in our view best coheres with the phenomenology of the sense of agency.

Chapter 8

Virtual actions and abstract attitudes

… each bodily stimulation for the normal subject awakens, not an actual movement, but a sort of "virtual movement"; the part of the body addressed escapes from anonymity, appears through a strange tension, and as a certain power for action within the frame of the anatomical apparatus. The normal subject's body is not merely ready to be mobilized by real situations that draw it toward themselves, it can also turn away from the world, apply its activity to the stimuli that are inscribed upon its sensory surfaces, lend itself to experiments and, more generally, be situated in the virtual. […] for the normal person each motor or tactile event gives rise in consciousness to an abundance of intentions that run from the body as a center of virtual action either toward the body itself or toward the object.

—*Maurice Merleau-Ponty (1945/2012, p. 111)*

8.1 What next?

As we have stated at the beginning of this book, our goal is to put on the table a series of theoretically rich proposals for understanding embodied action and perception. We want to show that it is possible to move beyond the predominantly critical stance toward representationalism and articulate, in complementary fashion, a series of positive ideas and see them at work. We have not dwelled too long on debating representationalism apart from our brief discussion in Chapter 2. Many others have done this job before, and the debates will surely continue. Our sights are rather set on the promise that a world-involving, enactive story can go far without representations in any of the broad senses used by philosophers and cognitive scientists. We think this goal can be achieved even if, in its vagueness, the word representation can be made to cover a lot of ground indeed.

We also think that the foregoing chapters deliver on this promise. They draw on enactive ideas, dynamical formalizations, modeling work, empirical evidence, and phenomenological insights to support, expand, and complement existing embodied accounts of action and perception and give them a world-involving interpretation, one in which agent and environment co-define themselves through a history of engagements that cannot be captured in terms of input/output informational exchanges. The sensorimotor approach to perceptual experience—one of the most developed and most debated of these embodied accounts—has served us as a guiding thread. We have demonstrated that the notions of sensorimotor contingencies on which this account is based can be made precise in dynamical systems terms and that the concept of mastery admits powerful non-representational interpretations. We have also responded to the worries raised by Thompson (2005) concerning the lack of a suitable account of subjectivity in sensorimotor contingencies theory by providing a compatible theory of agency and showing that it leads to a solid alternative account of the sense of agency.

Based on this, should we now expect an avalanche of embodied functionalists jumping ship to join the enactivist camp? Chances are that, despite the ground covered, in the mind of many we have not yet addressed the more pressing dilemmas facing non-representational accounts of action and perception. In part, this may be a typical "moving-target" situation. We should remember that many people still find it difficult to conceive of a non-representational account of mastery, let alone one of norms and goals as one should expect from a sound theory of agency, and that we have provided these here. But it is true that our account still needs to provide further answers to be more convincing; in particular, we should be able to decide whether it is feasible to explain more sophisticated forms of action with this approach. This feasibility, often denied to non-representational accounts, is the topic of this final chapter.

As we have seen in Chapter 2, representatives of embodied functionalism recognize the importance of the situated body and the benefits of dynamical explanations but they do not think these are sufficient for changing the general conception of cognition as a form of computation. They have felt secure all along because they rely on what they see as a stronghold beyond the reach of embodied/dynamical explanations: cases of so-called decoupled or "offline" cognition. At least some intelligent activity will not be explainable in terms of dynamical couplings with the environment because they demand the processing of information that is unavailable at the moment in the coupling.

What is wrong with this picture? We summarize some points already made in Chapter 2:

1. This is a sort of I-cannot-imagine-things-being-otherwise argument, an argument by default. These hardly ever provide a strong basis for deciding on scientific issues, especially in areas where our knowledge and conceptual categories are still rather limited and in flux. And by definition, it is impossible to question the adequacy of a theoretical framework by relying on these arguments.

2. Many criticisms of computationalism cast doubt on the information processing metaphor (e.g., the assumptions of near-decomposability of the human brain, the assumption of universal applicability of informational descriptions, the problems and limitations of a functionalist account of meaning, and so on). One cannot simply assume that certain experiences and mental phenomena (e.g., imagining unicorns) suffice to overthrow these criticisms, especially since in themselves these critical perspectives never deny those experiences (and in some cases they rely precisely on paying closer and more systematic attention to lived experience). In short, embodied functionalism is not problem-free and cannot be said to have successfully addressed the issues its defenders feel so confident it can explain.

3. The evidence of representational capacities in the human mind (imagination, symbolic thinking, mental imagery, etc.) is precisely that, evidence of something humans are

capable of *doing*, evidence of a type of activity, not in itself evidence that the processes underlying this activity are themselves representational.

4. If it is true that enactive accounts have not yet managed to explain the full spectrum of human cognition, it also remains true that computationalism (embodied or not) has not done so either.

Here we would like to gesture toward certain directions to further lay down our enactive, world-involving path, guided by some of these critical concerns. We will not describe a full enactive account of human action and perception (what this would even entail is not yet entirely clear), but we will show that, based on the ground covered in this book and the work of others, such an account is *feasible*, and we can even envision what it may look like in some of its aspects.

We start by addressing the already mentioned apparent limitation of the enactive account of action and perception: its inability to explain mental activity that is extended in time and scope beyond the current coupling with the environment, the so-called "representation-hungry" situations (Clark and Toribio 1994).

We will respond to arguments concerning the impossibility of enactive explanations of such cases. These arguments are usually based on a cocktail of (1) a misreading of the concept of sense-making and of the coupling between agent and environment, (2) the assumption that meaningful action relies on vehicles carrying content (i.e., applying a functionalist, not an enactivist, criterion for meaning) and that the dynamical coupling with the environment is an insufficiently versatile vehicle, and (3) an application of the "there-is-no-alternative" operator.

In the second part of the chapter, we propose a concept of virtual action based on the clustered network of schemes that constitute the sensorimotor agent introduced in Chapter 6. We suggest that the relations between schemes already activate a series of attitudes toward behavioral possibilities in addition to the ones that are enacted at the moment. This is done in accordance with the mastery of the lawful relations between schemes as well as the sensitivities that emerge from contextualization in activity genres (scheme subnetworks). The activation of these attitudes toward immediate and long-range behavioral possibilities opens the door for enactive explanations of more complex phenomena such as planning, rehearsing, and imagining.

Finally, we address the question of "perceptual attitudes," that is, the fact that human beings (at least) are able to switch between different modes of engagement with the world. From instrumental, goal-oriented attitudes to contemplative, abstract ones. The evidence in this case strongly supports the hypothesis that such attitude switching between instrumental and abstract forms of perception and action is enabled, and in some aspects even constituted, by social interactive experience and socio-linguistic skills. An enactive interpretation is sketched of how these enabling and constitutive relations work, thus suggesting the road for further work in world-involving, non-representational explanations of more abstract forms of cognition.

This ground will be covered as one sketches a rough map. It will provoke new questions, probably more questions than answers. But they will be interesting. The objective is to show that enactive explanations do not end in the embodied, complex dynamical stories of concrete agent-environment engagements. They do start there, in the concrete lived moment, and in a sense, they see all forms of cognitive activity, all forms of mind, as ever more sophisticated ways in which the concrete moment is *lived*. So they do also eventually come back to the concrete moment. It is possible that these explanations will eventually find their limitation. But we see no a priori reason not to continue the exploration of how far they can go.

8.2 **Modes of the mind**

In her book *Human Minds*, developmental psychologist Margaret Donaldson puzzles over the human capacity for constant invention of novel activities and goals. In some cases, these occupy a good part of our waking hours (obsessions with sports, hobbies, music, social networks, etc.). Some obsessions (the word still applies) involve high degrees of sophistication and demand high-level cognitive capacities (e.g., doing research, developing new technologies, understanding legal systems or financial markets) What is curious is that the study of human intelligence always focuses on how we learn to achieve ever more complex cognitive tasks, for instance, how we go from moving about without bumping into things to doing math, but hardly ever looks at the question of why we even bother to climb this complexity hill. There must be a motivation at the heart of every human action, but motivations will vary in kind. Of the different dimensions that we could use to describe the complexity of human activity and motivations, Donaldson proposes to look at a dimension of degrees of concreteness/abstraction in all aspects of the mind, including cognition, action, perception, and emotion. This is not the only interesting dimension, but it is quite useful in the current context.

This dimension is typically, and poorly, described as having two poles and many undefined cases in between. As the distinction between "higher" and "lower" forms of cognition can often be crude, embodied functionalists tend to speak of "offline" and "online" intelligence instead (Wheeler 2005). The first kind of intelligence is demonstrated by fluid embodied engagements with our surroundings (e.g., navigating a noisy train station full of moving objects and people without bumping into them). Offline intelligence, in one of Wheeler's examples, involves "mentally weighing up the pros and cons of moving into a new city" (p. 12). Such seemingly intuitive distinctions belie the inextricable Cartesian roots hiding in everyday language and the common sense from which they originate, the pervasive mediational epistemology Charles Taylor warned us about. The online/offline distinction mischaracterizes differences in cognitive complexity and reflexivity as levels of body involvement (the walking is done by the body; the "weighing of pros and cons" is done "mentally"). But I'm no less embodied and coupled to the world when I plan my holidays than when I ride a bike; I'm simply doing different things with my body and my coupling. To preconceive these differences in this way is simply to use a different name for the separation between body and mind. Adopting the online/offline and related terminology means implicitly buying into a dualistic perspective. And uncritically assuming this dichotomy in the framing of the question is unlikely to lead to answers that do not perpetuate its inherent dualism.

Fortunately, Donaldson provides us with a more discerning alternative. She distinguishes four modes in which our minds function (for our current purposes, we simplify some important elements in her more elaborate proposal). Instead of drawing directly on a scale of sophistication, she proposes that these modes depend on our focus of concern and motivation. The first mode is the *point mode*. It deals with here-and-now coping. The focus of concern is the currently engaged activity involving a concrete ongoing motivation. Most animal activity falls within the point mode, and a large part of everyday human activity does as well. The point mode is the mode we engage when we make a pause while eating lunch and have a drink of water, when we avoid bumping into other people at the station, when we get dressed in the morning, tie our shoelaces, and so on, but also when we need to make a split second decision, anticipate a change in mood in our interlocutor, or estimate the rate of currency exchange while paying a cab fare in a foreign country.

Next is the *line mode*. It expands the focus of concern to the immediate past and the possible immediate future. For this reason, it may also involve a larger spatial region than the space I currently occupy and can easily reach. Picking up the umbrella on a cloudy morning before leaving the house is an example of line-mode activity. The line mode expands the point mode, but it often involves it at the same time: for instance, line-mode activity may concern things and events I perceive in the distance *now*, (I see someone I would prefer to avoid who is waving at me from the opposite side of the road) but whose consequences (the expected boring conversation) are *not yet* directly manifested. These potential consequences nevertheless allow me to regulate my activity now.

Next is the *construct mode*, which involves a decentering of cognitive activity. The concern of the agent focuses on events that have happened or may happen somewhere at some point in time. These events do not necessarily involve the agent herself; they could be about other people or a generalized other. Construct mode activity includes the recognition of different perspectives and points of view, generalizations about events, motivations, and behavior, and practical reasoning and norms (e.g., "it is best not to drink and drive"). A concrete engagement in the construct mode may still involve elements of the line and the point modes ("since it is best not to drink and drive, I will say no to the offer of another round of drinks that I can see my friend is about to make").

Finally, the *transcendent mode* involves no locus of concern as such. It involves concepts and relations between ideas, abstract contemplation, thought, values, and emotion.

As already suggested, cognitive activity is probably most of the time not performed neatly within a single mode. While it is reasonable to think that there is a developmental progression and that cognitive skills move from the point mode toward the transcendent mode with age and education, most of the time we are involved in some complex concrete mixture of activity in the different modes. For instance, one may be trying to think of the best way to write a sentence that clearly encapsulates a complex idea, while at the same time being annoyed at some dog's loud barking and contemplating plans to move to a quieter neighborhood; all of this while exercising the skill of writing on an old laptop's keyboard.[1]

The idea of this classification is to have a relatively simple way of distinguishing between human activities that is nevertheless richer than the dichotomous, dualistic, and normatively biased opposition between offline and online or between high and low level mental activity.

8.3 **The poverty of the coupling?**

For embodied functionalism, in spite of important lessons learned from the embodied turn in cognitive science, representationalism remains the only game in town. We have already criticized this attitude, in part since it tends to be self-justificatory (define cognition as the manipulation of representations and there seems to be little alternative to representationalism) and in part because it is far from clear that the representational approach could produce a full, coherent and naturalized explanations of the mind.

But let us briefly consider what drives this kind of arguments, which often begin with an acknowledgment/dismissal combo ("yes, this is all fine; no, it doesn't change anything") of the lessons of enactive and dynamical approaches. As we have suggested earlier, the main drive behind these arguments is the supposed poverty of the coupling between agent and environment. If sense-making, as we have proposed, depends on adaptively regulating this dynamical coupling, it would seem that no sense-making could ever escape Donaldson's point mode. All sense-making, as defined by enactivists, would be limited to here-and-now sensorimotor engagements, at best.

It is true that the dynamical coupling between agent and environment plays a central role in sense-making. But why must this role be akin to that of a vehicle for carrying content? In fact, no one has ever made such a claim, or anything resembling it, except the embodied functionalists that build their arguments on its basis.

Embodied functionalists rely on the assumption that many cognitive problems are representation-hungry (Clark and Toribio 1994; Clark 1997), in other words, problems that deal with absent elements (not directly perceivable or realized) or abstract ones (including unspecified general goals). In Donaldson's categorization this would mean all mental activity in the construct and transcendent modes, and probably most activity in the line mode too. However, the idea that the class of such representation-hungry problems is defined univocally, or that, given a particular cognitive phenomenon, it is a straightforward matter to decide whether it is representation-hungry or not has come under considerable scrutiny and criticism (e.g., Chemero 2009; Di Paolo et. al 2010; Degenaar and Myin 2014).

---

[1] More examples showing how the different modes can be intertwined in everyday activities can be found in some of the few, but influential, studies that bring together cognitive science, ethnography, and anthropology (e.g., Agre 1988; Hutchins 1995; Lave 1988).

While Clark and Toribio (1994) have in mind what we could roughly describe as higher cognitive mental function when talking about representation-hungry problems, their proposed category would seem to include any kind of motor control problem for which direct sensorimotor feedback seems insufficient. For instance, performing smooth arm movements (or smooth movements in any other multijoint system) involves compensating for interaction torques that emerge on a given joint owing to the movement around another joint. Controlling for interaction torques would seem to be a representation-hungry problem because sensory feedback can be ambiguous (i.e., relevant information is absent). However, Buhrmann and Di Paolo (2014a) have demonstrated in a model implementing simple spinal circuit equilibrium-point control that the dynamics of the musculoskeletal system is able to compensate by itself for such interaction torques.

Similarly, Izquierdo and Di Paolo (2005) have provided a simple demonstration using a simulated agent of how some problems that could be deemed representation-hungry can be solved without the need (even without the possibility) of manipulating internal representations. They evolve the neural controller of a simulated robot able to move in one dimension using a simple array of distance sensors. The task is to approach or avoid a falling object according to its shape (a typical minimal cognition task in Randall Beer's style). However, without the agent having any way of knowing it, its sensor array may be reversed in the left-right direction. So if the falling object is to the right of the agent it stimulates the sensors on the right-hand side in the case of no reversion, or the sensors on the left-hand side in the case of reversion. In either case, the agent must approach the object. This task is a type-2 classification problem (one in which the instantaneous stimulus is ambiguous with respect to the correct moment to moment decision about the course of action to take, Clark and Thornton 1997). However, a robot controlled by a stateless machine can actually solve the task. Such a controller is incapable of handling any representation about the history of stimulation from which to extract relevant information to solve the task. Although from moment to moment, each instantaneous change in the movement of the simulated agent will be different depending on whether its sensors are reversed or not, the accumulated history of these changes still results in a trajectory that catches the object at the end. The key aspect of the non-representational solution is that the agent must be able to engage in a continuous coupling with the environment, and it is in this time-extended process that the ambiguities of the sensory input are resolved. This follows on from similar, and remarkable, demonstrations of simple embodied agents performing supposedly representation-hungry tasks, such as delayed, history-contingent choices, learning and conditional classification using the minimal cognition paradigm championed by Beer and colleagues (e.g., Beer 2003; Izquierdo et al. 2008; Williams et al. 2008).

In short, what is or is not a representation-hungry task is anything but a trivial or even a decidable matter. Perhaps the category should be dropped altogether (see also Degenaar and Myin 2014). The implication up to this point is that in explaining how we move beyond Donaldson's point mode into more sophisticated forms of action, non-representational explanations cannot yet be discarded. So far this seems a weak clarification, but it gains weight in the light of the fact that this is precisely what many people do: they assume that there is no other way to solve some cognitive tasks other than through internal representations.

Consider the problem of "tracking" targets that are momentarily absent or of anticipating behavioral outcomes before the relevant environmental features are perceived (or even actualized). It is profitable to see how such cases can be explained in dynamical systems terms. The lesson we aim at is that whenever we consider an agent in coupling with the environment, it is not just the instantaneous coupling that matters for sensemaking, but the history and dynamics of the agent and the environment as well. The very notion of anticipation has been considered from this dynamical perspective in the cases of ultrafast cognition mentioned in Chapters 2 and 7 (Wallot and Orden 2012). In these cases, fast cognitive performance (typically in go/no-go tasks) occurs reliably in times that are faster or of the same order as the estimated physiological limits of information transmission (to say nothing of processing). Since the explanation for this reason cannot involve any kind of anticipatory or predictive modeling (let alone more complex forms of representations), the problem must be approached differently. The agent is seen as a complex dynamical system, which is historically situated and able to move toward critical dynamical states. As the agent understands the possibilities of the go/no-go task, the coupled dynamics become "poised" in a metastable state, which can quite rapidly crystallize into the right cognitive performance. The stimulus acts as a trigger and not so much as bearing information that needs to be processed and decoded.

Evidence of ultrafast cognition supports the notion that cognitive systems can be strongly anticipative. According to Stepp and Turvey (2010), strong anticipation arises from the (ecologically situated) system itself via lawful regularities embedded in the system's ordinary mode of functioning in coupling with the environment. Weak anticipation, by contrast, arises from a model of the system via internal simulation. The distinction has been introduced in formal terms by Dubois (2003). A strongly anticipatory system is attuned to its situation by the integration of feedback, delays, and synergies into its overall dynamics. Before movement starts, the system is already in some form of attractor that will lead to the appropriate, context-sensitive behavior. Perturbations, like pulling on the chin while you are trying to articulate a phoneme (Kelso et al. 1984), simply result in taking a different transient toward the same attractor (equifinality with respect to the outcome, rather than to implementation of movement). In other words, the existence of the attractor means that the system already anticipates possible perturbations by allowing for different changes in its transients. And the setting up of the attractor is a sort of feedforward action, hence by the time the movement actually starts, nothing more needs to be "computed," not even the appropriate response to perturbations; the agent-environment coupled system just relaxes toward a stable state.

This relaxation can indeed be context and history sensitive as demonstrated in Buhrmann and Di Paolo, (2014b), where an arm model with a touch sensor at the tip is able to follow a surface in front of it at various orientations and positions and then continue moving along the same plane even when the sensor is turned off (as if predicting where the remainder of the surface "should be"). These trajectories, however, all lead as different transients toward the same attractor, and, in each case, it is the form of the transient that has been modulated by a history of coupling with the environment. Nothing in the system changes internally, nothing is represented.

The relative simplicity of these examples should not distract us from the important lessons they provide. The first lesson is that they show how complex action, perception, and cognition can emerge in an autonomous agent in coupling with the environment. There is no explanatory constraint from the enactive perspective to reduce the determinants of sense-making to anything less than the history-dependent coupled system as a whole. When we consider aspects of the coupled system and wish to understand their influence on the outcome of an action, we need always to make sure the question is properly contextualized. Meaningful activity, therefore, cannot be reduced to what may be found in a part of this whole coupled system, namely, the variables and parameters currently involved in sustaining the agent-environment coupling. These are *not* vehicles for content. The second lesson is that the agent is historically situated and able to change as a whole in terms of its history of interactions with the environment and other agents. It is precisely for these reasons that there is in general no such thing as "the function" of the coupling. The whole idea of a part or a process having a function is contingent on the current mode of stability of the system (in terms of the network of sensorimotor schemes introduced in Chapter 6, elements may or may not have a definite function depending on whether the agent is situated within an activity cluster or in transition between one cluster and another). Yet these historical changes in the agent are not required to take the form of representational traces, which are then available for inspection. Many other options are possible. Once we abandon the computer metaphor, we begin to realize that there are other ways of being affected by history and attuned to current and future situations. Yet a third lesson may be drawn, one that stresses the situatedness of the agent, one that may be easy to miss in the a-temporal economy of information processing. No matter how large the embedded scales of action, how far they stretch into the past or the future, or how far afield the decentered scope of cognitive activity is, in other words, no matter in which of Donaldson's modes action occurs, it always and *necessarily* occurs as the concrete acts of an embodied agent here and now. "The action itself is an event. It initiates novelty in the world. It is no longer a possibility moving toward the actual, but an aspect of the actual itself, the flesh of advancing process" (Ricoeur 1950/2007, p. 205).

If these lessons are true, then it must mean that sense-making is a very particular kind of activity, one that is grounded in the contemporaneous materiality of agency and the world and yet able to extend its sensitivities and its efficacies way beyond. And it is able to achieve this without the aid of representational traces or predictive models, but inherently in the sort of activity that it is (see also Haugeland 1995). To see this more clearly, let us consider the concept of sense-making again and, in the process, clarify some aspects about the relational ontology of enactivism.

8.4 **Sense-making and virtual fields**

As we have insisted, our account of mastery and sensorimotor agency is a world-involving one. It is an account where the world does not play the role assigned by functionalism in cognitive science: that of being a source of information. This sets the enactive approach apart from other traditions that have also stressed the importance of the relation between agent and environment, such as for instance the Gibsonian tradition in ecological psychology (Gibson 1979). There are many convergences between the ecological and enactive approaches. The ecological insistence on the richness of the environment as a source of information is a response to representationalists, for whom the purpose of perception is to do well within an economy (production, access, distribution, consumption) of a scarce commodity (information) which must be extracted, deduced, or even constructed by mental activity.

We agree with ecological psychologists when they highlight that real environments are rich enough to access directly their relevant meaningful aspects. We think they are in fact *too rich*, in that sense-making always involves a massive reduction of all the environmental energies that might affect the agent, to those within the dimensions of biological, sensorimotor, and social historically contingent meaning. However, the ecological response to the traditional view of the poverty of the environment is a move within the same conversation, one that does not question the nature of cognition as being in the business of managing information. They simply move from an economy of scarcity to one of abundance, but they do not switch discourse.

A world-involving account of a given cognitive phenomenon means precisely that the world is involved at more fundamental levels than as a source of information. In fact, it is one of the challenges of enactive cognitive science to provide clear theoretical reasons of what special circumstances allow us to speak of information processing as a limiting case, something that can happen, but under particular conditions.

What will happen, then, if we stop thinking of cognition as the management of content extracted from the environment that must be carried in vehicles from one place to another (see Hutto and Myin 2013)?

The first point to consider is that from an enactive perspective any form of action and any form of perception is a form of sense-making (Di Paolo 2005; Thompson 2007a). As we saw earlier, sense-making is the adaptive regulative activity of an autonomous and precarious agent. Through this activity, the agent is sensitive to tendencies in its states and its relations with the environment that would break through its boundary of viability. The agent responds to these negative tendencies by regulating their course so that such a breakdown may be avoided. No magic is involved in this idea as demonstrated by the various models of sense-making we have discussed earlier.

Let us expand on the notion of sensitivities and regulation of tendencies. The autonomous organism (the argument also applies to a sensorimotor agent) is capable of evaluating some interactions with the world as beneficial or not for its own continued self-individuation. In other words, the maintenance of systemic identity is a source of intrinsic normativity. The problem is that the formal condition of operational closure of a living system is in itself binary, not graded. Some form of mediation is missing in this picture, something that can link actual states with the tendencies and capacities involved in evaluating the consequences for the system's viability. Since sense-making is the regulation of the coupling with the world based on this evaluation, it requires not only an autonomous organization but also an adaptive one as discussed in Chapter 5. Adaptivity is the capability of an autonomous system to respond to tendencies in the trajectories of its states and its relations to the world, so that when these tendencies approach the boundary of its own viability the system modulates its coupling with the world in a way that tends to avert the crossing of this boundary. Successful adaptive responses thus regulate states and relations before the breakdown of closure (death) has occurred.

Adaptivity, therefore, works on the *virtual field* that surrounds the current dynamical configuration of the agent–world system. In two senses: first, the agent must be responsive to whether or not tendencies approach the boundary of viability—if the crossing of this boundary is actualized, it is simply too late—second, it must make use of its capacities to modulate the coupling with the world by introducing changes that alter the virtual field in the vicinity of the current states (modifying the direction of the negative tendencies).

What exactly do we mean by a *virtual field*? The notion of the virtual in the current context follows from the tradition of granting real ontological status to capacities and tendencies, especially those in the concrete vicinity of current actuality. This tradition includes Aristotle's concept of potentiality, Spinoza's concept of affect, and related (but not identical) usages of the term "virtual" by Bergson and Deleuze. For our purposes, we additionally use the rich conceptual resources of dynamical systems theory to speak of trajectories, traces, and tendencies. In a few words, in this tradition the virtual is that which is real but not actualized, such as the glass's capacity to hold liquids or the tendency of liquid water to become solid at freezing temperatures. Such capacities and tendencies need not be currently actualized (this glass is empty and in principle could remain empty forever but its water-holding capacity is still real).

Notably, these capacities and tendencies are always *relational*, unlike properties of an object that belong to it in all contexts. Hence, virtual capacities and tendencies can be potentially infinite in number (the glass can be used as a paperweight; water also has a tendency to slow down high-energy neutrons inside a nuclear reactor). This does not mean that we can manipulate circumstances so that any object could acquire *any* arbitrary capacity or tendency (the glass does not have the tendency to slow down fast neutrons, and liquid water makes for a rather ineffective paperweight). Actualization of virtual capacities and tendencies occurs always as an *event* or a happening, or an act. These actualizations are situated in history, have duration, and so forth.

By speaking of *fields*, we want to focus attention on the concrete situatedness of virtual tendencies and capacities, that is, not just any abstract counterfactual, but those that neighbor a current state of affairs. This adjacent structured field can be studied by examining the dynamical landscape around a current trajectory (this landscape need not be fixed in time). Such fields have been investigated in many simulation models of embodied cognition to different degrees of explicitness. It is common in such models to artificially travel back and forth along a recorded behavioral trajectory, manipulate circumstances, and rerun the model to ask "what-if" questions in order to map systematically a virtual field (e.g., Beer 2003; Iizuka and Di Paolo 2007; this is the method we have used to explore the sensorimotor habitat for our model in Chapter 3). The very concepts of stability and metastability in dynamical systems theory imply the idea of a virtual field since they point to structural aspects of adjacent flows that surround a currently actualized trajectory of states in a system. As we saw in Chapters 6 and 7, these ideas underlie complex concepts such as readiness-to-act and commitment.

We come to an intermediate conclusion: in a strict sense, all cases of sense-making, and therefore all action and perception, involve constitutively elements of virtuality.

In other words, sense-making, which even in its most basic forms entails adaptivity, always occurs in a "thick," extended here-and-now, not just in the actual changes of state of an agent but in its virtual field. By this we mean that, given the current situation, not only the actual states matter to the sense-maker but also the virtual traces and tendencies that surround these states, whether they become actualized or simply modified but not necessarily actualized in the course of events. Anchored in the here-and-now, sense-making always operates on a basis that transcends the present moment.

In the enactive analysis, making sense of a situation—evaluating its relevance to the sustaining of a precarious identity and responding accordingly—requires by definition both *sensitivity* to and *regulation* of the virtual field of possibilities. The sensitivity and capacity for regulation are present in all forms of life with large variations in complexity and refinement. For the average person, it is not the same to walk a straight course stepping along the curb at street level or stepping along the border of the terrace of a tall building, even if the sensorimotor trajectories that should be actualized are nearly identical. For the average pigeon, the two situations are similar. This is because both are differently sensitive to risks and have different capacities to respond to them. Apparently similar situations present to them radically different virtual fields.

We see then that, from an enactive perspective, normal action involves both sensitivity to the virtual adjacent field of possibilities and traces and a way of manipulating this field (for instance, by reducing risk or creating opportunities). From a sensorimotor standpoint, we have seen that this virtual field is modulated by the mastery of the laws of sensorimotor contingencies, which depends on the agent's history of adaptive equilibration, the structure of the sensorimotor relations between schemes, and its concrete situatedness within an activity.

8.5 **Virtual actions**

A useful concept that may help us shape future enactive explanations of how we move through Donaldson's four modes is that of virtual action. We take a virtual action to be an action that has not been fully actualized, and yet has real consequences for the agent and the world.

This idea inherits at least some of the explanatory burden of what in classical Artificial Intelligence goes under the name of planning. Whether in real robotic systems or in other contexts, planning in the traditional view involves maintaining and using an accurate enough representation of the world. Action-based approaches relax this requirement a little. For instance, Rick Grush, in his emulation theory (2004), suggests that internal models of the body and the environment, whose proposed primary purpose is to improve motor control, can be run "offline" to produce motor and visual imagery and explore simulated actions without actualizing them (see, also, Clark and Grush 1999).

Our proposal, predictably different, is that virtual actions involve the skill of extending sensitivities and regulations further along the virtual field and recursively invoke the virtual consequences of actions not yet taken.

This skill—for which we acknowledge that a more detailed operational account is not yet available in full—demands no new fundamental principle over and above the concept of sense-making. It does require, yes, more complexity. Consider how we have discussed the concept of mastery. This can be recast in terms of sensitivities to the virtual field conditioned on possible actions that are not immediately taken. We have mentioned in Chapter 7 the sense of being capable of actions and trusting the surrounding world, a sense of *I-can*, and speculated about its relation to the partial activation of several schemes in response to motivation and environmental solicitations. This is already a sort of virtual sensitivity. This sensitivity may become oriented toward a particular course. I see a cup in front of me and my perception involves constitutively my mastery of the virtual possibilities afforded by this cup and my body in interaction. I may move my head slightly or reach to grab the cup and my mastery of the relation between schemes in my sensorimotor network shapes my perception of the cup by the characteristics of these action possibilities, even those I do not take. We saw this in George Mead's and Margaret Floy Washburn's descriptions of the role of inhibition of viable actions in favor of a selected one (Chapter 2). To perceive in the sensorimotor approach is to master the influence of virtual actions (i.e., of possible adjacent actions, which are not taken).

What we need in order to explain more sophisticated forms of virtual action is an account that furthers this mastery into more complex sensitivities. Such mastery may only occur in organisms above a certain level of sensorimotor complexity. Such organisms act in the here-and-now but "with reference to a spread-out environment" linked to the spatially and temporally distant, "an environment both extensive and enduring [that] is immediately implicated in present behavior" (Dewey 1929, p. 279). As well as being a trace of recent events, the immediately present cup is like a "hub" for directly solicited virtual actions (move the head around it, grasp it, drink from it, etc.). A skillful sense-maker can regulate these immediate sensitivities to include virtual events projected further forward in time (grasp the cup now but then place it on top of another surface, near the desk at which I will sit to write a report).

There is no fundamental difficulty here, only the complexity involved in skillfully extending and navigating the virtual possibilities afforded by the here-and-now anchoring of sensorimotor activity. This complexity would be unmanageable if an agent were required to be sensitive to all possible virtual actions. However, I rarely grab a cup to put it on top of my head or to fill it with sand. As we saw in our account of sensorimotor agency, I live within activities that cluster sensorimotor schemes into functional and structural self-sustaining relations. Within these activities, I live in a situated normativity that shapes the space of my current affordances (see, also, Rietveld 2008b, Rietveld and Kiverstein 2014). Inside a cluster, schemes "make sense" in relation to each other and this permits not only the regulation of immediately adjacent consequences but of whole streams of virtual activity at the moment I enact a single particular scheme.

We can illustrate what this would look like using our network description of sensorimotor agency.

Within a cluster of sensorimotor schemes, within an activity or behavioral genre, the enactment of a scheme, as we saw in Chapters 6 and 7, is influenced by previously enacted schemes and by those schemes that are currently inhibited or activated to be enacted in parallel. The enactment of a scheme has virtual implications for the future propagation of the activity. These past, present, and future influences shape the enactment of the scheme here and now, and the enactment in turn modifies these virtual influences. I grasp the cup differently depending on whether I see tea in it and want to drink some or whether I will wash it in the kitchen sink or whether I will put it back in the cupboard. A skilled agent may be more sensitive to such influences, which may then involve activity further back or forward in time or space. I grab the cup differently whether, for example, (1) I aim to put it in the cupboard, close the cupboard door, turn around, and leave the kitchen or (2) I aim to empty the cold tea, rinse it, put the kettle on, choose a new tea bag, and wait till the water boils. The differences may be only slightly manifested in the actual movements and in postures that put me on course to the next events in my activity stream. In this way an agent regulates the relations between schemes within a cluster of activity. Eventually, these regulations may not take the form of an overt, enacted scheme that has been prepared by sensitivity and readiness to ever-longer series of projected actions. To act virtually is like any embodied skill, and practice may lead to progressive refinements. At some point the virtual act may simply be the act of regulating long-range sensitivity and readiness, even before acting, by a prolongation of the "onset" phase of an enactment. Most of the time, however, these regulations happen while we act. We are thus frequently *thinking in movement*, to use Maxine Sheets-Johnstone's phrase (Sheets-Johnstone 1981).

There may be different processes involved in these regulations and sensitivities, which count as virtual actions; some of them were discussed in Chapters 6 and 7. In Figure 8.1 we illustrate a hypothetical situation. On the left we see a part of a sensorimotor network. A scheme in the blue cluster (S_1) is currently being enacted. This facilitates the subsequent activation in the agent of the support structures for schemes S_2 and S_3. We assume that the environmental conditions allow either of these two to be enacted next. But S_2 and S_3 tend to inhibit each other and cannot be enacted together. In addition, each of them leads to the facilitation of schemes in different clusters (e.g., the activity of preparing a new cup of tea or the activity of leaving the kitchen after putting the cup away). Even before the competition between S_2 and S_3 is resolved, the activity in the support structures for

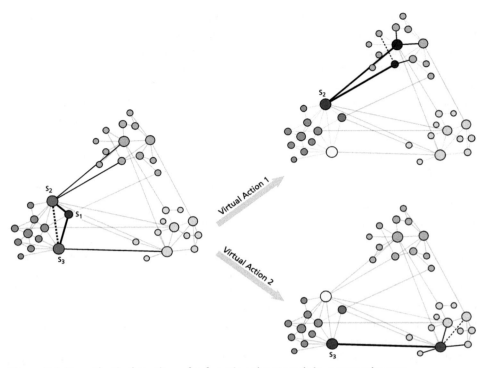

Figure 8.1 Hypothetical section of a functional network in a sensorimotor agent. **Left:** A single scheme (S_1) is currently enacted, as indicated by its black shading. At the same time, through the influence of S_1 as well as the potential contribution of environmental solicitations, two other schemes are already "primed" (S_2 and S_3, darker shade). This could involve, e.g., the "subthreshold" activation of the support structures in the agent required for these schemes. Though S_2 and S_3 have no overt effect yet, their latent activity may nevertheless already influence yet other schemes in the network (illustrated as thicker connections to nodes in different clusters), which may as a result become more likely to be enacted in the future. These "primed" schemes constitute a strongly anticipative context for the current action, which thereby extends its horizon of influence into the future. The two schemes S_2 and S_3 are also mutually inhibiting and thus they compete with each other for enactment. **Right:** activity shifts to either S_2 or S_3 and one of the two schemes is enacted (top and bottom). Several events in the agent can cause the tipping of the balance between the two competing schemes. This may be the influence of other active nodes in the network on S_2 or S_3 (not shown), an uneven influence of S_1 on the two schemes, or a more holistic act, such as the agent changing the relation to the environment with the result that one affordance becomes more salient than another. All these are examples of virtual actions, and their result is the shifting of activity in relevant parts of the network. Note that the agent-side structures involved in previously enacted schemes may continue to exhibit influence on other parts of the network, even when no longer having any overt effects. Such "lingering" activity extends a currently enacted scheme's horizon of influence into the past, which current activity can resignify. (See color plate.)

these schemes at "subthreshold level" may already pre-activate schemes in the green and yellow clusters. The situation is that of a system poised at a critical state. Moreover, the influence of S_1 and the way it is enacted (which is not entirely up to the agent since the environment is involved) may still "linger" in its support structures (e.g., neuromuscular processes, posture, inertia). Maybe the influence of other recently enacted schemes also lingers. A virtual action in this case may take the form of tipping the balance between S_2 and S_3. It may occur by the way S_1 is enacted in actuality, which already orients the flow of activity toward one of the options, or it may occur as an emergent down-regulation of the agent as a whole (e.g., through activity in other parts of the network that influence either S_2 or S_3). The result is that the functional network takes a different shape (Figure 8.1, right) depending on this "internal gesture." Note that the spreading of latent activity in the network, that is, the influence of schemes enacted in the past and the influence of currently enacted schemes on those that may be enacted in the future, not only biases *which* schemes are activated and when, but may also modulate the way in which they are enacted. This is what we mean when we talk about currently enacted schemes being "colored" by previous ones and in turn "coloring" future ones—or what we have referred to as the "thick" here-and-now.

This is one possible example showing how the agent may regulate a virtual flow, in this case through changes in the relation between the agent-side support structures for interlocked schemes. It is important to note, however, that the example should not be interpreted as suggesting that virtual actions are somehow "internal," which would take us close to a representational story. Firstly, they only count as actions because they are the regulations induced by a whole agent, even if what in some cases is regulated may be the relations between internal support structures. Secondly, virtual actions involve changes to the relations between schemes without necessarily activating all of the schemes involved; this is why we call them *virtual*. But they are *actions* because they have consequences for the relation between agent and environment, more precisely adaptive consequences for the ongoing individuation of a way of life. In this sense, virtual actions are like other forms of sense-making, an activity that must be enacted, and that may fail or succeed.

In addition to future oriented regulation of activity flows (what shall I do next and how?), virtual actions can also help us fill in gaps and overcome micro-breakdowns with an activity. I see a meaningful course of action solicited/enabled by the environment, but the activation of the most suitable sensorimotor scheme to embark on that course is not possible in the current situation (say, I cannot find the pen I need to take down a phone number, or the hand I normally use for writing is injured). Hence, I must fill in these gaps with schemes that enable the activity I aim to enact. The enabling relations between schemes guide this search for alternatives, which again is performed by the agent as an emergent regulation of how schemes relate to each other, in other words, as a virtual action.

It is important to stress that all action—all sense-making—contains elements of virtuality. Virtual actions are, in particular, actions that regulate virtual fields, that is, the relations between networked schemes within clusters of situated activity.[2] They take place in the shaping of activity flows before they are actualized, as well as in the reshaping of the virtual implications of *past* actions, which are then re-signified by current enactments as the agent orients her activity toward future possibilities.[3] In the sensorimotor realm, anticipation and memory can be aspects of one and the same thing.

Whether in cases of future orientation, or in the reshaping of the meaning of past virtualities, or while she finds alternative routes toward a goal in the form of sensorimotor planning, rehearsing or imagining, the sensorimotor agent is not required, under this view, to represent the relations between schemes in an activity (as *we* represent them in our network diagrams) to achieve any of these things. Since these relations are the same relations that constitute her as an agent, every reaffirmation that she enacts and every breakdown that she suffers and overcomes contributes toward developing her sensitivities to *herself* as an agent, and eventually, toward maturing a practical, embodied form of self-mastery. We see here yet another meaning of the concept of sensorimotor mastery, this time applied to the contingent "laws" and norms embodied in the precarious organization of a sensorimotor agent as a result of her personal history. This meaning makes explicit a connotation of mastery as *power* that is only implicit in the other meanings we examined before. The emerging and reflexive power and self-understanding of the sensorimotor body becomes systematically potentiated during development with the acquisition of language and, with it, of socially enabled modes of self-regulation and self-control (Cuffari, Di Paolo, and De Jaegher 2015).

A consequence of our view is that only a *sensorimotor* agent can be capable of virtual actions of this kind. A biological agent that has not achieved sensorimotor autonomy would then be unable to perform this kind of manipulation of the virtual relations between schemes because it lacks both (a) the normativity that is established by the networked individuation of precarious schemes, and (b) the capacity to modulate its scheme network activity in relation to these norms. This does not mean that all sensorimotor agents would necessarily be capable of virtual actions, only that virtual actions "occur" in the regulation of the various structural and functional influences subtended by schemes in a closed, self-sustaining network.

---

[2] We note that our proposal for a non-representational account of virtual actions is highly compatible with ecological approaches. Indeed the notion of affordance is a paradigmatic case of a virtual relation between organism and environment, which should be considered in its activity-dependent situatedness (Rietveld and Kiverstein 2014). A recent proposal by Brincker (2014) elaborates a similar perspective using the notion of affordance navigation to give an account of complex cognitive phenomena such as the passing of false-belief tests.

[3] The idea of current action reshaping the past is explored in the context of social interactions (Di Paolo 2015). When engaged in them it may be possible to own personal intentions retroactively once an action is completed socially (e.g., through the response of others). As noticed by Elias (2015) and Wallot (2015), this retroactive effect, here understood as a present reshaping of the virtual implications of past events, need not only occur in cases of social interaction but can happen more generally.

8.6 **The zero mode**

We noted in Chapter 6 that we sometimes find ourselves at the hinges between activities, those moments where we must begin to enact a new microworld. These transitions may be more or less noticeable, more or less dramatic, but they are frequent. They sometimes imply some kind of breakdown, as changes from one activity to the next do not always occur in a smooth fashion (activities are often interrupted, put on hold, changed on the fly, etc.)

These moments of breakdown are among the most mysterious and less studied aspects of the human mind. These are precisely the moments where functionalism itself breaks down (hence the prevailing focus on task-driven studies in psychology and neuroscience, with the exception of more recent studies of default brain activity and studies involving meditation). At these breakdown points, the coordinates that grant stability to functional descriptions are in flux. They are a blind spot for functionalist theories. However, although we may be momentarily world-less in a sense, we are not mindless. In fact, in coping with these moments our autonomy as organisms and as agents is reasserted. So it is indeed a way of being a human mind to be lost and in search for new coordinates, and eventually to find them. But this way of the mind does not fit precisely in any of Donaldson's four modes, since they are all defined in terms of the locus or focus of concern. What occurs during transitional breakdowns is that concern is precisely *unfocused*. Goals may be confused, actions misdirected, aims fruitlessly multiplied and abandoned without properly starting on a new activity course. We call this the *zero mode*.

According to Varela (1992), this is where enaction occurs, precisely in navigating oneself outside this nebulous state back into a microworld. Varela conceived these transitions as relatively fast (fractions of a second; in our network model this would also be the case as one needs just a few jumps to get from one cluster to another, while one can remain jumping between schemes within a cluster for longer periods). But in some cases, the reconstitution of meaningful concern can take longer. For instance, if we are engaged in an activity and someone comes to tell us some shockingly bad news, some of us may drop what we are doing and attend to the new and unexpected implications of this event. But others may find it difficult to disengage from what they are doing, partly because the shock may not have fully registered or partly because they cannot find what is the appropriate reaction. So for a while their previous activity continues in an uncanny manner.

Another situation in which the zero mode is persistent occurs when sensorimotor skills are systematically disrupted, as we have already seen in Kohler's visual inversion experiments. In such cases the clustered linkages between schemes are thrown into disarray and must be reconstituted almost entirely. The subject who is starting to wear visual inversion goggles is unable to engage fully in normal activities. For some time, she will always find herself back at the starting point of getting the first scheme right. And it is precisely how this scheme is conditioned by what should follow next that makes the adaptation process so difficult, because what should follow next no longer makes the same sense as before the visual inversion. Thus in order to grab a cup to place it in the cupboard the untrained subject needs to overcome first the overall aim of her arm and hand movements, but even when this is overcome the scheme will not be ready for what comes next as the cupboard may be perceived in the wrong spatial position. With training not only are schemes enacted with some success, but the relations between them begin to be reconstituted as well. During this period the zero mode of activity makes itself very apparent.

In the same way that concrete activity may involve more than one mode, elements of the zero mode may also be present, for instance, in a creative resolution to micro-breakdowns (breakdowns within an activity that can normally be recovered by the adaptive relation between schemes). This could lead to a reconfiguration of the network of schemes, even to the appearance of totally new ones. Activity in the zero mode is what drives spontaneous creativity in agents capable of sensorimotor open-endedness (as described in Chapters 4 and 6). We can expect then the zero mode to be active in situations of play as new elements, rules, people and situations are added to a game, and the self-defining purposes of playful engagements shift. Successfully traversing the zero mode is creative sense-making at its purest. However, this only occurs as the result of the precariousness of the material relations that constitute the agent and the environment. It is because of material precariousness that breakdowns will occur and adaptive activity must restlessly reconstitute meaning by finding states and relations that are more viable if the organization is to survive. And it is because of the materiality involved that this reconstitution of meaning can in principle become open-ended.

8.7 Abstraction and perceptual attitudes: the missing social dimension

Virtual actions as we have defined them, the active regulation of virtual fields that affect sense-making, are the counterexample to claims that non-representational approaches are incapable of explaining cognition across multiple scales, and beyond Donaldson's point mode. But they cannot be the whole story. We still need to understand how it is possible to train these extended sensitivities, and to adopt more reflective attitudes on the activities we engage in.

At various occasions throughout the preceding chapters, we have remarked on the need to include a social dimension in our enactive approach to action and perception. A proper treatment of the complexities brought into our proposals once we consider that sensorimotor agents are also, in general, social agents, is beyond the scope of this book. The interactions between biological and sensorimotor agency are complex and hard to untangle. Adding a social dimension to these interactions, and, as argued by Cuffari, Di Paolo, and De Jaegher (2015), including properly social forms of agency into the picture, will apparently make matters even more complex. This may be so. Then again, it may be the case that, as it often happens when dealing with multiple levels of analysis, several unresolved issues in one dimension can only be resolved once another dimension is taken into consideration. Be that as it may, in this section we will limit ourselves only to exploring one particular issue that is directly relevant to the current discussion regarding the feasibility of the enactive program: the question of how an agent is capable of adopting abstract perceptual attitudes. This again will show that the enactive approach can offer explanations for phenomena beyond Donaldson's point and line modes.

8.7.1 **The abstract object, a false starting point**

We said earlier that sensorimotor agents are generally involved in some concrete meaningful activity, a microworld. In general, this is implicitly acknowledged in embodied theories of perception. Yet, once we consider the starting points for most of these theories, they do not often involve the agent embedded in some concrete activity, but instead they consider a relatively demotivated agent confronted in the abstract with some external object. Most of the examples that illustrate the sensorimotor approach to perceptual experience are of this kind (see Chapters 3 and 4).

Like in representational theories (Marr 1982), it may seem that the simplest starting point for theorizing about perception is a situation involving a single agent confronted with a single static and neutral object. More elaborate cases incorporating multiple objects, in motion, available for different uses, or set within naturalistic scenes and social situations, must follow this prototypical case as so many degrees of increasing complexity. Everyday experience and experimental evidence tell a different story. I am hardly ever confronted with abstract objects, which I regard and inspect in their perceptual presence in a manner devoid of interest or social significance. Most of the time, I work my way around a world of social and biological purposes and norms, for which objects are seldom present to me as the bottles or tomatoes that populate the sensorimotor literature (O'Regan and Noë 2001, p. 945; Noë 2004, p. 62). I don't normally inspect a bottle with my fingers to perceive in it a Gestalt shape as opposed to disjointed sensations; I pour some beer out of it. I don't regard the red tomato and consider its unseen sides and its voluminous presence; I cut it and add it to the salad. The problem of how the object maintains its identity despite its manifold of appearances is not a problem for an agent whose attitude is pragmatic, but only for one whose attitude is abstract. Once I pick up a hammer to drive a nail, it is clear to me that it remains the same hammer between blows, as its identity is verified by its being part of the sensorimotor schemes that I enact.

Given our account of how sensorimotor agency is constituted as a self-sustaining network of schemes organized into clusters of activities, a pragmatic, instrumental attitude in the sensorimotor agent is always more primary that an abstract one.[4] But then we should ask how could the latter emerge from the former. How does the sensorimotor agent ever develop such a stance in regards to objects closer to Donaldson's construct mode (exploring the shape of the bottle or regarding the tomato as a soft red volume)?[5]

---

[4] We must not forget that the pragmatic attitude is itself a developmental outcome. Hence it is more primary than the abstract attitude, but not altogether primordial.

[5] While framed as a problem of abstract perception of objects, the same questions apply to the capability of producing abstract acts. In his classical studies of abstract and concrete behavior in patients

It has been suggested that the answer to this question must implicate the social world (Froese and Di Paolo 2009; Gallagher 2008; 2009; 2014). Shaun Gallagher aptly describes embodied theories of perception, and more traditional ones, as *philosophically autistic.* Their guiding assumption is that of a transparent, literal, and accessible objective reality that contrasts with an opaque, complex, and half-hidden social world. This assumption applies pretty much across the board in the sensorimotor approach literature. It is implicit in the unjustified supposition that one must first investigate how we perceive objects and only then move on to investigate how we perceive the social world (e.g., Engel et al. 2013).

Were it not for my co-inhabiting a world in which a myriad of perspectives, motivations, and interests flourish, I would never learn that an object has a presence beyond my own "solipsistic" concerns. It is because I can somehow see that I am not the only one for whom the object is meaningful, that I can also appreciate as an abstract object what remains at the intersection of those motivated perspectives—which "cancel out" like the summation of so many randomly oriented vectors.

Our approach to sensorimotor agency has filled in a gap in sensorimotor contingencies theory and addressed the criticisms by Thompson (2005) in a non-representational manner (as advocated by Hutto 2005). But our proposal leaves us with a problem lying, so to speak, at the opposite end: if perception is so constituted by *my* enactment of sensorimotor schemes, *my* deployment of *my* skillful mastery of the laws of the sensorimotor contingencies that correspond to *my* sensorimotor body, why is it that I perceive a world that transcends my activity? Why, in other words, do I perceive objects as being out there, publicly present to me and to others?

Even after attempting to resolve this issue through a story describing the complexification of a sensorimotor network through differentiation, integration, and clustering of schemes (Chapter 6), we will still find the problem unresolved. At the end of such an account, the agent would access the external object in multiple ways: the confluences of several modes of activity-dependent engagement between body and object. But is this sufficient for the object to have an existence that transcends the agent's concrete activity? The object has its own synergy with the agent's body and so it is manifested in a particular way, one that belongs to this object and not to another one. It is a synergy of the agent's modes of seizing the world, a set of organized relations of sensorimotor schemes. Two objects may be seen as different, but the agent sees both as emerging from her own activity. The object, distinct from other objects and in relation to them, remains

---

with neural lesions, Kurt Goldstein correlates the lack of a capability in dealing with abstraction (e.g., arranging objects in arbitrary geometrical relations and performance of meaningless movements) with impaired capacity in patients to put themselves in the place of another (Goldstein 1963 p. 55, based on lectures delivered in 1940). This is in line with what we will suggest in the rest of this chapter. The link between lack of understanding of abstract attitudes and lack of understanding of other perspectives is suggestive of a common capacity for decentering which is impaired in these patients. It also helps explain their socially awkward or misplaced emotional reactions as whole situations are understood only partially through disjointed concrete elements.

inescapably oriented toward the agent's interests and norms if we were to apply the equilibration story developed in Chapter 4. It could not be otherwise. What distinguishes an object is the achievement of the multiple equilibration events of the agent's sensorimotor schemes, a process whose norms are those of sensorimotor agency. Perceived objects therefore achieve only a relative independence; independence from each other, but not from acts regulated by biological and sensorimotor norms. A sensorimotor agent of this kind inhabits a "solipsistic" world.[6]

One solution to this problem, as indicated, is to go social, to include perceptual acts whose normativity transcends that of the constitution of precarious sensorimotor agency. Let us examine some of the evidence that support this move (see Di Paolo [2016] for further details).

8.7.2 **Social determinants of perception**

In a broad sense, perception is influenced by the social world because objects themselves, the situations they are embedded in, and the practices they involve are largely social in origin (Costall 1995; Heft 2007). Infants play with socially produced toys and wear socially produced clothes. To this we should add the social origins of various sensory standards (musical scales, prevalence of certain geometrical shapes and color palettes, etc.). This is also the case even for "natural" objects encountered by the developing child: they are subject to socially established norms and practices (you don't take insects to your mouth, you don't yank out the pretty flowers, etc.).

This is not a trivial observation. It points to the possibility of cultural factors affecting perception. These factors are indeed found to co-vary with perceptual "styles," even if we restrict ourselves only to the visual perception of object and scenes. Cultural influences are found in the perception of visual illusions (Segall, Campbell, and Herskovits 1963), perception of depth and orientation, and pictorial representations (Deregowski 1989). A well-researched cultural variation in visual perception is the contrast between holistic attention to context in East Asian cultures and analytic attention to parts in some Western cultures (Nisbett and Miyamoto 2005). Holistic viewers show greater skill in attending to

---

[6] The problem is an old one. Phenomenologically, the issue of object presence is closely related to the problem of the transcendence of the world with respect to subjective experience. The argument regarding the insufficiency of a single developing subject undergoing progressive equilibration of her sensorimotor organization for constituting an abstract, objective perceptual attitude parallels Husserl's argument against the solipsistic constitution of the objective world through the variations between normal and pathological or disturbed experiences (Husserl 1989; see, also, Gallagher 2008). Like Mead (1934, 1938) who viewed the meaning and objectivity of the world as a consequence of the socially acquired expertise at adopting various roles, the Husserl of *Ideen II* sees the transcendent as constitutively intersubjective, as does Merleau-Ponty (1964). Each of these philosophers (despite important differences) seems to appeal to a similar solution involving the transformative consequences for the subjective grasp on the world that depend on effectively coordinating perspectives with others.

background and relations between objects while analytic viewers excel at discriminating figures from backgrounds (Witkin and Berry 1975). Such differences in visual style are manifested from a certain age onwards, but not before (Duffy et al. 2009).

Of course, cultural variations in perceptual styles do not immediately imply an inter-personal influence on development. It is also possible that other factors are at play, for instance, the structure of typical visual scenes (Miyamoto, Nisbett, and Masuda 2006). Other relevant factors tend to co-vary with cultural ones (climate, diet, landscapes, etc.) and these could have effects on perceptual development. While they are suggestive of the influence of social practices, cultural variations are not definitive proof. We can at least claim that they play a contextual role in modulating perceptual styles, i.e., they affect the style of perception but do not seem to enable its possibility (see De Jaegher, Di Paolo, and Gallagher 2010). Conversely, lack of cultural variation is not by itself sufficient to discard social influences on perceptual development.

Social cues also affect visual attention and object salience. The gaze of others influences affective judgments with participants finding objects more appealing if they are congru-ent with perceived gaze direction (Bayliss et al. 2007; Hayes et al. 2008). Eye-tracking and event-related potential measurements indicate that 4-month-old infants are sensitive to the difference between directed and averted gaze with respect to an object (Hoehl et al. 2008) and can be similarly sensitive to other cues such as head movements. Again, such influences modulate object perception, but the evidence so far does not prove that these influences play a strong enabling role in the development of perceptual skills, only a con-textual one.

Evidence of stronger social influences on attention and manipulation capabilities in infants can be found in the development of triadic competencies during the first year of age. During this period infants become increasingly sensitive to joint-attention con-texts as they move from dyadic interactions with the caregiver to triadic interactions also involving objects (Striano and Stahl 2005). Primary intersubjectivity (i.e., early recipro-cal engagements, affective tuning, and mutual regulations between caregiver and infant) is progressively followed by co-regulated engagements involving shared attention and action on objects, or secondary intersubjectivity (Trevarthen and Hubley 1978). Dyadic interactions are established early on. Infants attempt to re-engage interactions that are paused artificially (still-face paradigm) already in the second month (Tronick et al. 1978). At around 6 months, objects dominate the infants' attention while they sometimes look up at their mother. Between 9 and 12 months, they engage in full triadic exploration of objects paying attention to both the adult and the object and focusing on the relation between the two.

There are close developmental links between dyadic and triadic competences (Striano and Reid 2006). This is confirmed in naturalistic longitudinal studies. Kaya de Barbaro and colleagues observe robust changes in the infant's sensorimotor organization that go together with changes in the organization of social engagement (de Barbaro, Johnson, and Deák 2013). At 4 months infants engage all the sensory modalities in the explora-tion of the object in front of them, grasping it, looking at it, taking it to the mouth (see, also, Smith, Yu, and Pereira 2011). This is followed in the second half of the first year by more frequent separation of activities allowing decoupled attentional engagement with the adult and other objects in synchrony with object manipulation. Mothers are active during this transition in engaging the infant's attention to the object (showing it, moving it in conjunction with smiles and vocalizations). They solicit eye contact and attempt to switch the infant's attention from one object to another. These solicitations result in the infant at 6 months being able to decouple attention to objects and parent through different modalities, holding one object in the hand while looking at another one or attending to the mother. Transitions between objects or holding two objects at once go together with a reduction in the mother's bids for attention to novel objects and more engagement in attending together to the same object, imitating, and facilitating the activity of the infant. At 12 months the infant displays bi-manual coordination in handling objects in elaborate sensorimotor schemes. Mothers at this stage perform manipulations of objects that are more complex, and the infant attends to these demonstrations and alters subsequent manipulations. At this age, interaction patterns organize in activity turns and involve partial acts such as giving or requesting.

These and other naturalistic observations (e.g., Nomikou, Rohlfing, and Szufnarowska 2013; Rossmanith et al. 2014) suggest that the infant's attention and sensorimotor skills are "tutored" by context-sensitive scaffolding resulting in a socially guided mastery of attention and object manipulation. These social influences are more than merely contextual, they are enabling in the sense that these experiences play a role in shaping a particular kind of sensorimotor network in the infant.

As the child grows, she progressively adopts decentered attitudes toward objects and others.[7] Interpersonal coordination, breakdowns, and recoveries of coordination that make up a social interaction (De Jaegher and Di Paolo 2007) seem to play an enabling role in learning to decenter. In the 1970s, a team of psychologists at the University of Geneva studied the effects of interactions while children attempted to solve a problem together (e.g., Doise, Mugny, and Perret-Clermont 1975; Doise and Mugny 1984). The problems were typical Piagetian tasks: conservation of length and volume, rearranging objects to a new location conserving their relative positions, as well as games that could be played alone or in collaboration.

---

[7] Adults are much better at decentering than children, but not infallible. They seem to achieve a decentered solution to a problem only after having considered and inhibited an egocentric one. For instance, adults instructed to move an object in a shared frame of reference often look at the egocentric solution first (a salient object they can see but is occluded for the instructor) and so do children (Epley, Morewedge, and Keysar 2004). However, adults inhibit this initial reaction and proceed to reach toward the correct object visible to all while children have more difficulty doing this. This is compatible with the notion of decentering involving an active regulation of one's own perspective from the perspective of a real or putative other, rather than an acquired and readily available novel kind of personal "objective" perspective, in which case one would not expect adult egocentric looking to occur at all. Decentering, it would seem, is a virtual *act*.

In these experiments, children are individually pre-tested on the task in question and ranked as low, intermediate, and high performers. Then they are allocated into pairs to solve a similar problem together, and finally tested on their own once more. Provided that certain conditions apply, there are reliable positive improvements in individual performance after the interaction. Doise and colleagues observe that individual progress depends on interpersonal conflicts during the joint condition and on how they are resolved. Improvements in performance do not correlate with the observation of the correct solution to a problem, as for instance when pairing a passive low with a dominant high performer. They correlate instead with the degree of participation of both children in working out the task together. These experiments show a clear enabling influence between interactive factors (conflict, participation, breakdowns, and agreement) and progression in sensorimotor/cognitive tasks that require a decentered attitude.

Finally, we should also consider the influence of language and play in shaping, enabling, and possibly also constituting the capability to adopt abstract perceptual attitudes. These effects are major and not fully understood. For Vygotsky (1978), linguistic mediation is the key factor that differentiates human from animal perception. Lexical categories structure the perceptual world (e.g., Brown and Lenneberg 1954; Luria 1976; Rosch et al. 1976). As we have suggested earlier, linguistic abilities endow an agent not only with tools for understanding the world but also with instruments of self-mastery, for regulation of virtual fields. Language enables the control of perception by permitting the voluntary shifting of attention, easier retention of past perceptual events, and action planning. These influences are too numerous to describe here in detail. Even if we limit ourselves to the non-overtly social aspect of perception, evidence shows that there is a deep influence of language on color perception (Thierry et al. 2009), object recognition (Boutonnet and Lupyan 2015), motion detection (Meteyard et al. 2007), and spatial understanding (Richardson et al. 2003).

Flexible relations to objects during mother–infant interaction underpin object substitutions in make-believe play, which are at the basis of symbolic function (Vygotsky 1978). Meaning substitutions work when the child adopts the right embodied gestures (a pencil becomes a rocket ship by re-enacting a countdown to lift-off and imitating the corresponding movements and soundtrack). This capacity becomes progressively abstract during the second year (Watson and Jackowitz 1984) allowing substitutions to depend less on similarities (a toy banana for a phone) and increasingly on linguistic/gestural schemes, usually in a context of social interaction. As single make-believe episodes turn into social dramatic play (Starks 1960; Bodrova and Leong 1998), children learn to not only self-regulate and adopt different social roles but also to perceive and use objects as the recipients of novel social meanings, no longer just centered on the immediate sensorimotor grasp (Sinha 2009).

The developmental relations between language-mediated play, language comprehension and use, the development of malleable object relations, and the self-regulation of attitudes and roles are not fully understood. But it would seem that several of these processes are of the essence for the capability to regulate perceptual attitudes. To be able to adopt a detached observational attitude toward an object, the child must be able to empty it of its most immediate meaning, to temporarily bracket its significance. This capability logically, not just empirically, implies the power to manipulate meanings and relations between objects, uses, and perspectives and this suggests that some linguistic and interactive abilities play a constitutive role in abstract object perception.

8.7.3 Abstraction, an enactive interpretation

From an enactive perspective, any explanation of the emergence of decentered perceptual attitudes must be based on how the agent's sensorimotor organization varies through processes of (metastable, precarious) equilibration. Dynamically, such processes rely on the emergence and subsequent resolution of conflicts: conflicts within a given scheme (adjusting the strength and precision of the grip upon realizing that the cup is a bit too full), conflicts between schemes (adjusting to driving on the opposite side of the road), and interpersonal conflicts. The latter seem to be at the root of the most compelling evidence briefly described in the previous section. We see this in the experiments by Doise and colleagues, where interpersonal conflicts that afford some degree of joint participatory resolution are the ones that most influence individual progress.

Conflict is also present implicitly at the phases of qualitative change in mother–infant dyadic and triadic interactions. In the observations by de Barbaro and colleagues, we see mothers insisting on contextually diversifying the infant's acts and attention when these are multimodally over-focused on a single object and later encouraging refocused attention and engagements as the infant learns to manipulate different objects at the same time. It is as if mothers were providing a context aimed first to enlarge the infant's set of sensorimotor schemes, let them accommodate to each other, only to follow this with a context aimed to refocus the now diverse modes of engagement so as to make them work together. This guidance is conflictive (though this is not to be understood as antagonistic) since the mother's influence tends to dis-equilibrate the infant in specific ways.

If we represented these transitions using the network description of sensorimotor agency, the stages would look as in Figure 8.2. A first stage (at around 4 months or less) shows highly synergistic schemes. The infant focuses on the object with full attention and multiple modalities. As the mother induces disequilibrium, these schemes begin to differentiate. At around 6 months, the infant is capable of manipulating and attending to more than one object and to the mother. Schemes begin to be grouped by different activities. Now the mother induces another disequilibrium by soliciting that the infant be more attentive to specific objects. This attention is shared and the object is manipulated together. The diversified schemes begin to accommodate to each other and form new links between them toward the end of the first year as shown in the figure.

These examples of *interpersonal* equilibration, as well as those described by Doise and colleagues during peer interactions in young children, are cases of participatory sense-making (De Jaegher and Di Paolo 2007). According to this concept, novel meaning is jointly achieved through a history of breakdowns and recoveries of interactive coordination. From a systemic perspective, this is a way of resolving the apparent contradiction

4 months 6 months 9–12 months

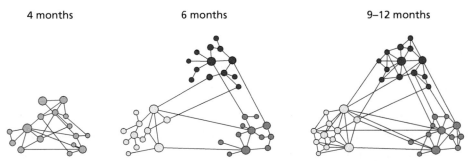

Figure 8.2 Schematic (hypothetical) representation of the scaffolding effects. These correspond to mother-infant interactions on the infant's sensorimotor network during the first year. The first panel shows the highly synergistic schemes that infants of about 4 months of age bring to an object. As the mother solicits bids for attention to herself and other objects, these schemes begin to diversify (middle panel) into new, loosely linked clustered schemes. The mother now encourages integration of these schemes during play by sustaining attention and manipulations on single objects and on relations between objects and persons. This results in the network growing increasingly consolidated as the schemes for attention, object manipulation, shared attention, social acts, etc., get progressively integrated (third panel). (See color plate.)

between the autonomy of the cognitive, living agent and the possibility of socially consti-tuted, incorporated skills. Social processes can intervene, but not in a determining fash-ion, within the self-constituted identity of the autonomous agent (otherwise she would cease to be autonomous). However, a history of co-specifications can lead to forms of individual agency that would be unreachable through the systemic transformations avail-able only to the isolated agent.

In order to understand how these social influences transform sensorimotor agency such that objects may be perceived in the abstract, in the construct mode, we need to ask what it means for a perceptual attitude to be constituted intersubjectively. What aspects of such an attitude (which undoubtedly belongs to a particular, individual embodied agent) can be said to be essentially social?

Social interactions tend to have a life of their own and involve phenomena that are irre-ducible to the sum of the individual actions and intentions of the participants (De Jaegher and Di Paolo 2007; De Jaegher, Di Paolo, and Gallagher 2010). The interactive order is what brings people together (there is a pull toward sustaining interaction), but it is also a source of heteronomy for each participant. Individual acts during the interaction are sub-ject to a double normative framework: they are, on the one hand part of individual sense-making and on the other, they are moves in the unfolding of the interactive encounter.

Cuffari and colleagues (2015) discuss this primordial tension between individual and interactive norms. Their dialectical model explores the concept of participatory sense-making, which moves from general coordination of intentional activity during inter-action, through various forms of complex social agency, to "languaging." One form of self-differentiation in participatory sense-making is the emergence of complementary social acts, acts that require interpersonal coordination to bring them to completion, such as a handshake or the act of offering/accepting an object. Each individual enacts a partial sensorimotor scheme, and to learn to perform social acts successfully implies interpersonal equilibration between these schemes. The ability to participate in such acts is an irreducibly social skill. Social interaction is not merely enabling of this skill (via mutual equilibration) but it is also constitutive because the social act can only be conceived as being bound by the normativity of the interactive order. I may perform a dexterous gesture offering you the object you have been looking for, but if I do it with bad social timing, say, when you turn to address someone else, the social act is awkward or it simply fails.

The partial schemes that jointly make a social act also tend to be reciprocal and not just mutually fitting, but also mutually addressed. At some point the child's sensorimotor repertoire is endowed with coherent partial schemes, easy to coordinate with one another (e.g., a scheme for offering an object and the scheme for receiving it), but which unlike other sensorimotor schemes, are inherently oriented toward an other (like one half of a handshake). In our sensorimotor network description, these could be represented as schemes with open arrows leading to complementary schemes in another person. But the child learns to act on both sides of the complementary relation. As the child learns to hand over a toy to the mother and then to accept it back from her, she is not too far from being able to "hand it over" to a resting place, while doing something else, and then "accepting" it again. This is different from simply putting the toy down and then picking it up. The attitude toward the toy is tinted with social normativity involving an implied other. The toy is partially objectified by the solitary combination of two complementary partial schemes enacted by a same individual. The resulting whole social—though solitary—act implies the object is being "held" and regarded by an absent viewpoint. In this way, the reciprocal sensorimotor skills involved in partial schemes invite the shifting of perspectives, which even in solitary play can be enacted, as in the above example, by the child "giving" the toy to an absent other whom she partially incarnates.

Phenomenologically, abstract object perception also points to a constitutively intersubjective connection. The abstract perceptual attitude involves perceiving the object with its properties and relations to the perceiver and its surroundings, but neutralized of any instrumental interest (as far as possible; we never reach a fully abstract attitude). Successful instrumental actions are complete in themselves. A contemplative attitude toward the object is open instead. We do not know in principle where it may lead, how long it will keep us occupied, and, if the object is unfamiliar, what we are likely to discover. The child experiences such unfinished, prolonged, sensorimotor engagements during interaction with others, in particular during play. Actions in the interactive context are hardly ever closed on themselves since they also serve in sustaining the encounter (when they fully close, they signal a pause or a break from the interaction). Playful interactions are especially open-ended; they engage the zero mode as meanings can break loose from direct instrumental concerns.

The contemplative perceptual attitude is phenomenologically afforded by the infant's experience in expecting actions not to finish with a clear goal. But this is not the only phenomenological clue. As the child regards an extraneous object, her attitude is one of curiosity, an interrogative attitude. The present object is addressed as a social other, one that is unable to complete fully the child's open partial acts. The object is interrogated with hands, eyes, mouth, and ears. And the child is open to being surprised (or disappointed or quickly bored). As she has learned in social encounters, she expects answers and solicitations to further action, but in this case, she has to facilitate them herself.

A similar interpretation is offered by Thomas Fuchs (2012) who presents an analysis of the development of first-, second-, and third-person perspectives in infancy focusing on the emergence of reflexive, meta-perspectival stances in the child that are constitutive of social and self-understanding. He highlights the developmental relevance of conflict during interpersonal interactions. Vasu Reddy (2008) suggests that there are primordial aspects of second-person interactions from which concepts of self and others develop. The infant experiences in herself what it means to be an object of attention for others and this experience serves her to understand that others may attend also to other objects and that objects can be made into social actors. This knowing oneself through another's actions and expressions fits with Merleau-Ponty's description of the difference between concrete and abstract movement. In concrete movement, the motor project is directed toward others and the world. In abstract movement, instead, it is directed toward the body itself (e.g., while trying to execute some prescribed series of postures or moves without immediate significance, the motor project aims at moving my limbs, my head, etc., as specified) (Merleau-Ponty 1945/2012, p. 114). But again, this reflexive orientation toward the body is achievable by first experiencing the body as an object toward which it is possible to hold motor projects, guided, as it were, by the attitudes of others as they direct their attention and actions to myself.

According to our enactive interpretation, as decentered attitudes develop, the child approaches objects starting from a second-person perspective; one, however, that never quite reaches the full engagement of this perspective as in a real social interaction, or does so partially (personalization of toys, animism). This perspective does not have all the elements of a proper second-person attitude as the object shows itself not to be an autonomous other. This invokes the child's use of complementary partial schemes to sustain the encounter in a quasi-interactive mode involving perspective shifting. The noninstrumental object invites and at the same time resists a second-person engagement. By eliciting in the child a second-person stance that cannot succeed, but invites creative completion in the zero mode through the adoption of multiples roles, the interrogative engagement with the object gives rise to a novel viewpoint. Out of a breakdown of the second-person attitude, a third-person perspective emerges.

8.8 **Perceptual mastery ... once again**

The perspective on virtual actions and on intersubjective factors that constitute some perceptual attitudes sketched in the previous sections will require further elaboration.

Ideally, we would like to take these ideas to the operational/dynamical descriptions we have developed for other proposals in this book. These future developments will help us clarify the precise relations between the different modes of the human mind. Nevertheless, we can already re-examine some of the central aspects of the sensorimotor approach to perception based on these incomplete proposals.

We return to Noë's (2004) account according to which I experience a tomato as having an unseen side and volume, because of my "knowledge" of how sensory and motor changes would co-vary if I were to look at the tomato from a different perspective. Put in these terms, this unperformed act is just abstractly counterfactual and invites representational explanations. Gallagher (2008) rightly points out that if it were a question of imagining these non-actualized moves, I would not then perceive the hidden aspects of the object as coexisting at this moment with my view of it. I would *know* the hidden sides "as a necessary consequence of a certain law of the development of my perception" (Merleau-Ponty 1964, p. 14). But what is given in perceptual experience are not necessary truths, but concrete presences (Merleau-Ponty 1964). In other words, presence itself *cannot* be constituted by any knowledge about lawful co-variations.

Gallagher seeks the solution to the phenomenological conundrum in open intersubjectivity. According to the Husserlian account, "if the absent profiles cannot be correlated with my possible but non-actualized perceptions, then the absent profiles may be correlated with the possible perceptions that others could currently have" (Gallagher 2008, p. 172). Possible truths, however, have the same inconvenience as necessary truths: they are intellectual notions that differ from the concrete experience of presence. Those "possible perceptions" belonging to others may not be actualized either at the moment I decide to stare at the tomato. And if they were, the right synergy between my body and theirs could still be absent. They also remain abstract. In the absence of direct actualizations, these possible perceptions of others must also be imagined, remembered, or deduced. The same criticism Merleau-Ponty would make to Noë seems to apply to Gallagher's proposal. Moreover, if there is an apperceptive horizon of possible experiences subtended by the object, which includes mine and those of others, this phenomenological insight does not yet provide an explanation of how this quality of perception could come about not only in the absence of co-present alternative acts, but also without co-present others.

Gallagher is correct to invoke the intersubjective dimension so as to add degrees of freedom to the problem and aim for a closer phenomenological match. And as we have discussed, empirical evidence strongly backs this move. But he is too quick in ruling out virtual aspects of sense-making. These are not temporally disjointed from the perceptual act; in fact, we have already seen that they always accompany it. Dynamically speaking, alternative engagements with the object can be virtually co-present if we see perception as time-extended (Froese and Di Paolo 2009). However, and importantly, this occurs *in a concrete context-dependent manner*, not as conveniently invoked counterfactual acts. "The hidden side [of an object] is present in its own way. It is in my *vicinity*" (Merleau-Ponty 1964, p. 14, emphasis added). Virtualities, unlike abstract counterfactuals, are adjacent to actuality; they are in the concrete "vicinity" of the actual. They can take the form of sensitivities to proximate situations (the risks and opportunities I confront). They also take the form of inner tensions as certain evoked sensorimotor schemes are inhibited in favor of others. These tensions are contextual in that they follow the internal relations of schemes clustered around a given activity. They may be resolved through full-fledged virtual actions, as we illustrated in Figure 8.1.

The problem is that these virtualities may or may not involve those precise sensori-motor co-variations required to reveal specific aspects of the object, such as having an unseen side. This will depend on the concrete activity. As we have said in the discussion on mastery at the end of Chapter 6, the visual presence of the tomato does not seem to be informed by arbitrary possible schemes in which it could be involved, but mostly by those schemes enacted in the context of "tomato-related" activities. Many of these involve grasping the tomato, moving it from one place to another, and so on. Graspable objects such as tomatoes and bottles involve a field of virtual actions that allow us to see them as having volumes, solidity, and hidden sides. But this is not the case with objects we engage differently. We do not see, say, keys on a computer keyboard in this way. A computer user rarely observes keys outside the context of a keyboard, and seldom accesses their hidden side. The virtualities present in normal engagements with the keyboard create a percep-tual horizon that extends beyond the immediate visual and tactile sensations, but they do not suffice to constitute keys as having an underside. Keys in a keyboard are present as more than surfaces, but less than volumes, and this difference is related to the clustered schemes one typically uses to engage with them.

Of course, we *know* keys have a hidden side (and this is proof that knowledge itself cannot constitute presence). Following the line we have been pursuing, we do not need to invoke a representational story to account for this aspect should we decide to contem-plate keys abstractly, outside the context of typing. Instead, we need a story, like the one we have sketched above, that explains how I am able to shift perspectives and adopt an abstract perceptual attitude. We have argued that the skill behind this act is intersubjec-tively constituted. I can see keys as objects not by mentally disassembling the keyboard, or by invoking a flat imaginary being who lives inside my laptop. I do it by manipulating my sensitivities to the virtualities of my situation, in other words, by enacting the skill of perspective shifting that allows me to establish a dialogical, even playful, relation to the object. As a result, I may now regard the keys no longer instrumentally, but by invok-ing a different sensorimotor repertoire, for instance, the one used when looking at other loose small flat objects such as buttons or coins. Then I can invoke a series of manipula-tory schemes applicable to *these* other objects and influence my visual perception of keys as things I could virtually hold between my fingers and turn around. Such perspective-shifting skills come in degrees and relate to personal experience, e.g. photographers and painters appreciate complex patterns of color, texture, and light in ordinary objects more easily, in general, than other people. They can bring into play richer sensorimotor reper-toires in their perspective shifts.

It would seem therefore that object presence is always activity-dependent according to the only non-representational candidate explanation we have so far, which relies on the influence that sensorimotor schemes exert on each other. Each enactment (in particular, the enactment of schemes involved in perceiving an object) is colored by the regional influences in the cluster of schemes that make up an activity. The only way to alter how an object is perceived is to adopt a different perceptual attitude, that is, through the virtual act that allows it to be seen from the perspective of a different activity, or if possible, in the case of linguistic sensorimotor agents, to see it in a decentered manner.

The future challenge for the enactive approach will be to provide dynamical explanations of what is required for achieving these perspectival shifts so that a lived situation changes its meaning. Our proposal is that this requires *enacting* alterations to our current field of virtualities (i.e., altering that which we are sensitive to). This sophisticated skill, we propose, is constituted—not just learned—intersubjectively, and in this direction lies the future of the enactive approach to action and perception. A detached perceptual attitude is not so much an individual skill that I learn socially, but a social skill that I enact individually.

Chapter 9

Epilogue

9.1 On the continuity between life and mind

What is the relation between mind, body, and world? In broad strokes, this is what we have set out to explore in this book. We have approached this not as a general philosophical question, but through one of its multiple facets and in the form of specific contributions to an enactive theory of action and perception. These contributions orbit around a concept that gives this book its title: *sensorimotor life*. Of the multiple connotations of this concept, we want to highlight, as a closing reflection, the way it exemplifies what enactivists mean when they talk about the continuity between life and mind.

William James' (1878) remarks against Herbert Spencer's characterization of the evolution of both mental and bodily life as a kind of correspondence or adjustment of inner and outer relations, serve as an early statement against a naive understanding of life-mind continuity. Today we would describe Spencer's idea as an evolutionary and functionalist account of agency and normativity. According to him, mental evolution is best understood on the model of biological evolution, and his view of correspondence is interpreted as the establishment of inner subjective relations in the organism that match external objective relations in the environment and help the organism survive (a view not dissimilar to the mediational epistemology criticized by Charles Taylor and discussed earlier in the book).

In a way, James's criticisms of this correspondence view could not have been more enactive. He ridicules the notion that mind should be driven only by a passion for survival. The bases for human action, he postulates, are the wealth of diverse interests and concerns, the new truths that human beings create, and the richness of lived experience, aesthetic impulses, spiritual emotions, and personal affections. In words Margaret Donaldson would probably agree with, the life of the mind is not an adjustment to the actual relations with the world, but an ongoing creation of virtual relations, goals, and values guided by personal and societal interests and norms. This is the mental life not captured by a simple extension of an idea that may work in biology but proves limited when applied to comprehending the mind.

But are not enactivists champions of the continuity between life and mind? Should they not side with Spencer here? Most of what we have said in this book should point to a negative answer to this last question.

The term continuity can admit trivially right and trivially wrong interpretations, none of which is the one intended by enactivists and used in this book. On the "trivially right" side, continuity is merely seen as an affirmation that minds are empirically instantiated as biological organisms and that the kind of capabilities they exhibit will be constrained by biological processes. On the "trivially wrong" side, there is the complaint that enactivists are in danger of reducing mind to life and psychology to biology.

From the outset James sees these two aspects. He recognizes both the relational nature of mental life—"minds inhabit environments which act on them and on which they in turn react" (1890, p. 6), and that in this it resembles organic life. Yet, he warns against Spencer's functionalist flavor of life-mind continuity, according to which mental phenomena constitute simply another set of evolutionary adaptive functions. For James, the mark of the mental is an intrinsic teleology that cannot be reduced to the evolutionary fixation of such functions. Survival is too poor a condition for the richness of mental life to spring from it. In other words, what he criticizes is a reductive notion of continuity, one that fails to acknowledge the autonomy of mental life.

This intrinsic activity of the mind, that seeks and creates meaningful relations as well as adjusts to pre-existing external relations, is clearly manifested in the emergence of an open space of human interests. These are the true *a prioris* of cognition, according to James (1878). But where do they come from? How do they relate to the interest in survival ("*primus inter pares,* perhaps")? For James, this remained a mystery. On the question of how the mind—with its intrinsic teleology that cannot be reduced to outward facts—relates to the brain and the physical world, he advocated that a safe and verifiable psychology should content itself with studying the mere correspondence between mental and physical phenomena, in other words, with an "empirical parallelism" (James 1890, p. 182).

The enactive insistence on the continuity between life and mind is motivated precisely by these questions and promises a conceptual bridge connecting the two domains. Between triviality and nonsense lies a deeper meaning of the continuity thesis. In the fewest possible words: mental phenomena constitutively demand explanations of individuality, agency, and subjectivity, and the principles and conceptual categories for these explanations are the same as those required to explain the phenomenon of life. It is not— as James interprets the evolutionary continuity hypothesis—"that consciousness in some shape must have been present at the very origin of things" (an idea he mockingly refers to as "primordial mind-dust"), but, the idea that what he understands to be the mark of the mental, namely the "pursuance of future ends and the choice of means for their attainment" (1890, p. 8), or agency, in other words, can indeed be traced back to the organization of life at its origin.

This is comparable with the attitude adopted by John Dewey in his naturalistic theory of logic. According to him, the primary postulate of such a theory is the "continuity of the lower (less complex) and the higher (more complex) activities and forms" (Dewey 1938/1991, p. 30). Dewey maps the contour of the concept of continuity by making explicit what it excludes: a "complete rupture on one side and mere repetition of identities on the other; it precludes reduction of the 'higher' to the 'lower' just as it precludes complete breaks and gaps" (Dewey 1938/1991, p. 30). We have seen, in addition, compatible perspectives expressed elsewhere: apart from Francisco Varela and his colleagues, it is worth repeating in this respect the names of Hans Jonas, Maurice Merleau-Ponty, and Jean Piaget, all of them life-mind continuity supporters in their own ways.

Life-mind continuity, in this non-reductionist sense, entails a rejection of the sudden appearance of fully independent novel levels of description—for instance, the realm of human social normativity—without an account of how their emergence and relative autonomy are grounded on (understandable in terms of and in interaction with) phenomena at other levels. This is not only a historical or causal enquiry, for it is through the processes that give rise and constrain the new forms of autonomy that we can understand how these new forms are constituted, in short, what they are, at their most fundamental.

The continuity thesis therefore proposes the need for a theoretical path that links organic, mental, and social phenomena, that is, the three dimensions of human bodies proposed by Thompson and Varela. The project, however, remains non-reductionist for these three reasons: (1) it seeks explanations of emergent phenomena through theoretical and experimental investigations of self-organization and complex multiscale interactions; (2) it replaces the notion of an independence of levels of enquiry (e.g., biology, psychology, and sociology) with a concept of relative autonomy and postulates the conditions by which this autonomy can be investigated; and (3) it advances the possibility of various kinds of interactions between levels leading potentially to evolving forms of cross-level dependencies and transformations.

It is telling in this context that James often makes frequent use of the term *mental life* ("Psychology is the Science of Mental Life, both of its phenomena and conditions," he writes in capital letters as the opening sentence of his *Principles of Psychology*). As with sensorimotor life, this phrase is rich in connotations. It points to the mental realm as one that is *lived through*, one we inhabit and participate in. It also suggests that what characterizes this realm is that it is a *form of life*, a life that sustains itself, with its own values and norms, which is also fragile and may find itself threatened, get sick, or stop operating if the appropriate conditions disappear. We have described in this book how literal this idea can be once we understand that the principles for organic life (precarious self-individuation, sense-making, and agency) not only contribute to explaining the properties of sensorimotor life but also find their own instantiations in that domain. In this important and all-pervading region of mental life, we have seen how it is possible to answer the question left open by James's criticisms of biological reductionism: Where do human interests come from?

In this book, we have provided a partial answer to this question and a method to search further. Sensorimotor life is an autonomous dimension of the body, constrained and enabled by organic and by social life, but not entirely determined by them. The way sensorimotor life affirms itself is through the processes we have described in the preceding chapters, by creating organized and dynamic linkages between sensorimotor engagements, habits, dispositions, sensitivities, and skills that allow a sensorimotor agent to emerge as a distinct entity and sustain active and potentially open-ended meaningful relations to the world, relations she seeks in order to reassert herself.

The explanations we have provided do not start from the assumptions of functionalism, as we have stated repeatedly. We have criticized these assumptions but we have not devoted our energies to extend or refine these criticisms. Instead, we have provided positive proposals that are operational and therefore open to be tested, improved, or rejected. Our approach to questions about the organization of action and perception, about sensorimotor agency and experience, in short about mental life, stems from a deeply embodied, non-representational, and world-involving perspective.

The more we move further into an enactive understanding of mental life, as we go from the metabolic constitution of organisms to the individuation of sensorimotor acts, activities, and whole agents, the more we are confronted with fundamental matters that need to surface explicitly. In this book in particular, we have had to explore pivotal ontological and epistemological issues as we followed our lines of inquiry: the relational nature of the mind, the co-emergence of identity and interaction, the virtual fields that constitute the material to be sensed and transformed by agency, and the notion of the world as involved in all forms of sense-making beyond being a source of information.

Each time we "climb up" the ladder of the complexity of mental phenomena, we seem to make a corresponding move "downward" into a better understanding of the material, relational, and organic bases of life and mind. Like growth that pushes the branches upward and digs the roots deeper, the continuity of life and mind demands to be transited in both directions. This is a good sign. It confirms the enactive approach as a non-reductionist yet scientifically engaged philosophy of nature.

Bibliography

Abney, D. H., Paxton, A., Dale, R., and Kello, C. T. (2014). Complexity matching in dyadic conversation. *Journal of Experimental Psychology General* 143, 2304–15. doi:10.1037/xge0000021.

Adler, J. (1966). Chemotaxis in bacteria. *Science* 153(3737), 708–16.

Agmon, E. and Beer, R. D. (2014). The dynamics of action switching in an evolved agent. *Adaptive Behavior* 22, 3–20. doi: 10.1177/1059712313511649.

Agre, P. (1988). *The dynamic structure of everyday life*. MIT AI Lab Technical Report 1085.

Agre, P. E. (1997). *Computation and human experience*. Cambridge: Cambridge University Press.

Aguilera, M. (2015). *Interaction dynamics and autonomy in cognitive systems, from sensorimotor coordination to collective action* (PhD Thesis). Universidad de Zaragoza, Spain. Retrieved from http://maguilera.net/phdthesis.

Aguilera, M., Barandiaran, X. E., Bedia, M. G., et al. (2015). Self-organized criticality, plasticity and sensorimotor coupling. Explorations with a neurorobotic model in a behavioural preference task. *PLoS ONE* 10(2), e0117465. doi: 10.1371/journal.pone.0117465.

Aguilera, M., Bedia, M. G., Santos, B. A., and Barandiaran, X. E. (2013). The situated HKB model: how sensorimotor spatial coupling can alter oscillatory brain dynamics. *Frontiers in Computational Neuroscience* 7, 117. doi: 10.3389/fncom.2013. 00117.

Alexander, B. K., Coambs, R. B., and Hadaway, P. F. (1978). The effect of housing and gender on morphine self-administration in rats. *Psychopharmacology* 58, 175–9.

Alexandre, G. (2010). Coupling metabolism and chemotaxis-dependent behaviours by energy taxis receptors. *Microbiology* 156(8), 2283–93.

Alexandre, G. and Zhulin, I. B. (2001). More than one way to sense chemicals. *Journal of Bacteriology* 183(16), 4681–6. doi:10.1128/JB.183.16.4681-4686.2001.

Alon, U., Surette, M. G., Barkai, N., and Leibler, S. (1999). Robustness in bacterial chemotaxis. *Nature* 397(6715), 168–71. doi:10.1038/16483.

Alsmith, A. J. T. (2015). Mental activity and the sense of ownership. *Review of Philosophy and Psychology* 6, 881–96. doi: 10.1007/s13164-014-0208-1.

Apperly, I. (2010). *Mindreaders: the cognitive basis of "Theory of Mind."* Hove, East Sussex: Psychology Press.

Arbib, M.A., Érdi P., and Szentágothai J. (1998). *Neural organization: structure, function, and dynamics*. Cambridge, MA: MIT Press.

Ashby, W. R. (1960). *Design for a brain*. 2nd ed. New York: John Wiley.

Auvray, M., Hanneton, S., and O'Regan, J. K. (2007). Learning to perceive with a visuo—auditory substitution system: localization and object recognition with "The vOICe." *Perception* 36, 416–30. doi: 10.1068/p5631.

Auvray, M. and Myin, E. (2009). Perception with compensatory devices. From sensory substitution to sensorimotor extension. *Cognitive Science* 33, 1036–58. doi: 10.1111/j.1551-6709.2009.01040.x

Aytekin, M., Moss, C.F., and Simon, J. Z. (2008). A sensorimotor approach to sound localization. *Neural Computation* 20, 603–35. doi: 10.1162/neco.2007.12-05-094.

Bach-y-Rita, P. (1967). Sensory plasticity. *Acta Neurologica Scandinavica* 43, 417–26. doi: 10.1111/j.1600-0404.1967.tb05747.x.

Bach-y-Rita, P. (1972). *Brain mechanisms in sensory substitution*. New York: Academic Press.

Bach-y-Rita P. (2004). Tactile sensory substitution studies. *Annals of New York Academy of Sciences* 1013, 83–91.

Bandura, A. (2006). Towards a psychology of human agency. *Perspectives on Psychological Science* 1(2), 164–80.

Barandiaran, X. E. (2007). Mental life: conceptual models and synthetic methodologies for a post-cognitivist psychology. In: **B. Wallace, A. Ross, J. Davies,** and **T. Anderson** (eds.), *The world, the mind and the body: psychology after cognitivism*. Exeter: Imprint Academic, pp. 49–90.

Barandiaran, X. E. (2008). *Mental Life: A Naturalized Approach to the Autonomy of Cognitive Agents*. (Unpublished PhD Thesis). University of the Basque Country (UPV–EHU), Donostia–San Sebastián, Gipuzkoa, Spain. http://www.barandiaran.net/phdthesis/

Barandiaran, X. E. (2016). Autonomy and enactivism: towards a theory of sensorimotor autonomous agency. *Topoi* 1–22. (in press). doi:10.1007/s11245-016-9365-4.

Barandiaran, X. E. and **Di Paolo, E. A.** (2008). Artificial mental life. In: **S. Bullock, J. Noble, R. A. Watson,** and **M. A. Bedau** (eds.), *Artificial life XI: Proceedings of the Eleventh International Conference on the Simulation and Synthesis of Living Systems*. Cambridge, MA: MIT Press, p. 747.

Barandiaran, X. E. and **Di Paolo, E. A.** (2014). A genealogical map of the concept of habit. *Frontiers in Human Neuroscience* 8, 522. doi: 10.3389/fnhum.2014.00522.

Barandiaran, X. E., **Di Paolo, E.,** and **Rohde, M.** (2009). Defining agency: Individuality, normativity, asymmetry, and spatio-temporality in action. *Adaptive Behavior* 17(5), 367–86.

Barandiaran, X. and **Egbert, M.** (2014). Norm-establishing and norm-following in autonomous agency. *Artificial Life* 20, 5–28.

Barandiaran, X. E. and **Moreno, A.** (2006). On what makes certain dynamical systems cognitive: A minimally cognitive organization program. *Adaptive Behavior* 14(2), 171–85.

Barandiaran, X. E. and **Moreno, A.** (2008). Adaptivity: from metabolism to behavior. *Adaptive Behavior* 16(5), 325–44. doi:10.1177/1059712308093868.

Barker, R. G. (1968). *Ecological psychology: concepts and methods for studying the environment of human behavior*. Palo Alto, CA: Stanford University Press.

Barrett, N. (2015). The normative turn in enactive theory: An examination of its roots and implications. *Topoi* (online first), doi:10.1007/s11245-015-9355-y.

Barsalou, L. (2008). Grounded cognition. *Annual Review of Psychology* 59, 617–45.

Bayliss, A. P., Frischen, A., Fenske, M. J., and **Tipper, S. P.** (2007). Affective evaluations of objects are influenced by observed gaze direction and emotional expression, *Cognition* 104, 644–53.

Bayne, T. (2011). The sense of agency. In: **F. Macpherson** (ed.), *The senses: classic and contemporary philosophical perspectives*. Oxford: Oxford University Press, pp. 355–74.

Beaton, M. (2013). Phenomenology and embodied action. *Constructivist Foundations* 8(3), 298–313.

Bedia, M. G., Aguilera, M., Gomez, T., Larrode, D. G., and **Seron, F.** (2014). Quantifying long-range correlations and 1/f patterns in a minimal experiment of social interaction. *Frontiers in Psychology* 5, 1281. doi:10.3389/fpsyg.2014.01281.

Beer, R. D. (1990). *Intelligence as adaptive behaviour: an experiment in computational neuroethology (perspectives in artificial intelligence)*. San Diego, CA: Academic Press.

Beer, R. D. (1997). The dynamics of adaptive behavior: a research program. *Robotics and Autonomous Systems* 20, 257–89. doi:10.1016/S0921-8890(96) 00063-2.

Beer, R. D. (2003). The dynamics of active categorical perception in an evolved model agent. *Adaptive Behavior* 11, 209–43. doi:10.1016/j.neunet.2009.03.002.

Beer, R. D. and **Williams, P. L.** (2015). Information processing and dynamics in minimally cognitive agents. *Cognitive Science* 39, 1–38. doi: 10.1111/cogs.12142.

Bermúdez, J. L. (2015). Bodily ownership, bodily awareness and knowledge without observation. *Analysis* **75**, 37–45. doi:10.1093/analys/anu119.

Bernstein, N. A. (1967). *The coordination and regulation of movement*. London: Pergamon.

Bernstein, N. A. (1996). *Dexterity and its development*. Ed. **M. L. Latash** and **M. T. Turvey**. Hillsdale, NJ: Erlbaum.

Berwick, R. C., Okanoya, K., Beckers, G. J., and Bolhuis, J. J. (2011). Songs to syntax: the linguistics of birdsong. *Trends in Cognitive Sciences* **15(3)**, 113–21.

Bennett, M. R. and Hacker, P. M. S. (2003). *Philosophical foundations of neuroscience*. Oxford: Blackwell.

Bermejo, F., Di Paolo, E. A., Hüg, M. X., and Arias, C. (2015). Sensorimotor strategies for recognizing geometrical shapes: a comparative study with different sensory substitution devices. *Front Psychol* **6**, 679. doi: 10.3389/fpsyg.2015.00679.

Bird, J. and Di Paolo, E. A., (2008). Gordon Pask and his maverick machines. In: **P. Husbands, M. Wheeler,** and **O. Holland** (eds.), *The mechanical mind in history*. Cambridge, MA: MIT Press, pp. 185–211.

Bird J. and Layzell, P. (2002). The evolved radio and its implications for modelling the evolution of novel sensors. In: **D. B. Fogel, M. A. El-Sharkawi, X. Yao, G. Greenwood, H. Iba, P. Marrow,** and **M. Shackleton** (eds.), *Proceedings of the 2002 Congress on Evolutionary Computation CEC2002*. Piscataway, NJ: IEEE Press, pp.1836–41.

Bloodgood, R. A. (2010). Sensory reception is an attribute of both primary cilia and motile cilia. *Journal of Cell Science* **123**, 505–09.

Boden, M. (1972). *Purposive explanations in psychology*. Cambridge, MA: Harvard University Press.

Boden, M. (2006). *Mind as machine: a history of cognitive science*. 2 vols. Oxford: Oxford University Press.

Bodrova, E. and Leong, D. J. (1998). Development of dramatic play in young children and its effects on self-regulation: the Vygotksian approach. *Journal of Early Childhood Teacher Education* **19(2)**, 115–24.

Boom, J. (2009). Piaget on equilibration. In: **U. Müller, J. I. M. Carpendale,** and **L. Smith** (eds.), *The Cambridge companion to Piaget*. Cambridge: Cambridge University Press, pp. 132–49.

Boutonnet, B. and Lupyan, G. (2015). Words jump-start vision: a label advantage in object recognition. *Journal of Neuroscience* **35(25)**, 9329–35.

Braitenberg, V. (1984). *Vehicles: experiments in synthetic psychology*. Cambridge, MA: MIT Press.

Brincker, M. (2014). Navigating beyond "here & now" affordances—on sensorimotor maturation and "false belief" performance. *Frontiers in Psychology* **5**, 1433. doi: 10.3389/fpsyg.2014.01433.

Brooks, R. (1991). Intelligence without representation. *Artificial Intelligence* **47**, 139–59. doi:10.1016/0004-3702(91)90053-M.

Brown, R. W. and Lenneberg, E. H. (1954). A study in language and cognition. *Journal of Abnormal Psychology* **49**, 454–62.

Bruner, J. (1990). *Acts of meaning*. Cambridge, MA: Harvard University Press.

Buhrmann, T., Di Paolo, E. A., and Barandiaran, X. (2013). A dynamical systems account of sensorimotor contingencies. *Frontiers in Psychology* **4**, 285. doi: 10.3389/fpsyg.2013.00285.

Buhrmann, T. and Di Paolo, E. A. (2014a). Spinal circuits can accommodate interaction torques during multijoint limb movements. *Frontiers in Computational Neuroscience* **8**, 144. doi: 10.3389/fncom.2014.00144.

Buhrmann, T. and Di Paolo, E. A. (2014b). Non-representational sensorimotor knowledge. In: **A. P. del Pobil, E. Chinellato, E. Martinez-Martin, J. Hallam, E. Cervera,** and **A. Morales** (eds.), *From* *animals to animats 13, Proceedings of the 13th International Conference on Simulation of Adaptive Behavior*. Heidelberg: Springer, pp. 21–31.

Buhrmann, T. and Di Paolo, E. A. (2015). The sense of agency: A phenomenological consequence of enacting sensorimotor schemes. *Phenomenology and the Cognitive Sciences*, doi: 10.1007/ s11097- 015- 9446- 7. (Online first).

Bullmore, E. and Sporns, O. (2009). Complex brain networks: graph theoretical analysis of structural and functional systems. *Nature Reviews Neuroscience* 10(3), 186–98.

Buytendijk, F. J. J. (1958). *Mensch und Tier. Ein Beitrag zur vergleichenden Psychologie*. Reinbek: Rowohlt Verlag.

Buytendijk, F. J. J. (1970). Some aspects of touch. *Journal of Phenomenological Psychology* 1(1), 99–122.

Buzsáki, G. (2010). Neural syntax: cell assemblies, synapsembles, and readers. *Neuron* 68(3), 362–85.

Byrge, L., Sporns, O., and Smith, L. B. (2014). Developmental process emerges from extended brain-body–behavior networks. *Trends in Cognitive Sciences* 18(8), 395–403.

Calvo, F. (2008). Towards a general theory of antirepresentationalism. *The British Journal for the Philosophy of Science* 59(3), 259–92.

Canguilhem, G. (1966/1991). *The normal and the pathological*. New York: Zone Books.

Caracciolo, M. (2012). Narrative, meaning, interpretation: an enactivist approach. *Phenomenology and the Cognitive Sciences* 11, 367–84.

Cariani, P. (1993). To evolve an ear: epistemological implications of Gordon Pask's electrochemical devices. *Systems Research* 10, 19–33.

Carlisle, C. (2014). *On habit*. London: Routledge.

Cassirer, E. (1944). The concept of group and the theory of perception. *Philosophy and Phenomenological Research* 5(1), 1–36.

Castellini, M. A., Davis, R. W., and Kooyman, G. L. (1992). Annual cycles of diving behavior and ecology of the Weddell seal. *Bulletin of the Scripps Institution of Oceanography* 28, 1–54.

Čejková, J., Novák, M., Štěpánek, F., and Hanczyc, M. M. (2014). Dynamics of chemotactic droplets in salt concentration gradients. *Langmuir* 30, 11937–44. doi: 10.1021/la502624f.

Chambon, V., Sidarus, N., and Haggard, P. (2014). From action intentions to action effects: how does the sense of agency come about? *Front Hum Neurosci* 8, 320. doi: 10.3389/fnhum.2014.00320.

Chapman, S. (1968). Catching a baseball. *Am J Phys* 36, 868–70. doi:10.1119/1.1974297.

Chapman, M. (1992). Equilibration and the dialectics of organization. In: H. Beilin and P. B. Pufall (eds.), *Piaget's theory: prospects and possibilities*. Hillsdale, New Jersey: Lawrence Erlbaum, pp. 39–59.

Chemero, A. (2009). *Radical embodied cognitive science*. Cambridge, MA: MIT Press.

Chiel, H. J. and Beer, R. D. (2009). Computational neuroethology. In L. Squire (ed.), *Encyclopedia of neuroscience, Vol. 3*. Amsterdam: Elsevier, pp. 23–28.

Christensen, W., Bicknell, K., McIlwain, D., and Sutton, J. (2015). The sense of agency and its role in strategic control for expert mountain bikers. *Psychology of Consciousness: Theory, Research, and Practice* 2(3), 340–53.

Cisek, P. (2007). Cortical mechanisms of action selection: the affordance competition hypothesis. *Philosophical Ttransactions of the Royal Society of London B: Biological Sciences* 362(1485), 1585–99.

Cisek, P. and Kalaska, J. F. (2010). Neural mechanisms for interacting with a world full of action choices. *Annual Review of Neuroscience* 33, 269–98.

Clark, A. (1996). Happy couplings: emergence and explanatory interlock. In: M. A. Boden (ed.), *The philosophy of artificial life*. Oxford: Oxford University Press, pp. 262–81.

Clark, A. (1997). *Being there: putting brain, body and world together again*. Cambridge: MA: MIT Press.

Clark, A. (2013). Whatever next? Predictive brains, situated agents, and the future of cognitive science. *Behavioral and Brain Sciences* **36**(3), 181–204.

Clark, A. and Chalmers, D. (1998). The extended mind. *Analysis* **58**(1), 7–19.

Clark A. and Grush, R. (1999). Towards a cognitive robotics. *Adaptive Behavior* **7**(1), 5–16.

Clark A. and Thornton, C. (1997). Trading spaces: computation, representation and the limits of uninformed learning. *Behavioral and Brain Sciences* **20**, 57–90.

Clark, A. and Toribio, J. (1994). Doing without representing? *Synthese* **101**, 401–431.

Cliff, D. (1991). Computational neuroethology: a provisional manifesto. In: J.-A. Meyer and S. W. Wilson (eds.), *From animals to animats. Proceedings of the First International Conference on Simulation of Adaptive Behaviour*. Cambridge MA: MIT Press, pp. 29–39.

Colombetti, G. (2014). *The feeling body: affective science meets the enactive mind*. Cambridge, MA: MIT Press.

Colombetti, G. and Torrance, S. (2009). Emotion and ethics: an inter-(en)active approach. *Phenomenology and the Cognitive Sciences* **8**, 505–26.

Cornilleau-Pérès, V. and Droulez, J. (1994). The visual perception of 3D shape from self-motion and object-motion. *Vision Research* **34**, 2331–36. doi: 10.1016/0042-6989(94)90279-8.

Costall, A. (1995). Socializing affordances. *Theory & Psychology* **5**, 467–81.

Craik, K. J. W. (1943). *The nature of explanation*. Cambridge: Cambridge University Press.

Csikszentmihályi, M. (1990). *Flow: the psychology of optimal experience*. New York: Harper & Row.

Cuffari, E., Di Paolo, E. A., and De Jaegher, H. (2015). From participatory sense-making to language: there and back again. *Phenomenology and the Cognitive Sciences* **14**(4), 1089–25.

Cummins, F. (2013). Towards an enactive account of action: speaking and joint speaking as exemplary domains. *Adaptive Behavior* **13**(3), 178–86.

David, N., Newen, A., and Vogeley, K. (2008). The "sense of agency" and its underlying cognitive and neural mechanisms. *Consciousness and Cognition* **17**(2), 523–34.

Dehaene, S., Kerszberg, M., and Changeux, J.-P. (1998). A neuronal model of a global workspace in effortful cognitive tasks. *Proceedings of the National Academy of Sciences USA* **95**, 14529–34.

de Barbaro, K., Johnson, C. M., and Deák, G. O. (2013). Twelve-month "social revolution" emerges from mother-infant sensorimotor coordination: A longitudinal investigation. *Human Development* **56**, 223–48.

De Jaegher H. (2013). Embodiment and sense-making in autism. *Frontiers in Integrative Neuroscience* **7**, 15.

De Jaegher, H. and Di Paolo, E. A. (2007). Participatory sense-making: an enactive approach to social cognition. *Phenomenology and the Cognitive Sciences* **6**, 485–507.

De Jaegher, H, Di Paolo, E. A., and Gallagher, S. (2010). Can social interaction constitute social cognition? *Trends in Cognitive Sciences* **14**, 441–7.

De Jaegher, H. and Froese, T. (2009). On the role of social interaction in individual agency. *Adaptive Behavior* **17**(5), 444–60. doi:10.1177/1059712309343822.

de Vignemont, F. (2013). The mark of bodily ownership. *Analysis* **73**, 643–51. doi:10.1093/analys/ant080.

de Vignemont, F. and Fourneret, P. (2004). The sense of agency: a philosophical and empirical review of the "who" system. *Consciousness and Cognition* **13**, 1–19. doi:10.1016/S1053-8100(03)00022-9.

Degenaar, J. and Myin, E. (2014). Representation-hunger reconsidered. *Synthese* **191**, 3639–48.

Delevoye-Turrell, Y., Giersch, A., and Danion, J.-M. (2003). Abnormal sequencing of motor actions in patients with schizophrenia: evidence from grip force adjustments during object manipulation. *The American Journal of Psychiatry* **160**, 134–41.

Dennett, D. C. (1993). Review of F. Varela, E. Thompson and E. Rosch, *The embodied mind*. *American Journal of Psychology* **106**, 121–6.

Dennett, D. C. (2013). *Intuition pumps and other tools for thinking*. New York: W. W. Norton & Company.

Deregowski, J. B. (1989). Real space and represented space: cross-cultural perspectives. *Behavioral and Brain Sciences* **12**, 51–119.

Dewey, J. (1896). The reflex arc concept in psychology. *Psychological Review* **3**, 357–70.

Dewey, J. (1929). *Experience and nature*. London: George Allen & Unwin.

Dewey, J. [1938] (1991). *Logic: the theory of inquiry*.Carbondale: Southern Illinois University Press

Di Paolo, E. A. (2000a). Behavioral coordination, structural congruence and entrainment in a simulation of acoustically coupled agents. *Adaptive Behavior* **8**, 25–46. doi: 10.1177/105971230000800103.

Di Paolo, E. A. (2000b). Homeostatic adaptation to inversion of the visual field and other sensorimotor disruptions. In: **J.-A. Meyer, A. Berthoz, D. Floreano, H. Roitblat**, and **S. W. Wilson** (eds.), *From animals to animats 6. Proceedings of the Sixth International Conference on Simulation of Adaptive Behavior*. Cambridge, MA: MIT Press, pp. 440–9.

Di Paolo, E. A. (2003). Organismically-inspired robotics: homeostatic adaptation and teleology beyond the closed sensorimotor loop. In: **K. Murase** and **Asakura** (eds.), *Dynamical systems approaches to embodiment and sociality*. Adelaide: Advanced Knowledge International, pp. 19–42.

Di Paolo, E. A. (2005). Autopoiesis, adaptivity, teleology, agency. *Phenomenology and the Cognitive Sciences* **4**, 429–52. doi:10.1007/s11097-005-9002-y.

Di Paolo, E. A. (2009). Extended life. *Topoi* **28**, 9–21. doi: 10.1007/s11245-008- 9042-3.

Di Paolo, E. A. (2010). Robotics inspired in the organism. *Intellectica* **53/54**, 129–62.

Di Paolo, E. A. (2014). The worldly constituents of perceptual presence. *Frontiers in Psychology* **5**, 450. doi: 10.3389/fpsyg.2014.00450.

Di Paolo, E. (2015). Interactive time-travel: on the intersubjective retro-modulation of intentions. *Journal of Consciousness Studies* **22**, 49–74.

Di Paolo, E. A. (2016). Participatory object perception. *Journal of Consciousness Studies* **23(5–6)**, 228–58.

Di Paolo, E. A. (2017) (forthcoming). The enactive conception of life. In: **A. Newen, S. Gallagher**, and **L. de Bruin** (eds.), *The Oxford handbook of cognition: embodied, embedded, enactive and extended*. Oxford: Oxford University Press.

Di Paolo, E. A., Barandiaran, X. E., Beaton, M. and Buhrmann, T. (2014). Learning to perceive in the sensorimotor approach: Piaget's theory of equilibration interpreted dynamically. *Frontiers in Human Neuroscience* **8**, 551

Di Paolo, E. A. and De Jaegher, H. (2012). The interactive brain hypothesis. *Frontiers in Human Neuroscience* **6**, 163, doi: 10.3389/fnhum.2012.00163.

Di Paolo, E. A. and Iizuka, H. (2008). How (not) to model autonomous behaviour. *Biosystems* **91**, 409–23. doi: 10.1016/j.biosystems.2007.05.016.

Di Paolo, E. A., Noble, J., and Bullock, S. (2000). Simulation models as opaque thought experiments. In: **M. A. Bedau, J. S. McCaskill, N. H. Packard**, and **S. Rasmussen** (eds.), *Artificial life VII: The Seventh International Conference on the Simulation and Synthesis of Living Systems, Reed College, Portland, Oregon, USA, 1–6 August, 2000*. Cambridge, MA: MIT Press, pp. 497–506.

Di Paolo, E. A., Rohde, M., and De Jaegher, H. (2010). Horizons for the enactive mind: values, social interaction and play. In: **J. Stewart, O. Gapenne**, and **E. A. Di Paolo** (eds.), *Enaction: towards a new paradigm of cognitive science*. Cambridge, MA: MIT Press, pp. 33–87.

Di Paolo, E. A. and Thompson, E. (2014). The enactive approach. In: **L. Shapiro** (ed.), *The Routledge handbook of embodied cognition*. London: Routledge Press, pp. 68–78.

Doise, W., Mugny, G., and Perret-Clermont, A-N. (1975). Social interaction and the development of cognitive operations. *European Journal of Social Psychology* **5(3)**, 367–83.

Doise, W. and **Mugny, G.** (1984). *The social development of the intellect.* Oxford: Pergamon.

Donaldson, M. (1992). *Human minds: an exploration.* London: Penguin Books.

Dorogovtsev, S. N., Goltsev, A. V., and **Mendes, J. F. F.** (2008). Critical phenomena in complex networks. *Review of Modern Physics* **80**, 1275.

Dotov, D. G., Nie, L., and **Chemero, A.** (2010). A demonstration of the transition from ready-to-hand to unready-to-hand. *PLoS ONE* **5**, e9433. doi: 10. 1371/journal.pone.0009433.

Dow, J. M. (2015). Just doing what I do: on the awareness of fluent agency. *Phenomenology and the Cognitive Sciences* (online first) doi: 10.1007/s11097-015-9445-8.

Dretske, F. (1981). *Knowledge and the flow of information.* Cambridge, MA: MIT Press.

Dretske, F. (2003). Experience as representation. *Philosophical Issues* **13**, 67–82. doi:10.1111/ 1533-6077.00005.

Dreyfus, H. L. (1965). *Alchemy and artificial intelligence.* The RAND Corporation Paper P-3244.

Dreyfus, H. L. (1967). Why computers must have bodies in order to be intelligent. *Review of Metaphysics* **21**, 13–32.

Dreyfus, H. L. (1972). *What computers can't do: a critique of artificial reason.* New York: Harper and Row.

Dreyfus, H. L. (2006). Overcoming the myth of the mental. *Topoi* **25(1–2)**, 43–9.

Dreyfus, H. L. (2007a). The return of the myth of the mental. *Inquiry* **50(4)**, 352–65.

Dreyfus, H. L. (2007b). Response to McDowell. *Inquiry* **50(4)**, 371–7. doi:10.1080/00201740701489401.

Dreyfus, H. L. and **Taylor, C.** (2015). *Retrieving realism.* Cambridge, MA: Harvard University Press.

Dubois, D. M. (2003). Mathematical foundations of discrete and functional systems with strong and weak anticipations. In: **M. V. Butz, O. Sigaud,** and **P. Gérard.** *Anticipatory behavior in adaptive learning systems.* Berlin: Springer. pp. 110–32.

Duff, A., Fibla, M. S., and **Verschure, P. F. M. J.** (2011). A biologically based model for the integration of sensory-motor contingencies in rules and plans: a prefrontal cortex based extension of the distributed adaptive control architecture. *Brain Research Bulletin* **85**, 289–304. doi: 10.1016/ j.brainresbull.2010. 11.008.

Duffy, S., Toriyama, R., Itakura, S., and **Kitayama, S.** (2009). Development of cultural attention strategies in young children in North America and Japan. *Journal of Experimental Child Psychology* **102**, 351–9.

Dumas, G., Kelso, J. A. S., and **Nadel, J.** (2014). Tackling the social cognition paradox through multi-scale approaches. *Frontiers in Psychology* **5**, 882. doi:10.3389/fpsyg.2014.00882.

Dupuy, J-P. (2000). *The mechanization of the mind: on the origins of cognitive science.* Princeton University Press.

Edelman, G. M. (1987). *Neural Darwinism: the theory of neuronal group selection.* New York: Basic Books.

Egbert M. D. and **Barandiaran X. E.** (2014). Modeling habits as self-sustaining patterns of sensorimotor behavior. *Frontiers in Human Neuroscience* **8**, 590. doi: 10.3389/fnhum.2014.00590.

Egbert, M. and **Di Paolo, E. A.** (2009). Integrating behavior and autopoiesis: an exploration in computational chemo-ethology. *Adaptive Behavior* **17**, 387–401.

Egbert, M. D., Barandiaran, X. E., and **Di Paolo, E. A.** (2010). A minimal model of metabolism-based chemotaxis. *PLoS Computational Biology* **6**, e1001004.

Elias, J. Z. (2015). Tensions theoretical and theorized: comment on Di Paolo's "Interactive time-travel." *Journal of Consciousness Studies* **22(1–2)**, 75–9.

Engel, A. K., Fries, P., and **Singer, W.** (2001). Dynamic predictions: oscillations and synchrony in top–down processing. *Nature Reviews Neuroscience* **2(10)**, 704–16.

Engel, A. K., Maye, A., Kurthen, M., and **König, P.** (2013). Where's the action? The pragmatic turn in cognitive science. *Trends in Cognitive Sciences* **17(5)**, 202–9.

Epley, N., Morewedge, C. K., and Keysar, B. (2004). Perspective taking in children and adults: equivalent egocentrism but differential correction. *Journal of Experimental Social Psychology* **40**, 760–8.

Erlhagen, W. and Schöner, G. (2002). Dynamic field theory of movement preparation. *Psychological Review* **109**, 545–72.

Farina, M. (2013). Neither touch nor vision: sensory substitution as artificial synaesthesia? *Biology and Philosophy* **28**, 639–55. doi: 10.1007/s10539-013-9377-z.

Farrer, C. and Frith, C. D. (2002). Experiencing oneself vs another person as being the cause of an action: the neural correlates of the experience of agency. *Neuroimage* **15(3)**, 596–603.

Farrer, C., Franck, N., Georgieff, N., Frith, C. D., Decety, J., and Jeannerod, M. (2003). Modulating the experience of agency: a positron emission tomography study. *Neuroimage* **18(2)**, 324–33.

Farrer, C., Frey, S. H., Van Horn, J. D., Tunik, E., Turk, D., Inati, S., and Grafton, S. T. (2008). The angular gyrus computes action awareness representations. *Cerebral Cortex* **18(2)**, 254–61.

Feinberg, I. (1978). Efference copy and corollary discharge: implications for thinking and its disorders. *Schizophrenia Bulletin* **4(4)**, 636–40.

Fentress, J. C. (1983). The analysis of behavioral networks. In: J-P Ewert, R. R. Capranica, and D. J. Ingle (eds.), *Advances in vertebrate neuroethology*. New York: Plenum Press, pp. 939–68.

Fentress, J. C. (1984). The development of coordination. *Journal of Motor Behavior* **16(2)**, 99–134.

Fentress, J. C. and Gadbois, S. (2001). The development of action sequences. In: E.M. Blass (ed.), *Handbook of behavioral neurobiology, Vol. 13: developmental psychobiology, developmental neurobiology and behavioral ecology: mechanisms and early principles*. New York: Kluwer Academic, pp. 393–431.

Fernando, C., Szathmáry, E., and Husbands, P. (2012). Selectionist and evolutionary approaches to brain function: a critical appraisal. *Frontiers in Computational Neuroscience* **6**, 24. doi: 10.3389/fncom.2012.00024.

Fischer, K. W. (1980). A theory of cognitive development: the control and construction of hierarchies of skills. *Psychological Review* **87**, 477–531.

Fischer, K. W. and Bidell, T. R. (1998). Dynamic development of psychological structures in action and thought. In: R. M. Lernerand and W. Damon (eds.), *Handbook of child psychology in theoretical models of human development (Vol. 1)*. 5th ed., New York: Plenum Press, pp. 117–21.

Fischer, K. W. and Bidell, T. R. (2006). Dynamic development of action, thought, and emotion. In: W. Damon and R. M. Lerner (eds.), *Theoretical models of human development. handbook of child psychology Vol. 1* (6th ed.). New York: Wiley. pp. 313–99.

Fischer, K. W. and van Geert, P. (2014). Dynamic development of brain and behavior. In: P. C. M. Molenaar, R. M. Lerner, and K. M. Newell (eds.) *Handbook of developmental systems theory and methodology*. New York: Guilford Press, pp. 287–315.

Fischer, K. W., Yan, Z. and Stewart, J. (2003). Adult cognitive development: Dynamics in the developmental web. In: J. Valsiner and K. Connolly (eds.), *Handbook of developmental psychology*. Thousand Oaks, CA: Sage, 491–516.

Fodor, J. (1981). *Representations*. Cambridge, MA: MIT Press.

Foucher, J., Vidailhet, P., Chanraud, S., Gounot, D., Grucker, D., Pins, D., et al. (2005). Functional integration in schizophrenia: too little or too much? Preliminary results on fMRI data. *Neuroimage* **26(2)**, 374–88.

Frankfurt, H. G. (1978). The problem of action. *American Philosophical Quarterly* **15(2)**, 157–62.

Fridland, E. (2014). They've lost control: reflections on skill. *Synthese* **191(12)**, 2729–50.

Friston, K. J. (1999). Schizophrenia and the disconnection hypothesis. *Acta Psychiatrica Scandinavica* **99(s395)**, 68–79.

Friston, K. J. (2010). The free-energy principle: a unified brain theory? *Nature Reviews Neuroscience* **11(2)**, 127–38.

Frith, C. (2005). The self in action: lessons from delusions of control. *Consciousness and Cognition* **14**, 752–70. doi:10.1016/j.concog.2005.04.002.

Frith, C. D., Blakemore, S. J., and **Wolpert, D. M.** (2000a). Abnormalities in the awareness and control of action. *Philosophical Transactions of the Royal Society of London. Series B, Biological Sciences* **355**, 1771–88. doi:10.1098/rstb.2000.0734.

Frith, C. D., Blakemore, S., and **Wolpert, D. M.** (2000b). Explaining the symptoms of schizophrenia: abnormalities in the awareness of action. *Brain Research Reviews* **31(2–3)**, 357–63. doi:10.1016/S0165-0173(99)00052-1.

Froese, T. (2014). Steps toward an enactive account of synesthesia. *Cognitive Neuroscience*, 5(2), 126–127.

Froese, T. and **Di Paolo, E. A.** (2009). Sociality and the life-mind continuity thesis. *Phenomenology and the Cognitive Sciences* **8(4)**, 439–63.

Froese, T. and **Cappuccio, M.** (eds.) (2014). *Enactive cognition at the edge of sense-making: making sense of non-sense.* Palgrave-Macmillan.

Froese, T. and **Ikegami, T.** (2013). The brain is not an isolated "black box," nor is its goal to become one. *Behavioral and Brain Sciences* **36**, 213–14. doi:10.1017/S0140525X12002348.

Froese, T., **Lenay, C.,** and **Ikegami, T.** (2012). Imitation by social interaction? Analysis of a minimal agent-based model of the correspondence problem. *Frontiers in Human Neuroscience* **6**, 202.

Froese, T., **Virgo, N.,** and **Ikegami, T.** (2014). Motility at the origin of life: its characterization and a model. *Artificial Life* **20**, 55–76.

Froese, T., **Woodward, A.,** and **Ikegami, T.** (2013). Turing instabilities in biology, culture, and consciousness? On the enactive origins of symbolic material culture. *Adaptive Behavior* **21**, 199–214.

Froese, T. and **Ziemke, T.** (2009). Enactive artificial intelligence: Investigating the systemic organization of life and mind. *Artificial Intelligence* **173**, 466–500.

Fuchs, T. (2009). Embodied cognitive neuroscience and its consequences for psychiatry. *Poiesis & Praxis* **6(3–4)**, 219–33.

Fuchs, T. (2011). The brain—a mediating organ. *Journal of Consciousness Studies* **18**, 196–221.

Fuchs, T. (2012). The phenomenology and development of social perspectives. *Phenomenology and the Cognitive Sciences* **12**, 655–83.

Gallagher, S. (2000). Philosophical conceptions of the self: implications for cognitive science. *Trends in Cognitive Sciences* **4**, 14–21.

Gallagher, S. (2008). Intersubjectivity in perception. *Continental Philosophy Review* **41(2)**, 163–78.

Gallagher, S. (2007). The natural philosophy of agency. *Philosophy Compass* **2**, 347–57.

Gallagher, S. (2009). Two problems of intersubjectivity. *Journal of Consciousness Studies* **16**, 289–308.

Gallagher, S. (2012). Multiple aspects in the sense of agency. *New Ideas in Psychology* **30**, 15–31. doi:10.1016/j.newideapsych.2010.03.003.

Gallagher, S. (2014). In the shadow of the transcendental: social cognition in Merleau-Ponty and cognitive science. In: **K. Novotny, P. Rodrigo, J. Slatman,** and **S. Stoller** (eds.), *Corporeity and affectivity.* Leiden: Brill, pp. 149–58.

Gallagher, S. (2015a). Invasion of the body snatchers: how embodied cognition is being disembodied. *The Philosophers' Magazine* **(April, 2015)**, 96–102.

Gallagher, S. (2015b). The problem with 3-year olds. *Journal of Consciousness Studies* **22**, 160–82.

Gallagher, S., **Hutto, D., Slaby, J.,** and **Cole, J.** (2013). The brain as part of an enactive system. *Behavioral and Brain Sciences* **36**, 421–22.

Gallagher, S. and Miyahara, K. (2012). Neo-pragmatism and enactive intentionality. In: J. Schulkin (ed.), *Action, perception and the brain: adaptation and cephalic expression*. Basingstoke: Palgrave Macmillan, pp. 117–46.

Gendlin, E. (1962/1997). *Experiencing and the creation of meaning: a philosophical and psychological approach to the subjective*. Evanston, IL: Northwestern University Press.

Gibson, J. J. (1979). *The ecological approach to visual perception*. London: Routledge.

Goldman, A. (2006). *Simulating minds: the philosophy, psychology, and neuroscience of mindreading*. Oxford: Oxford University Press.

Goldman, A. and de Vignemont, F. (2009). Is social cognition embodied? *Trends in Cognitive Sciences* 15, 154–9.

Goldstein, K. (1963). *Human nature in the light of psychopathology*. New York: Shocken.

Goldstein, K. (1934/1995). *The organism. a holistic approach to biology derived from pathological data in man*. New York: Zone Books.

Gottlieb, G. (1992). *The genesis of novel behavior: individual development and evolution*. New York: Oxford University Press.

Graham, G. and Stephens, G. L. (1993). Mind and mine. In: G. Graham and G. L. Stephens (eds.), *Philosophical psychopathology*. Cambridge: MIT Press, pp. 91–109.

Grene, (1966). *The knower and the known*. London: Faber and Faber.

Grush, R. (2004). The emulation theory of representation: motor control, imagery, and perception. *Behavioral and Brain Sciences* 27, 377–96.

Guarniero, G. (1974). Experience of tactile vision. *Perception* 3, 101–4. doi: 10.1068/p030101.

Haggard, P. (2005). Conscious intention and motor cognition. *Trends in Cognitive Sciences* 9(6), 290–5.

Hanczyc M. M. (2014). Droplets: unconventional protocell model with life-like dynamics and room to grow. *Life* 4, 1038–49; doi:10.3390/life4041038.

Hanczyc M. M., Toyota T., Ikegami T., Packard N., and Sugawara T. (2007). Fatty acid chemistry at the oil–water interface: self-propelled oil droplets. *Journal of the American Chemical Society* 129, 9386–91. doi:10.1021/ja0706955.

Hanna, R. and Maiese, M. (2009). *Embodied minds in action*. Oxford: Oxford University Press.

Hari, R. and Kujala, M. V. (2009). Brain basis of human social interaction: From concepts to brain imaging. *Physiological Reviews* 89, 453–79.

Harvey, I. (2008). Misrepresentations. In: S. Bullock, J. Noble, R. A. Watson, and M. A. Bedau (eds.), *Artificial life XI: proceedings of the Eleventh International Conference on the Simulation and Synthesis of Living Systems*. Cambridge, MA: MIT Press, pp. 227–33.

Harvey, I., Husbands, P., Cliff, D., Thompson, A. and Jakobi, N. (1996). Evolutionary robotics: the Sussex approach. *Robotics and Autonomous Systems* 20, 205–24. doi:10.1016/S0921- 8890(96)00067-X.

Harvey, I., Di Paolo, E. A., Tuci, E., Wood, R. and Quinn, M. (2005). Evolutionary robotics: A new scientific tool for studying cognition. *Artificial Life* 11, 79–98. doi: 10.1162/1064546053278991.

Haugeland, J. (1991). Representational genera. In: W. Ramsey, S. Stich, and D. Rumelhart, (eds.), *Philosophy and connectionist theory*. Hillsdale, NJ: Erlbaum.

Haugeland, J. (1995). Mind embodied and embedded. In: L. Haaparanta and S. Heinamaa (eds.), *Mind and cognition: philosophical perspectives on cognitive science and artificial intelligence*, Acta Philosophica Fennica, 58, pp. 233–67.

Hausdorff, J. M., Zemany, L., Peng, C.-K., and Goldberger, A. L. (1999). Maturation of gait dynamics: stride-to-stride variability and its temporal organization in children. *Journal of Applied Physiology* 86(3), 1040–7.

Hayes, A.E., Paul, M. A., Beuger, B., and Tipper, S.P. (2008). Self produced and observed actions influence emotion: the roles of action fluency and eye gaze. *Psychological Research* 72(4), 461–72.

Hebb, D. O. (1955). Drives and the C. N. S. (conceptual nervous system). *Psychological Review* 62, 243–54.

Hebb, D. O. (1980). *An essay on mind*. New York: Psychology Press.

Heft, H. (2007). The social constitution of perceiver-environment reciprocity. *Ecological Psychology* 19(2), 85–105.

Helmholtz, H. (1867). *Handbuch der physiologischen Optik*. Leipzig: Voss.

Hendriks-Jansen, H. (1996). *Catching ourselves in the act: situated activity, interactive emergence, evolution, and human thought*. Cambridge, MA: MIT Press.

Hochberg, J. (1968). In the mind's eye. In: R. N. Haber (ed.), *Contemporary theory and research in visual perception*. London: Holt, Rinehart & Winston, pp. 309–31.

Hoehl, S., Reid, V. M., Mooney, J., and Striano, T. (2008). What are you looking at? Infants' neural processing of an adult's object-directed gaze. *Developmental Science* 11, 10–16.

Hoffding, S. (2014). What is skilled coping? Experts on expertise. *Journal of Consciousness Studies* 21(9–10), 49–73.

Hoffmann, M., Schmidt, N., Pfeifer, R., Engel, A., and Maye, A. (2012). Using sensorimotor contingencies for terrain discrimination and adaptive walking behavior in the quadruped robot Puppy. In: T. Ziemke, C. Balkenius, and J. Hallam, (eds.) *From animals to animats 12. Proceedings of the 12th International Conference on Simulation of Adaptive Behavior, SAB 2012, Odense, Denmark, August 27–30, 2012*. Berlin: Springer, pp. 54–64.

Hordijk, W., Steel, M., and Kauffman, S. (2012). The structure of autocatalytic sets: evolvability, enablement, and emergence. *Acta Biotheoretica* 60(4), 379–92.

Hordijk, W., Steel, M., and Kauffman, S. (2013). The origin of life, evolution, and functional organization. In P. Pontarotti (ed.), *Evolutionary biology: exobiology and evolutionary mechanisms*. Berlin: Springer, pp. 49–60.

Hordijk, W. and Steel, M. (2015). Autocatalytic sets and boundaries. *Journal of Systems Chemistry* 6, 1. doi: 10.1186/s13322-014-0006-2.

Hurley, S. (1998). *Consciousness in action*. Cambridge, MA: Harvard University Press.

Hurley, S. and Noë, A. (2003). Neural plasticity and consciousness. *Biology and Philosophy* 18, 131–68. doi:10.1023/A:1023308401356

Husbands, P., Holland, O., and Wheeler, M. (eds.) (2008). *The mechanical mind in history*. Cambridge, MA: MIT Press.

Husserl, E. (1989). *Ideas pertaining to a pure phenomenology and a pure phenomenological philosophy, second book*. Dordrecht: Kluwer.

Hutchins, E. (1995). *Cognition in the wild*. Cambridge, MA: MIT Press

Hutto, D. (2005). Knowing what? Radical versus conservative enactivism. *Phenomenology and the Cognitive Sciences* 4, 389–405. doi:10.1007/s11097-005-9001-z.

Hutto, D. D. and Myin, E. (2013). *Radicalizing enactivism: basic minds without content*. Cambridge, MA: MIT Press.

Huyck, C. R. and Passmore, P. J. (2013). A review of cell assemblies. *Biological Cybernetics* 107(3), 263–88.

Iizuka, H. and Di Paolo, E. A. (2007). Toward Spinozist robotics: exploring the minimal dynamics of behavioral preference. *Adaptive Behavior* 15, 359–76. doi:10.1177/1059712307084687.

Iizuka, H. and Di Paolo, E. A. (2008). Extended homeostatic adaptation: improving the link between internal and behavioural stability. In: M. Asada, J. C. T. Hallam, J.-A. Meyer, and J. Tani (eds.),

From animats to animals 10: the Tenth International Conference on the Simulation of Adaptive Behavior. Heidelberg: Springer, pp. 1–11.

Izhikevich, E. M. (2007). Solving the distal reward problem through linkage of STDP and dopamine signaling. *Cerebral Cortex* **17**, 2443–52.

Izquierdo, E. J., Aguilera, M., and Beer, R. D. (2013). Analysis of ultrastability in small dynamical recurrent neural networks. In: **P. Lio, O. Miglino, G. Nicosia, S. Nolfi,** and **M. Pavone** (eds.), *Advances in artificial life, Proceedings of the 12th European Conference of Artificial Life, ECAL 2013,* Cambridge, MA: MIT Press, pp. 51–58.

Izquierdo, E. J. and **Beer, R D.** (2013). Connecting a connectome to behavior: An ensemble of neuroanatomical models of C. elegans klinotaxis. *PLoS Computational Biology* **9(2)**, e1002890.

Izquierdo, E. J. and **Buhrmann, T.** (2008). Analysis of a dynamical recurrent neural network evolved for two qualitatively different tasks: walking and chemotaxis. In: **S. Bullock, J. Noble, R. A. Watson,** and **M. A. Bedau** (eds.), *Artificial life XI: Proceedings of the Eleventh International Conference on the Simulation and Synthesis of Living Systems.* Cambridge, MA: MIT Press, pp. 257–64.

Izquierdo, E. and **Di Paolo, E. A.,** (2005). Is an embodied system ever purely reactive? In: **M. S. Capcarrère, A. A. Freitas, P. J. Bentley, C. G. Johnson,** and **J. Timmis** (eds.), *Advances in artificial life, Proceedings 8th European Conference, ECAL 2005, Canterbury, UK, September 5–9, 2005.* Berlin: Springer. pp. 252–261.

Izquierdo, E., Harvey, I. and **Beer, R. D.** (2008). Associative learning on a continuum in evolved dynamical neural networks. *Adaptive Behavior* **16**, 361–84. doi: 10.1177/1059712308097316.

Jacob, F. and **Monod, J.** (1961). Genetic regulatory mechanisms in the synthesis of proteins. *Journal of Molecular Biology* **3**, 318–56.

James, W. (1878). Remarks on Spencer's definition of mind as correspondence. *Journal of Speculative Philosophy* **12(1)**, 1–18.

James, W. (1890). *The principles of psychology. Vol I.* London: Macmillan.

Johnson, M. (1987). *The body in the mind: the bodily basis of meaning, imagination and reason.* Chigago: University of Chicago Press.

Johnson, M. (2007). *The meaning of the body: aesthetics of human understanding.* Chicago: University of Chicago Press.

Jonas, H. (1966). *The phenomenon of life: toward a philosophical biology.* New York: Harper & Row.

Jonas, H. (1968). Biological foundations of individuality. *International Philosophical Quarterly* **8(2)**, 231–51.

Jonas, H. (1984). *The imperative of responsibility: in search of an ethics for the technological age.* Chicago: Chicago University Press.

Jonas, H. (1996). Evolution and freedom: on the continuity among life-forms. In: **H. Jonas,** *Mortality and morality. a search for the good after Auschwitz.* **L. Vogel,** (ed.), Evanston, IL: Northwestern University Press.

Juarrero, A. (1999). *Dynamics in action. intentional behavior as a complex system.* Cambridge, MA: MIT Press.

Juarrero Roqué A. (1985). Self-organization: Kant's concept of teleology and modern chemistry. *Review of Metaphysics* **39**, 107–35.

Kauffman, S. A. (1986). Autocatalytic sets of proteins. *Journal of Theoretical Biology* **119(1)**, 1–24.

Kauffman, S. A. (2002). *Investigations.* Oxford: Oxford University Press.

Kauffman, S. (2003). Molecular autonomous agents. *Philosophical Transactions of the Royal Society A: Mathematical, Physical and Engineering Sciences* **361(1807)**, 1089–99.

Kaye, K. (1979). The development of skills. In: **G. J. Whitehurst** and **B. J. Zimmerman** (eds.), *The functions of language and cognition.* New York: Academic Press, pp. 23–55.

Keijzer, F. (2001). *Representation and behavior*. Cambridge, MA: MIT Press.

Kelso, J. A. S. (1995). *Dynamic patterns: the self-organization of brain and behavior*. Cambridge, MA: MIT Press.

Kelso, J. A. S. (2009). Synergies: atoms of brain and behavior. In: D. Sternad (ed.), *A multidisciplinary approach to motor control*. Heidelberg: Springer, pp. 83–91.

Kelso, J. A. S. (2016). On the self-organizing origins of agency. *Trends in Cognitive Sciences* 20(7), 490–9.

Kelso, J. A. S., Dumas, G., and Tognoli, E. (2013). Outline of a general theory of behavior and brain coordination. *Neural Networks* 37, 120–31. doi:10.1016/j.neunet.2012.09.003.

Kelso, J. A. S., Tuller, B., Vatikiotis-Bateson, E., and Fowler, C. A. (1984). Functionally specific articulatory cooperation following jaw perturbations during speech: evidence for coordinative structures. *Journal of Experimental Psychology: Human Perception and Performance* 10(6), 812–32.

Kirkby, L., Sack, G., Firl, A., and Feller, M. B. (2013). A role for correlated spontaneous activity in the assembly of neural circuits. *Neuron* 80(5), 1129–44.

Kirsh, D. (1991). Today the earwig tomorrow man? *Artificial Intelligence* 47, 161–84.

Kirsh, D. (1995). The intelligent use of space. *Artificial Intelligence* 73, 31–68.

Kirsh, D. (1996). Adapting the environment instead of oneself. *Adaptive Behavior* 4, 415–52.

Knill, D. C. and Pouget, A. (2004). The Bayesian brain: the role of uncertainty in neural coding and computation. *Trends in Neurosciences* 27, 712–19. doi: 10.1016/j.tins.2004.10.007.

Kohler, I. (1964). The formation and transformation of the perceptual world. *Psychological Issues* 3(4), Monograph 12.

Konvalinka, I., Xygalatas, D., Bulbulia, J., Schjødt, U., Jegindø, E-M., Wallot, S., Van Orden, G., and Roepstorff, A. (2011). Synchronized arousal between performers and related spectators in a fire- walking ritual. *Proceedings of the National Academy of Sciences USA* 108, 8514–19. doi:10.1073/pnas. 1016955108.

Korsgaard, C. (2009). *Self-constitution: agency, identity, and integrity*. Oxford: Oxford University Press.

Kostrubiec, V., Zanone, P.-G., Fuchs, A., and Kelso, J. A. S. (2012). Beyond the blank slate: routes to learning new coordination patterns depend on the intrinsic dynamics of the learner-experimental evidence and theoretical model. *Frontiers in Human Neuroscience* 6, 222. doi: 10.3389/fnhum.2012.00222.

Kugler, P., Kelso, J. S., and Turvey, M. (1982). On the control and coordination of naturally developing systems. In: J. S. Kelso and J. Clark (eds.), *The development of movement control and coordination*. New York: Wiley, pp. 5–78.

Kuniyoshi, Y. and Sangawa, S. (2006). Early motor development from partially ordered neural-body dynamics: experiments with a cortico-spinal-musculo-skeletal model. *Biological Cybernetics* 95(6), 589–605. doi:10.1007/s00422-006-0127-z

Kuo, Z.-Y. (1967). *The dynamics of behavior development: an epigenetic view*. New York: Random House.

Kuo, Z.-Y. (1970). The need for coordinated efforts in developmental studies. In: L. R. Aronson, E. Tobach, D. S. Lehrman, and J. S. Rosenblatt (eds.), *Development and evolution of behavior: essays in memory of T. C. Schneirla*. San Francisco: Freeman, pp. 181–93.

Kyselo, M. (2015). The enactive approach and disorders of the self—the case of schizophrenia. *Phenomenology and the Cognitive Sciences* (online first), doi: 10.1007/s11097-015-9441-z.

Kyselo, M. and Di Paolo, E. A. (2015). Locked-in syndrome: a challenge for embodied cognitive science. *Phenomenology and the Cognitive Sciences* 14, 517–42, doi: 10.1007/s11097-013-9344-9.

Lafargue, G., Paillard, J., Lamarre, Y., and Sirigu, A. (2003). Production and perception of grip force without proprioception: is there a sense of effort in deafferented subjects? *European Journal of Neuroscience* 17(12), 2741-9.

Lagzi, I., Siowling, S., Paul, J. W., Kevin, P. B. and Bartosz, A. G. (2010). Maze solving by chemotactic droplets. *Journal of the American Chemical Society* **132**, 1198–9.

Lakoff, G. (1987). *Women, fire, and dangerous things: what categories reveal about the mind.* Chicago: University of Chicago Press.

Lakoff, G. and Johnson, M. (1980). *Metaphors we live by.* Chicago: University of Chicago Press.

Laflaquière, A., O'Regan, J. K., Argentieri, S., Gas, B., and Terekhov, A. V. (2015). Learning agent's spatial configuration from sensorimotor invariants. *Robotics and Autonomous Systems* **71**, 49–59. doi:10.1016/j.robot.2015.01.003.

Langer, S. K. (1967). *Mind: an essay on human feeling. Vol 1.* Baltimore: The Johns Hopkins Press.

Latash, M. L. (2008). *Synergy.* Oxford: Oxford University Press.

Lave, J. (1988). *Cognition in practice: mind, mathematics and culture in everyday life.* Cambridge: Cambridge University Press.

Lenay, C., Canu, S., and Villon, P. (1997). Technology and perception: the contribution of sensory substitution systems. In: *Proceedings of the Second International Conference on Cognitive Technology, 1997. "Humanizing the Information Age," Aizu, Japan.* Los Alamitos: IEEE, pp. 44–53.

Leont'ev, A. N. (1978). *Activity, consciousness, and personality.* Englewood Cliffs, NJ: Prentice-Hall.

Leslie, A. M., Friedman, O., and German, T. P. (2004). Core mechanisms in "theory of mind." *Trends in Cognitive Sciences* **8**, 528–33. doi:10.1016/j.tics.2004.10.001.

Lewis, M. D. (2000). The promise of dynamic systems approaches for an integrated account of human development. *Child Development* **71**(1), 36–43.

Luria, A. (1976). *Cognitive development. its cultural and social foundations.* Cambridge, MA: Harvard University Press.

Lynall, M., Bassett, D. S., Kerwin, R., McKenna, P. J., Kitzbichler, M., Müller, U., et al. (2010). Functional connectivity and brain networks in schizophrenia. *The Journal of Neuroscience* **30**(**28**), 9477–87.

MacKay, D. M. (1962). Theoretical models of space perception. In: C. A. Muses (ed.), *Aspects of the theory of artificial intelligence.* New York: Plenum Press. pp. 83–103.

Maes, P. (ed.) (1991). *Designing autonomous agents: theory and practice from biology to engineering and back.* Cambridge, MA: MIT Press.

Maiese, M. (2015). Transformative learning, enactivism, and affectivity. *Studies in Philosophy and Education* (online first) doi:10.1007/s11217-015-9506-z.

Malafouris, L. (2007). Before and beyond representation: towards an enactive conception of the Palaeolithic image. In: C. Renfrew and I. Morley (eds.), *Image and imagination: a global history of figurative representation.* Cambridge: McDonald Institute for Archaeological Research, pp. 289–302.

Malafouris, L. (2013). *How things shape the mind: a theory of material engagement.* Cambridge, MA: MIT Press.

Manicka, S. and Di Paolo, E. A. (2009). Local ultrastability in a real system based on programmable springs. In: G. Kampis, I. Karsai, and E. Szathmáry (eds.), *Advances in artificial life, Proceedings of the 10th European Conference on Artificial Life, ECAL09.* Heidelberg: Springer, pp. 91–8.

Marcel, A. J. (2003). The sense of agency: awareness and ownership of action. In: J. Roessler and N. Eilan (eds.), *Agency and self-awareness: issues in philosophy and psychology.* Oxford: Clarendon, pp. 48–93.

Marchetti, C. and Della Sala, S. (1998). Disentangling the alien and anarchic hand. *Cognitive Neuropsychiatry* **3**, 191–207. doi:10.1080/135468098396143.

Marler, P. (1981). Birdsong: the acquisition of a learned motor skill. *Trends in Neurosciences* **4**, 88–94.

Marques, H. G., Imtiaz, F., Iida, F., and Pfeifer, R. (2013). Self-organization of reflexive behavior from spontaneous motor activity. *Biological Cybernetics* **107**(1), 25–37.

Marr, D. (1982). *Vision*. New York: W. H. Freeman and Company.

Maturana, H. and Varela, F. J. (1980). *Autopoiesis and cognition: the realization of the living*. Dordrecht: D. Reidel Publishing.

Maturana, H. and Varela, F. J. (1987). *The tree of knowledge: the biological roots of human understanding*. Boston, MA: Shambhala.

Matyja J. R. and Schiavio A. (2013). Enactive music cognition: background and research themes. *Constructivist Foundations* **8**, 351–57.

Maye, A. and Engel, A. K. (2011). A discrete computational model of sensorimotor contingencies for object perception and control of behavior. In: *IEEE International Conference on Robotics and Automation (ICRA) 2011, Shanghai*, pp. 3810–15.

Maye, A. and Engel, A. K. (2013). Extending sensorimotor contingency theory: prediction, planning and action generation. *Adaptive Behavior* **21**, 423–36. doi: 10. 1177/1059712313497975.

McDowell, J. (2007a). What myth? *Inquiry* **50(4)**, 338–51.

McDowell, J. (2007b). Response to Dreyfus. *Inquiry* **50(4)**, 366–70.

McGann, M. (2010). Perceptual modalities: modes of presentation or modes of interaction? *Journal of Consciousness Studies* **17**, 72–94.

McGann, M. (2014). Enacting a social ecology: radically embodied intersubjectivity. *Frontiers in Psychology* **5**, 1321.

McGann, M. and De Jaegher, H. (2009). Self–other contingencies: enacting social perception. *Phenomenology and the Cognitive Sciences* **8**, 417–37. doi:10.1007/s11097-009-9141-7.

McGeer, T. (1990). Passive dynamic walking. *The International Journal of Robotics Research* **9**, 62–82.

McGregor, S. and Virgo, N. (2011). Life and its close relatives. In: G. Kampis, I. Karsai, and E. Szathmáry (eds.), *Advances in Artificial Life. Darwin Meets von Neumann. 10th European Conference, ECAL 2009, Budapest, Hungary, September 13–16, 2009, Revised Selected Papers, Part II*. Heidelberg: Springer, pp. 230–37.

McKenzie, S., Robinson, N. T., Herrera, L., Churchill, J. C., and Eichenbaum, H. (2013). Learning causes reorganization of neuronal firing patterns to represent related experiences within a hippocampal schema. *Journal of Neuroscience* **33(25)**, 10243–56.

McCoy, A. (2006). *A question of torture. CIA interrogation, from the Cold War to the war on terror*. New York: Henry Holt.

Mead, G. H. (1932). The physical thing. In: G. H. Mead, *The philosophy of the present*. A. E. Murphy (ed.). LaSalle, IL: Open Court, pp. 119–39.

Mead, G. H. (1934). *Mind, self, and society from the standpoint of a social behaviorist*. Chicago: University of Chicago Press.

Mead, G. H. (1938). *The philosophy of the act*. Chicago: University of Chicago Press.

Medaglia, J. D., Ramanathan, D. M., Venkatesan, U. M., and Hillary, F. G. (2011). The challenge of non-ergodicity in network neuroscience. *Network* **22**, 148–53. doi: 10.3109/09638237.2011.639604.

Merleau-Ponty, M (2010). *Child psychology and pedagogy: the Sorbonne lectures 1949-1952*. T. Welsh (trans.). Evanston, IL: Northwestern University Press.

Merleau-Ponty, M. (1942/1963). *The structure of behavior*. A. L. Fisher (trans.). Boston: Beacon Press.

Merleau-Ponty, M. (1945/2012). *Phenomenology of perception*. 2nd ed. D. Landes (trans.). London: Routledge.

Merleau-Ponty, M. (1964). *The Primacy of Perception*. Evanston, IL: Northwestern Press.

Meteyard, L., Bahrami, B., and Vigliocco, G. (2007). Motion detection and motion verbs: language affects low-level visual perception. *Psychological Science* **18(11)**, 1007–13.

Metzinger, T. (2000). The subjectivity of subjective experience: A Representationist analysis of the first-person perspective. In: T. Metzinger (ed.), *Neural correlates of consciousness*. Cambridge, MA: MIT Press, pp. 285–306.

Millikan, R. G. (1989). Biosemantics. *Journal of Philosophy* 86, 281–97.

Miyamoto, Y., Nisbett, R., and Masuda, T. (2006). Culture and the physical environment: holistic versus perceptual affordances. *Psychological Science* 17, 113–19.

Molenaar, P. C. M. and Campbell, C. G. (2009). The new person-specific paradigm in psychology. *Current Directions in Psychological Science* 18, 112–17. doi: 10.1111/j.1467-8721. 2009.01619.x.

Moran, G., Fentress, J. C., and Golani, I. (1981). A description of relational patterns of movement during "ritualized fighting" in wolves. *Animal Behaviour* 29(4), 1146–65.

Moreno, A. and Etxeberria, A. (2005). Agency in natural and artificial systems. *Artificial Life* 11(1–2), 161–75.

Moreno, A. and Lasa, A. (2003). From basic adaptivity to early mind. The origin and evolution of cognitive capacities. *Evolution and Cognition* 9, 12–24.

Moreno, A. and Mossio, M. (2015). *Biological autonomy: a philosophical and theoretical enquiry.* Berlin: Springer.

Mossio, M. and Bich, L. (2014). What makes biological organisation teleological? Synthese (online first) doi:10.1007/s11229-014-0594-z.

Mossio, M. and Moreno, A. (2010). Organisational closure in biological organisms. *History and Philosophy of the Life Sciences* 32, 269–88.

Mossio, M. and Taraborelli, D. (2008). Action-dependent perceptual invariants: from ecological to sensorimotor approaches. *Consciousness and Cognition* 17, 1324–40. doi:10.1016/ j.concog.2007.12.003.

Muchisky, M., Gershkoff-Stowe, L., Cole, E., and Thelen, E. (1996). The epigenetic landscape revisited: a dynamic interpretation. In: C. Rovee-Collier and L. P. Lipsitt (eds.), *Advances in infancy research, Vol. 10.* Norwood, NJ: Ablex, pp. 121–59.

Myin, E. (2003). An account of color without a subject? *Behavioral and Brain Sciences* 26, 42–3. doi:10.1017/S0140525X03440016.

Myin, E. and O'Regan, J. K. (2002). Perceptual consciousness, access to modality and skill theories: a way to naturalize phenomenology? *Journal of Consciousness Studies* 9, 27–45.

Nahmias, E. (2005). Agency, authorship, and illusion. *Consciousness and Cognition* 14, 771–85. doi:10.1016/j.concog.2005.07.002.

Neisser, U. (1967). *Cognitive psychology*. New York: Appleton-Century Crofts.

Newell, A. and Simon, H. A. (1976). Computer science as empirical inquiry: symbols and search. *Commun. ACM* 19(3), 113–26. doi:10.1145/360018.360022.

Newell, K. M., Liu, Y., and Mayer-Kress, G. (2003). A dynamical systems interpretation of epigenetic landscapes for infant motor development. *Infant Behavior and Development* 26(4), 449–72.

Nicolis, G. and Prigogine, I. (1977). *Self-organization in non-equilibrium systems: From dissipative structures to order from fluctuations.* New York: Wiley.

Nisbett, R. E. and Miyamoto, Y. (2006). The influence of culture: holistic vs. analytic perception. *Trends in Cognitive Sciences* 9, 467–73.

Noë, A. (2004). *Action in perception*. Cambridge, MA: MIT Press.

Nolfi S. and Floreano D. (2000). *Evolutionary robotics: the biology, intelligence, and technology of self-organizing machines.* Cambridge, MA: MIT Press.

Nomikou, I., Rohlfing, K. J., and Szufnarowska, J. (2013). Educating attention: recruiting, maintaining, and framing eye contact in early natural mother–infant interactions, *Interaction Studies* 14(2), 240–67.

Okubo, T. S., Mackevicius, E. L., Payne, H. L., Lynch, G. F., and Fee, M. S. (2015). Growth and splitting of neural sequences in songbird vocal development. *Nature* **528(7582),** 352–7.

O'Regan, J. K. and Noë, A. (2001). A sensorimotor account of vision and visual consciousness. *Behavioral and Brain Sciences* **24,** 939–1031. doi: 10.1017/s0140525x01000115.

Oyama, S. (1985). *The ontogeny of information: developmental systems and evolution.* Cambridge: Cambridge University Press.

Pacherie, E. (2007a). The anarchic hand syndrome and utilization behavior: a window onto agentive self-awareness. *Functional Neurology* **22,** 211–17.

Pacherie, E. (2007b). The sense of control and the sense of agency. *Psyche* **13(1),** 1–30.

Panksepp, J. (1998). The periconscious substrates of consciousness: affective states and the evolutionary origins of the SELF. *Journal of Consciousness Studies* **5(5–6),** 566–82.

Pask, G. (1959). Physical analogues to the growth of a concept. In: *Proceedings of Symposium No. 10 of the National Physical Laboratory: Mechanisation of Thought Processes, National Physical Laboratory, 24–27 November 1958. Volume 2.* London: Her Majesty's Stationery Office.

Pfeifer, R. and Scheier, C. (1997). Sensory-motor coordination: the metaphor and beyond. *Robotics and Autonomous Systems* **20,** 157–78.

Pfeifer, R. and Scheier, C. (1999). *Understanding intelligence.* Cambridge, MA: MIT Press.

Phattanasri, P., Chiel, H. J., and Beer, R. D. (2007). The dynamics of associative learning in evolved model circuits. *Adaptive Behavior* **15,** 377–96. doi: 10.1177/1059712307084688.

Philipona, D. L. and O'Regan, J. K. (2006). Color naming, unique hues, and hue cancellation predicted from singularities in reflection properties. *Visual Neuroscience* **23,** 331–9. doi:10.1017/S0952523806233182.

Philipona, D. and O'Regan, J. K. (2010). The sensorimotor approach in CoSy: The example of dimensionality reduction. In: H. I. Christensen, G.-J. M. Kruijff, and J. L. Wyatt (eds.), *Cognitive systems, Vol. 8,* Berlin: Springer, pp. 95–130.

Philipona, D., O'Regan, J. K., and Nadal, J.-P. (2003). Is there something out there? Inferring space from sensorimotor dependencies. *Neural Computation* **15,** 2029–49. doi:10.1162/089976603322297278.

Phillips, D. P., Danilchuk, W., Ryon, J., and Fentress, J. (1990). Food-caching in timber wolves, and the question of rules of action syntax. *Behavioural Brain Research* **38(1),** 16.

Piaget, J. (1936). *La Naissance de L'intelligence Chez L'enfant.* Neuchâtel, Paris: Delachaux et Niestlé.

Piaget, J. (1947). *The psychology of intelligence.* London: Routledge.

Piaget, J. (1969). *Biology and knowledge.* Chicago: University of Chicago Press.

Piaget, J. (1975). *L'Équilibration des Structures Cognitives: Problème Central du Développement.* Paris: Presses Universitaires de France.

Plessner, H. (1970). *Laughing and crying: a study of the limits of human behavior.* Evanston, IL: Northwestern University Press.

Poincaré, H. (1902). *La science et l'hypothèse.* Paris: Flammarion.

Polanyi, M. (1966). *The tacit dimension.* Chicago: University of Chicago Press.

Popova, Y. B. (2015). *Stories, meaning, and experience: narrativity and enaction.* London: Routledge.

Powers, W. T. (1973). *Behavior: the control of perception.* Chicago: Aldine de Gruyter.

Quinn, M. (2001). Evolving communication without dedicated communication channels. In: J. Kelemen and P. Sosík, (eds.), *Advances in artificial life. Proceedings of the 6th European Conference, ECAL 2001 Prague, Czech Republic, September 10–14, 2001.* Berlin: Springer, pp. 357–66.

Raibert, M. (1986). *Legged robots that balance.* Cambridge, MA: MIT Press.

Ramachandran, V. S. and Rogers-Ramachandran, D. (1996). Synaesthesia in phantom limbs induced with mirrors. *Proceedings of the Royal Society of London B,* **263,** 377–86. doi:10.1098/rspb.1996.0058.

Ratcliffe, M. (2008). *Feelings of being: phenomenology, psychiatry and the sense of reality.* Oxford: Oxford University Press.

Ravaisson, F. (1838/2008). *Of habit.* London: Continuum.

Reddy, V. (2008). *How infants know minds.* Cambridge, MA: Harvard University Press.

Reigl, M., Alon, U., and Chklovskii, D. B. (2004). Search for computational modules in the *C. elegans* brain. *BMC Biology* 2, 25.

Richardson, D.C., Spivey, M.J., Barsalou, L.W., and McRae, K. (2003). Spatial representations activated during real-time comprehension of verbs. *Cognitive Science* 27, 767–80.

Ricoeur, P. (1950/2007). *Freedom and nature: the voluntary and the involuntary.* E. Kohak (trans.). Evanston, IL: Northwestern University Press.

Rietveld, E. (2008a). The skillful body as a concernful system of possible actions: Phenomena and neurodynamics. *Theory & Psychology* 18(3), 341–63.

Rietveld, E. (2008b). Situated normativity: the normative aspect of embodied cognition in unreflective action. *Mind* 117(468), 973–1001.

Rietveld, E. and Kiverstein, J. (2014). A rich landscape of affordances. *Ecological Psychology* 26, 325–52.

Riley, M. A., Richardson, M., Shockley, K., and Ramenzoni, V. C. (2011). Interpersonal synergies. *Frontiers in Psychology* 2, 38. doi:10.3389/fpsyg.2011.00038.

Rosch, E., Mervis, C. B., Gray, W., Johnson, D., and Boyes-Braem, P. (1976). Basic objects in natural categories. *Cognitive Psychology* 8, 382–439.

Roskies, A. L. (2008). A new argument for nonconceptual content. *Philosophy and Phenomenological Research* 76(3), 633–59.

Rock, I. (1983). *Logic of perception.* Cambridge, MA: MIT Press.

Rose, J. K. and Rankin, C. H. (2001). Analyses of habituation in Caenorhabditis elegans. *Learning & Memory* 8, 63–9.

Rosenblueth, A., Wiener, N., and Bigelow, J. (1943). Behavior, purpose and teleology. *Philosophy of Science* 10, 18–24. doi:10.1086/286788.

Rosslenbroich, B. (2014). *On the origin of autonomy: a new look at the major transitions in evolution.* Berlin: Springer.

Rossmanith, N., Costall, A., Reichelt, A.F., López, B., and Reddy, V. (2014). Jointly structuring triadic spaces of meaning and action: book sharing from 3 months on. *Frontiers in Psychology* 5, 1390.

Ruiz-Mirazo, K. and Mavelli, F. (2008). On the way towards "basic autonomous agents": Stochastic simulations of minimal lipid-peptide cells. *BioSystems* 91(2), 374–87. doi:10.1016/ j.biosystems.2007.05.013.

Rupert, R. (2016). Embodied functionalism and inner complexity: Simon's 21st-century mind. In: L. Marsh and R. Frantz (eds.), *Minds, models and milieux: commemorating the centennial of the birth of Herbert Simon.* Palgrave Macmillian, pp. 7–33.

Ryle, G. (1949). *The concept of mind.* London: Hutchinson.

Sampaio, E., Maris, S. and Bach-y-Rita, P. (2001). Brain plasticity: "visual" acuity of blind persons via the tongue. *Brain Research* 908, 204–7. doi: 10.1016/s0006-8993(01)02667-1.

Sasahara, K., Cody, M. L., Cohen, D., and Taylor, C. E. (2012). Structural design principles of complex bird songs: a network-based approach. *PloS ONE* 7(9), e44436.

Sato Y., Iizuka H., and Ikegami T. (2013). Investigating extended embodiment using a computational model and human experimentation. *Constructivist Foundations* 9(1), 73–84.

Scheerer E. (1984). Motor theories of cognitive structure: a historical review. In: W. Prinz and W. Sanders (eds.), *Cognition and motor processes.* Berlin: Springer-Verlag, pp. 77–98.

Schmidt, N. M., Hoffmann, M., Nakajima, K., and **Pfeifer, R.** (2012). Bootstrapping perception using information theory: case studies in a quadruped robot running on different grounds. *Advances in Complex Systems* **16**, 1250078, doi:10.1142/S0219525912500786.

Schoggen, P. (1989). *Behavior settings: a revision and extension of Roger G. Barker's ecological psychology.* Stanford, CA: Stanford University Press.

Schöner, G. and **Dineva, E.** (2007). Dynamic instabilities as mechanisms for emergence. *Developmental Science* **10**, 69–74. doi:10.1111/j.1467-7687.2007.00566.x.

Schurger, A., Sarigiannidis, I., Naccache, L., Sitt, J. D., and **Dehaene, S.** (2015). Cortical activity is more stable when sensory stimuli are consciously perceived. *Proceedings of the National Academy of Sciences USA* **112**, E2083–92.

Schutz, W. F. (ed.) (2007). *The phenomenon of torture: readings and commentary.* Philadelphia: University of Pennsylvania Press.

Searle, J. (1980). Minds, brains and programs. *Behavioral and Brain Sciences* **3**, 417–57.

Searle, J. (1983). *Intentionality.* Cambridge: Cambridge University Press.

Segall, M. H., Campbell, D. T., and **Herskovits, J. M.** (1963). Differences in perception of geometric illusions. *Science* **139**, 769–71.

Seth, A. K. (2014). A predictive processing theory of sensorimotor contingencies: explaining the puzzle of perceptual presence and its absence in synesthesia. *Cognitive Neuroscience* **5**, 97–118. doi: 10.1080/17588928.2013.877880.

Seung, H. S. (2003). Learning in spiking neural networks by reinforcement of stochastic synaptic transmission. *Neuron* **40(6),** 1063–73.

Sheets-Johnstone, M. (1981). Thinking in movement. *Journal of Aesthetics and Art Criticism* **39(4),** 399–407.

Sheets-Johnstone, M. (2003). Child's play: a multidisciplinary perspective. *Human Studies* **26(4),** 409–30.

Sheets-Johnstone, M. (2011). *The primacy of movement.* Expanded 2nd ed. Philadelphia: John Benjamins Publishing Company

Shockley, K., Santana, M.-V., and **Fowler, C. A.** (2003). Mutual interpersonal postural constraints are involved in cooperative conversation. *Journal of Experimental Psychology: Human Perception and Performance* **29**, 326–32.

Shusterman, R. (2008). *Body consciousness: a philosophy of mindfulness and somaesthetics.* Cambridge: Cambridge University Press.

Simon, H. A. (1969). *The sciences of the artificial.* Cambridge, MA: MIT Press.

Simondon, G. (2005). *L'Individuation à la Lumière des Notions de Forme et d'Information.* Grenoble: Millon.

Simons, D. J., Wang, R. X. F., and **Roddenberry, D.** (2002). Object recognition is mediated by extraretinal information. *Perception and Psychophysics* **64**, 521–30.

Sinha, C. (2009). Objects in a storied world. Materiality, normativity, narrativity. *Journal of Consciousness Studies* **16(6–8),** 167–90.

Skewes, J. and **Hooker, C.** (2009). Bio-agency and the problem of action. *Biology and Philosophy* **24(3),** 283–300.

Sloman, A. (2009). Some requirements for human-like robots: why the recent over-emphasis on embodiment has held up progress. In: **B. Sendhoff, E. Körner, O. Sporns, H. Ritter,** and **K. Doya** (eds.) *Creating brain-like intelligence.* Berlin: Springer, pp. 248–77.

Smith, L. B. and **Thelen, E.** (2003). Development as a dynamic system. *Trends in Cognitive Sciences* **7(8),** 343–48.

Smith, L. B., Yu, C., and Pereira, A. F. (2011). Not your mother's view: the dynamics of toddler visual experience. *Developmental Science* **14(1)**, 9–17.

Sporns, O. (2013). Structure and function of complex brain networks. *Dialogues in Clinical Neuroscience* **15(3)**, 247–62.

Sporns, O. and Edelman, G. M. (1993). Solving Bernstein's problem: a proposal for the development of coordinated movement by selection. *Child Development* **64**, 960–81.

Starks, E. B. (1960). Dramatic play. *Childhood education* **37(4)**, 163–7.

Steels, L. (1993). The artificial life roots of artificial intelligence. *Artificial Life* **1(1–2)**, 75–110.

Steinbuch, J. G. (1811). *Beitrag Zur Physiologie Der Sinne*. Nürnberg: Johann Leonard Schragg.

Stephens, G. L. and Graham, G. (2000). *When self-consciousness breaks: alien voices and inserted thoughts*. Cambridge, MA: MIT Press.

Stepp, N. and Turvey, M. T. (2010). On strong anticipation. *Cognitive Systems Research* **11**, 148–64. doi:10.1016/j.cogsys.2009.03.003.

Stern, D. (2010). *Forms of vitality. Exploring dynamic experience in psychology, arts, psychotherapy, and development*. Oxford: Oxford University Press.

Stewart, J. and Varela, F. J. (1991). Morphogenesis in shape-space. Elementary meta-dynamics in a model of the immune network. *Journal of Theoretical Biology* **153(4)**, 477–98.

Straus, E. (1966). *Phenomenological psychology*. London: Tavistock Publications.

Streeck, J., Goodwin, C., and LeBaron, C. D. (eds.). (2011). *Embodied interaction: language and body in the material world*. Cambridge: Cambridge University Press.

Striano, T. and Reid, V. M. (2006). Social cognition in the first year. *Trends in Cognitive Science* **10(10)**, 471–76.

Striano, T. and Stahl, D. (2005). Sensitivity to triadic attention in early infancy. *Developmental Science* **8**, 333–43.

Suchman, L. (1987). *Plans and situated actions: the problem of human-machine communication*. Cambridge: Cambridge University Press.

Sugar, T. G., McBeath, M. K., and Wang, Z. (2006). A unified fielder theory for interception of moving objects either above or below the horizon. *Psychonomic Bullettin &. Review* **13**, 908–17. doi:10.3758/BF03194018.

Sumino, Y. and Yoshikawa, K. (2014). Amoeba-like motion of an oil droplet. *The European Physical Journal Special Topics* **223**, 1345–52.

Sutton, R. S. and Barto, A. G. (2009). *Reinforcement learning: an introduction*. Cambridge, MA: MIT Press.

Synofzik, M., Vosgerau, G., and Newen, A. (2008a). Beyond the comparator model: a multifactorial two-step account of agency. *Consciousness and Cognition* **17**, 219–39. doi:10.1016/j.concog.2007.03.010.

Synofzik, M., Vosgerau, G., and Newen, A. (2008b). I move, therefore I am: a new theoretical framework to investigate agency and ownership. *Consciousness and Cognition* **17**, 411–24. doi:10.1016/j.concog. 2008.03.008.

Taylor, C. (1964). *The explanation of behaviour*. London: Routledge.

Taylor, C. (1977). What is human agency? In: T. Mischel (ed.), *The self: psychological and philosophical issues*. Oxford: Blackwell, pp. 103–35.

Taylor, C. (2002). Foundationalism and the inner-outer distinction. In: Smith, H. N. (ed.), *Reading McDowell: on mind and world*. London: Routledge, pp. 106–19.

Taylor, C. (2006). Merleau-Ponty and the epistemological picture. In: T. Carman and M. B. N. Hansen (eds.), *The Cambridge Companion to Merleau-Ponty*. Cambridge University Press, pp. 26–49.

Thelen, E., **Corbetta, D.**, and **Spencer, J. P.** (1996). The development of reaching during the first year: the role of movement speed. *Journal of Experimental Psychology: Human Perception and Performance* **22**, 1059–76.

Thelen, E. and **Fisher, D. M.** (1983). From spontaneous to instrumental behavior: kinematic analysis of movement changes during very early learning. *Child Development* **54**, 129–40.

Thelen, E. and **Smith, L. B.** (1994). *A dynamic systems approach to the development of cognition and action.* Cambridge, MA: MIT Press.

Thierry, G, **Athanasopoulos, P.**, **Wiggett, A.**, **Dering, B.**, and **Kuipers, J. R..** (2009). Unconscious effects of language-specific terminology on preattentive color perception. *Proceedings of the National Academy of Sciences USA* **106**, 4567–70.

Thom, R. (1983). *Mathematical models of morphogenesis.* New York: Wiley.

Thompson, A. (1997). An evolved circuit, intrinsic in silicon, entwined with physics. In: T. **Higuchi**, **M. Iwata** and **L. Weixin**, (eds.) *Evolvable systems: from biology to hardware. First International Conference on Evolvable Systems (ICES'96) Tsukuba, Japan, October 7–8, 1996.* Heidelberg: Springer., pp. 390–405.

Thompson, E. (2005). Sensorimotor subjectivity and the enactive approach to experience. *Phenomenology and the Cognitive Sciences* **4(4)**, 407–27.

Thompson, E. (2007a). *Mind in life: biology, phenomenology and the sciences of mind.* Cambridge, MA: Harvard University Press.

Thompson, E. (2007b). Representationalism and the phenomenology of mental imagery. *Synthese* **160**, 397–415.

Thompson, E., **Palacios, A.**, and **Varela, F. J.** (1992). Ways of coloring: comparative color vision as a case study for cognitive science. *Behavioral and Brain Sciences* **15**, 1–26.

Thompson, E. and **Varela, F. J.** (2001). Radical embodiment: neural dynamics and consciousness. *Trends in Cognitive Sciences* **5**, 418–25.

Tognoli, E. and **Kelso, J. S.** (2014). The metastable brain. *Neuron* **81(1)**, 35–48.

Tolman, E. C. (1939). Prediction of vicarious trial and error by means of the schematic sowbug. *Psychological Review* **46**, 318–36.

Tolman, E. C. (1941). Discrimination vs. learning and the schematic sowbug. *Psychological Review* **48**, 367–82.

Tolman, E. C. (1948). Cognitive maps in rats and men. *Psychological Review* **55(4)**, 189–208.

Tononi, G. and **Edelman, G. M.** (2000). Schizophrenia and the mechanisms of conscious integration. *Brain Research Reviews* **31(2)**, 391–400.

Trevarthen, C. and **Hubley, P.** (1978), Secondary intersubjectivity: confidence, confiding and acts of meaning in the first year. In **A. Lock** (ed.), *Action, gesture, and symbol.* London: Academic Press., pp. 183–229.

Tronick, E. Z., **Als, H.**, **Adamson, L.**, **Wise, S.**, and **Brazelton, T.** (1978). The infant's response to entrapment between contradictory messages in face-to-face interaction. *Journal of the American Academy of Child Psychiatry* **17**, 1–13.

Tsakiris, M. and **Haggard, P.** (2005). Experimenting with the acting self. *Cognitive Neuropsychology* **22**, 387–407. doi:10.1080/02643290442000158.

Tschacher, W., **Rees, G. M.**, and **Ramseyer, F.** (2014). Nonverbal synchrony and affect in dyadic interactions. *Personality and Social Psychology* **5**, 1323. doi:10.3389/fpsyg.2014.01323.

Turner, J. S. (2000). *The extended organism: the physiology of animal-built structures.* Cambridge, MA: Harvard University Press.

Turvey, M.T. (1990). Coordination. *American psychologist* **45**, 938–53.

Turvey, M. and **Carello, C.** (1995). Some dynamical themes in perception and action. In: T. **Van Gelder** and R. **Port** (eds),. *Mind as motion: explorations in the dynamics of cognition.* Cambridge, MA: MIT Press.

Turvey, M. T. and **Carello, C.** (1996). Dynamics of Bernstein's level of synergies. In: M. **Latash** and M. T. **Turvey** (eds.), *Dexterity and its development.* Hillsdale, NJ: Erlbaum, pp. 339–76.

Urban, P. (2014). Toward an expansion of an enactive ethics with the help of care ethics. *Frontiers in Psychology* 5, 1354. doi: 10.3389/fpsyg.2014.01354.

van der Maas, H. L. J., and **Molenaar, P. C. M.** (1992). Stagewise cognitive development: an application of catastrophe theory. *Psychological Review* 99, 395–417.

van der Schyff, D. (2015). Music as a manifestation of life: exploring enactivism and the "Eastern perspective" for music education. *Frontiers in Psychology* 6, 345. doi: 10.3389/fpsyg.2015.00345.

van Geert, P. (1998). A dynamic systems model of basic developmental mechanisms: Piaget, Vygotsky, and beyond. *Psychological Review* 105(4), 634–77.

Van Gelder, T. and **Port, R.** (1995). *Mind as motion: explorations in the dynamics of cognition.* Cambridge, MA: MIT Press.

Van Orden, G. C., **Holden, J. G.,** and **Turvey, M. T.** (2003). Self-organization of cognitive performance. *Journal of Experimental Psychology* 132, 331–50.

Van Orden, G., **Kloos, H.,** and **Wallot, S.** (2011). Living in the pink: intentionality, wellbeing, and complexity. In: C. A. **Hooker** (ed.), *Handbook of the philosophy of science, Vol. 10: philosophy of complex systems.* Amsterdam: Elsevier, pp. 639–82.

Varela, F. J. (1979). *Principles of biological autonomy.* New York: North Holland.

Varela F. J. (1991), Organism: a meshwork of selfless selves. In: **Tauber A. I.** (ed.), *Organism and the origins of self.* Dordrecht: Kluwer, 79–107.

Varela, F. J. (1992). Making it concrete: before, during and after breakdowns. *Revue Internationale de Psychopathologie* 4, 435–50.

Varela, F. J. (1995). Resonant cell assemblies: a new approach to cognitive functions and neuronal synchrony. *Biological Research* 28, 81–95

Varela, F. J. (1996a). Neurophenomenology: a methodological remedy for the hard problem. *Journal of Consciousness Studies* 3, 330–49.

Varela, F. J. (1996b). The early days of autopoiesis: Heinz and Chile. *Systems Research* 13, 407–16.

Varela, F. J. (1997). Patterns of life: intertwining identity and cognition. *Brain and Cognition* 34, 72–87.

Varela, F. J. (1999a). *ethical know-how: action, wisdom, and cognition.* Stanford, CA: Stanford University Press.

Varela, F. J. (1999b). The specious present: a neurophenomenology of time consciousness. In: J. **Petitot,** F. J. **Varela,** B. **Pacoud,** and J.-M. **Roy** (eds.), *Naturalizing phenomenology.* Stanford, CA: Stanford University Press. pp. 266–314.

Varela, F., **Coutinho, A., Dupire, B.,** and **Vaz, N.** (1988). Cognitive networks: immune, neural and otherwise. *Theoretical Immunology* 2, 359–75.

Varela, F. J. and **Frenk, S.** (1987). The organ of form: towards a theory of biological shape. *Journal of Social and Biological Structures* 10(1), 73–83.

Varela, F., **Lachaux, J., Rodriguez, E.,** and **Martinerie, J.** (2001). The brainweb: phase synchronization and large-scale integration. *Nature Reviews Neuroscience* 2(4), 229–39.

Varela, F. J., **Thompson, E.,** and **Rosch, E.** (1991). *The embodied mind: cognitive science and human experience.* Cambridge, MA: MIT Press.

Vargas, P., **Di Paolo, E. A., Harvey, I.,** and **Husbands, P.** (eds.) (2014). *The horizons of evolutionary robotics.* Cambridge, MA: MIT Press.

Vaughan, E., Di Paolo, E. A., and **Harvey, I.** (2014). Incremental evolution of an omni-directional biped for rugged terrain. In: **P. Vargas, E. A. Di Paolo, I. Harvey, and P. Husbands.** *Horizons for evolutionary robotics.* Cambridge, MA: MIT Press, pp. 237–78.

Vera, A. H. and **Simon, H. A.** (1993). Situated action: a symbolic interpretation. *Cognitive Science* **17(1),** 7–48.

Vernon, D. (2010). Enaction as a conceptual framework for developmental cognitive robotics. *Paladyn Journal of Behavioral Robotics* **1(2),** 89–98.

Vuillemin, J. (1972). Poincaré's philosophy of space. *Synthese* **24,** 161–79.

Von Holst, E. and **Mittelstaedt, H.** (1950). Das Reafferenzprinzip. *Naturwissenschaften* **37(20),** 464–76. doi:10.1007/BF00622503.

Vygotsky, L. S. (1978). *Mind and society: the development of higher mental processes.* Cambridge, MA: Harvard University Press.

Waddington, C. H. (1977). *Tools for thought.* New York: Basic Books.

Wallot, S. (2015). Intentions and synergies: the cases of control and speed. *Journal of Consciousness Studies* **22,** 80–3.

Wallot, S. and **Van Orden, G.** (2012). Ultrafast cognition. *Journal of Consciousness Studies* **19,** 141–60.

Walter, W. G. (1950). An imitation of life. *Scientific American* **182(5),** 42–5.

Walter, W. G. (1953). *The living brain.* London: G. Duckworth.

Washburn, M. F. (1914). The function of incipient motor process. *Psychological Review* **21,** 376–90.

Washburn, M. F. (1916). *Movement and mental imagery: outlines of a motor theory of the higher mental processes.* Boston: Houghton Mifflin.

Washburn, M. F. (1924). Gestalt psychology and motor psychology. *American Journal of Psychology* **37,** 516–20.

Watson, M. W. and **Jackowitz, E. R.** (1984). Agents and recipient objects in the development of early symbolic play,. *Child Development* **55(3),** 1091–7.

Watson, R. A. and **Szathmáry, E.** (2016). How can evolution learn? *Trends in Ecology & Evolution* **31(2),** 147–57.

Webb, B. (1995). Using robots to model animals: a cricket test. *Robotics and Autonomous Systems* **16(2–4),** 117–34.

Weber, A. and **Varela, F.** (2002). Life after Kant: natural purposes and the autopoietic foundations of biological individuality. *Phenomenology and the Cognitive Sciences* **1,** 97–125. doi: 10.1023/A:1020368120174.

Wegner, D. M. (2003). The mind's best trick: how we experience conscious will. *Trends in Cognitive Sciences* **7(2),** 65–9.

Wegner, D. M. and **Sparrow, B.** (2004). Authorship processing. In: **Gazzaniga M.** (ed.), *The Cognitive Neurosciences.* 3rd ed. Cambridge, MA: MIT Press. pp. 1201–9.

Wegner, D. M. and **Wheatley, T.** (1999). Apparent mental causation: sources of the experience of will. *American Psychologist* **54,** 480–92.

Weiss, M., Hultsch, H., Adam, I., Scharff, C., and **Kipper, S.** (2014). The use of network analysis to study complex animal communication systems: a study on nightingale song. *Proceedings of the Royal Society of London B: Biological Sciences* **281(1785),** 20140460.

Wenke, D., Fleming, S. M., and **Haggard, P.** (2010). Subliminal priming of actions influences sense of control over effects of action. *Cognition* **115(1),** 26–38.

Wexler, M. and **van Boxtel, J. J. A.** (2005). Depth perception by the active observer. *Trends in Cognitive Sciences* **9,** 431–38. doi:10.1016/j.tics.2005.06.018.

Wheeler, M. (2005). *Reconstructing the cognitive world: the next step.* Cambridge, MA: MIT Press.

Wijnants, M. L., Hasselman, F., Cox, R. F. A., Bosman, A. M. T., and G. Van Orden, G. (2012). An interaction-dominant perspective on reading fluency and dyslexia. *Annals of Dyslexia* **62**, 100–19.

Wijnants, M. L., Bosman, A. M., Hasselman, F., Cox, R. F., and Van Orden, G. C. (2009). 1/f scaling in movement time changes with practice in precision aiming. *Nonlinear Dynamics, Psychology, and Life Sciences* **13(1)**, 79–98.

Williams, P. L., Beer, R. D., and Gasser, M. (2008). An embodied dynamical approach to relational categorization. In: B. C. Love, K. McRae, and V. M. Sloutsky (eds.), *Proceedings of the 30th Annual Conference of the Cognitive Science Society*. Washington, DC: Cognitive Science Society, pp. 223–8.

Wilson, M. (2002). Six views of embodied cognition. *Psychonomic Bulletin & Review* **9**, 625–36.

Winograd, T. and Flores, F. (1986). *Understanding computers and cognition: a new foundation for design*. Norwood, NJ: Ablex.

Witkin, M. A. and Berry, J. W. (1975). Psychological differentiation in cross-cultural perspective. *Journal of Cross-Cultural Psychology* **6**, 4–78.

Wolpert, D. M., Doya, K., and Kawato, M. (2003). A unifying computational framework for motor control and social interaction. *Philosophical Transactions of the Royal Society of London. Series B, Biological Sciences* **358**, 593–602. doi:10.1098/rstb.2002.1238.

Yamauchi, B. and Beer, R. D. (1994). Sequential behavior and learning in evolved dynamical neural networks. *Adaptive Behavior* **2**, 219–46. doi: 10.1177/105971239400200301.

Young, I. M. (2005). *On female body experience: "Throwing like a girl" and other essays*. Oxford: Oxford University Press.

Yu, C. and Smith, L. B. (2013). Joint attention without gaze following: human infants and their parents coordinate visual attention to objects through eye-hand coordination. *PLoS ONE* **8(11)**, e79659. doi:10.1371/journal.pone.0079659.

Zahavi, D. (2013). Mindedness, mindlessness and first-person authority. In: J. Schear (ed.), *Mind, reason and being-in-the-world: the McDowell-Dreyfus debate*. London: Routledge, pp. 320–43.

Zhang, Y., Lu, H., and Bargmann, C. I. (2005). Pathogenic bacteria induce aversive olfactory learning in Caenorhabditis elegans. *Nature* **438**, 179–84. doi:10.1038/nature04216.

Zuckerman, M. (1979). *Sensation seeking. beyond the optimal level of arousal*. Hillsdale, NJ: Lawrence Erlbaum.

Zuckerman, M. and Haber, M. M. (1965). Need for stimulation as a source of stress response to perceptual isolation. *Journal of Abnormal Psychology* **70**, 371–7.

Index